Experimental Methods for Engineers

Experimental Methods for Engineers

Second Edition

J. P. Holman
Professor of Mechanical Engineering
Southern Methodist University

McGraw-Hill Book Company

New York   St. Louis   San Francisco   Düsseldorf   Johannesburg
Kuala Lumpur   London   Mexico   Montreal   New Delhi   Panama
Rio de Janeiro   Singapore   Sydney   Toronto

Contents

Preface   xi

List of Selected Books and Periodicals   xv

Chapter 1   Introduction   1

Chapter 2   Basic Concepts   6

2-1   Introduction   6
2-2   Definition of terms   6
2-3   Calibration   7
2-4   Standards   8
2-5   The generalized measurement system   12
2-6   Basic concepts in dynamic measurements   14
2-7   System response   19
2-8   Distortion   20
2-9   Impedance matching   21
2-10 Experiment planning   23
Review questions   31
Problems   31
References   32

Chapter 3   Analysis of Experimental Data   33

3-1   Introduction   33
3-2   Causes and types of experimental errors   34
3-3   Error analysis on a commonsense basis   36
3-4   Uncertainty analysis   37
3-5   Statistical analysis of experimental data   44
3-6   Probability distributions   47
3-7   The gaussian or normal error distribution   49
3-8   Probability graph paper   57
3-9   The chi-square test of goodness of fit   59
3-10 Method of least squares   65
3-11 Standard deviation of the mean   68
3-12 Graphical analysis and curve fitting   69
3-13 General considerations in data analysis   72
Review questions   73
Problems   74
References   77

Chapter 4   Basic Electrical Measurements and Sensing
            Devices   78

4-1   Introduction   78
4-2   The measurement of current   79

4-3   Voltmeters   85
4-4   Basic input circuitry   87
4-5   Bridge circuits   92
4-6   Amplifiers   100
4-7   Filter circuits   102
4-8   The vacuum-tube voltmeter (VTVM)   107
4-9   Digital voltmeters   109
4-10 The oscilloscope   112
4-11 Oscillographs   116
4-12 Counters, time, and frequency measurements   117
4-13 Transducers and electric sensing devices   124
4-14 The variable-resistance transducer   124
4-15 The differential transformer (LVDT)   124
4-16 Capacitive transducers   127
4-17 Piezoelectric transducers   129
4-18 Photoelectric effects   130
4-19 Photoconductive transducers   131
4-20 Photovoltaic cells   134
4-21 Ionization transducers   135
4-22 Magnetometer search coil   135
4-23 Hall-effect transducers   137
Review questions   138
Problems   139
References   142

Chapter 5   Displacement and Area Measurements   143

5-1   Introduction   143
5-2   Dimensional measurements   144
5-3   Gage blocks   145
5-4   Optical methods   146
5-5   Pneumatic displacement gage   149
5-6   Area measurements   151
5-7   The planimeter   152
5-8   Graphical and numerical methods for area measurement   155
5-9   Surface areas   157
Problems   158
References   159

Chapter 6   Pressure Measurement   160

6-1   Introduction   160
6-2   Dynamic response considerations   162
6-3   Mechanical pressure-measurement devices   165
6-4   Dead-weight tester   167
6-5   Bourdon-tube pressure gage   169
6-6   Diaphragm and bellows gages   169
6-7   The Bridgman gage   174

6-8 Low-pressure measurement 174
6-9 The McLeod gage 174
6-10 Pirani thermal-conductivity gage 176
6-11 The Knudsen gage 178
6-12 The ionization gage 179
6-13 The Alphatron 180
6-14 Summary 180
Review questions 180
Problems 181
References 184

Chapter 7   Flow Measurement   185

7-1 Introduction 185
7-2 Positive-displacement methods 186
7-3 Flow-obstruction methods 188
7-4 Practical considerations for obstruction meters 192
7-5 The sonic nozzle 201
7-6 Flow measurement by drag effects 203
7-7 Hot-wire anemometer 206
7-8 Magnetic flowmeters 210
7-9 Flow-visualization methods 211
7-10 The shadowgraph 213
7-11 The schlieren 215
7-12 The interferometer 217
7-13 The laser anemometer 219
7-14 Smoke methods 222
7-15 Pressure probes 224
7-16 Impact pressure in supersonic flow 229
7-17 Summary 231
Review questions 231
Problems 231
References 234

Chapter 8   The Measurement of Temperature   237

8-1 Introduction 237
8-2 Temperature scales 238
8-3 The ideal-gas thermometer 239
8-4 Temperature measurement by mechanical effects 240
8-5 Temperature measurement by electrical effects 244
8-6 Temperature measurement by radiation 259
8-7 Effect of heat transfer on temperature measurement 265
8-8 Transient response of thermal systems 273
8-9 Thermocouple compensation 273
8-10 High-speed temperature measurements 277
8-11 Summary 280
Review questions 280

Problems   280
References   284

Chapter 9 Thermal- and Transport-property Measurements 286

9-1   Introduction   286
9-2   Thermal-conductivity measurements   287
9-3   Thermal conductivity of liquids and gases   291
9-4   Measurement of viscosity   393
9-5   Gas diffusion   299
9-6   Calorimetry   303
9-7   Convection heat-transfer measurements   307
9-8   Humidity measurements   310
9-9   Heat-flux meters   313
Review questions   316
Problems   317
References   320

Chapter 10 Force, Torque, and Strain Measurements 321

10-1 Introduction   321
10-2 Mass balance measurements   322
10-3 Elastic elements for force measurements   326
10-4 Torque measurements   329
10-5 Stress and strain   331
10-6 Strain measurements   333
10-7 Electrical-resistance strain gages   334
10-8 Measurement of resistance strain-gage outputs   338
10-9 Temperature compensation   339
10-10 Strain-gage rosettes   340
10-11 The unbonded-resistance strain gage   343
Review questions   344
Problems   345
References   347

Chapter 11 Motion and Vibration Measurement 348

11-1 Introduction   348
11-2 Two simple vibration instruments   349
11-3 Principles of the seismic instrument   351
11-4 Practical considerations for seismic instruments   357
11-5 Sound measurements   360
Review questions   372
Problems   372
References   375

Chapter 12 Thermal- and Nuclear-radiation Measurements 376

12-1 Introduction   376
12-2 Detection of thermal radiation   376

12-3   Measurement of emissivity   382
12-4   Reflectivity and transmissivity measurements   384
12-5   Nuclear radiation   386
12-6   Detection of nuclear radiation   386
12-7   The Geiger-Müller counter   387
12-8   Ionization chambers   388
12-9   Photographic-detection methods   389
12-10 The scintillation counter   390
12-11 Neutron detection   390
12-12 Statistics of counting   391
Review questions   395
Problems   396
References   397

Chapter 13   Data Acquisition and Processing   399

13-1   Introduction   399
13-2   The general acquisition system   400
13-3   Data processing   402
13-4   Summary   402

Appendix   405

Index   415

Preface

Experimental measurements can be vexatious, and a textbook about experimental methods cannot alleviate all the problems which are perplexing to the experimental engineer. Engineering education has placed an increased emphasis on the ability of an individual to perform a theoretical analysis of a problem. Experimental methods are not unimportant; but analytical studies have, at times, seemed to deserve more emphasis. Laboratory work has also become more sophisticated in the modern engineering curricula. The conventional laboratory courses have consistently been modernized to include experiments with rather elaborate electronic instrumentation. Surprisingly enough, however, many engineering graduates do not exhibit the capability to perform simple engineering measurements with acceptable precision. Furthermore, they are amazingly inept when asked the question: How good is the measurement?

This book represents a first survey of experimental methods for undergraduate students. As such, it covers a broad range of topics and may be lacking in depth on certain of these topics. In these instances the reader is referred to more detailed treatments in specialized monographs for further information.

It is important that engineers be able to perform successful experiments, and it is equally important that they know or be able to estimate the accuracies of their measurements. This book proposes to discuss a rather broad range of instruments and experimental measurement techniques. Strong emphasis is placed on problem solving, and the importance of accuracy, error, and uncertainty in experimental measurements is stressed throughout all the discussions. The book is generally suitable for a one-semester course at the junior or senior level in conventional engineering curricula. Such a course would be expected to devote one laboratory period a week to the material in this book. To supplement this work, it would be desirable to cover some of the text material in a lecture session. The lectures would be concerned with the principles of instrumentation, whereas the laboratory periods would afford the student an opportunity to use some of the devices discussed in this text and others which may be available to faculty planning the course. The particular experiments, or the instruments used in the laboratory periods, will depend on the facilities available and the objectives set by each curriculum. A mathematical background through ordinary differential equations is assumed for the developments, and it is expected that basic courses in thermodynamics, engineering mechanics, and electric-circuit theory will precede a course based on this text.

Whatever the course arrangement for which this text is applied, it is strongly recommended that the problems at the end of each chapter receive careful attention. These problems force the student to examine several instruments in order to determine their accuracy and the uncertainties which might result from faulty measurement techniques. In many instances the problems are very similar to numerical examples in the text. Other problems require the student to extend the text material through derivations, design of experiments, etc. The selection of problems for a typical course will depend, naturally, on the types of experiments and laboratory facilities available for use with the course.

A few remarks concerning the arrangement of the text material are in order. A brief presentation of all topics was desired, so that a rather broad range of experimental methods could be discussed within the framework of a book suitable in length for a one-semester course. Chapters 1 and 2 provide initial motivational remarks and some brief definitions of important terms common to all measurement systems. Next, a simple presentation of some of the principles of statistical data analysis is given in Chapter 3. Some of the concepts in Chapter 3 are used in almost every subsequent chapter in the book, particularly the concept of experimental uncertainty.

Chapter 4 discusses several simple electrical-measurement circuits and the principles of operation of typical electric transducers. Many of these transducers are applicable to measurement problems described in later chapters. Chapters 5 and 6, concerning dimensional and pressure measurements, offer fairly conventional presentations of their subject matter, except that numerical examples and problems are included to emphasize the importance of experimental uncertainty in the various devices. Flow measurement is discussed in Chapter 7 in a rather conventional manner. A notable feature of this chapter is the section on flow-visualization techniques. Again, the examples and problems illustrate some of the advantages and shortcomings of the various experimental techniques. Chapter 8 is quite specific in its discussion of temperature-measurement devices. Strong emphasis is placed on the errors which may arise from conduction, convection, and radiation heat transfer between the temperature-sensing device and its thermal environment. Methods are given to correct for these effects.

Chapter 9 is brief but furnishes the reader with an insight into the problems associated with transport-property measurements. The material in this chapter is dependent on the measurement techniques discussed in Chapters 6, 7, and 8. It may be noted that the material in Chapter 9 could be dispersed through the three previous chapters and still achieve a balanced presentation; however, it was believed best to bring transport properties and thermal measurements into sharper focus by grouping them together in one chapter.

Static force, torque, and strain measurements are discussed in Chapter 10. The strain measurements are related to some elementary principles of experimental stress analysis, and the operation of the electrical-resistance strain gage is emphasized.

Some of the elementary principles of motion- and vibration-measurement devices are discussed in Chapter 11. Included in this presentation is a discussion of sound waves, sound-pressure level, and acoustic measurements. The inclusion of the acoustics material in Chapter 11 is somewhat arbitrary since this material would be equally pertinent in Chapter 6.

Chapter 12 discusses thermal- and nuclear-radiation measurements which are finding increasingly wide application in a number of industries. The presentation is brief; but some of the more important detection techniques are examined, and examples are given to illustrate the important principles. A short presentation of the statistics of counting illustrates the importance of background activity in nuclear-radiation detection. The thermal-radiation measurements are properly related to the material in Chapter 8.

Electronic data-acquisition and processing systems are being developed at a very rapid rate—so rapid, in fact, that the presentation in Chapter 13 had to be given in a very general manner. A discussion of specific systems would be outdated in short order; nevertheless, it is believed that this brief discussion will alert the reader to the importance of data-acquisition systems in modern experimental programs.

Some remarks concerning material which has been added in the second edition are appropriate at this point. After considerable thought, the decision was made to retain the broad, but brief, format of the first edition. The objective of the book remains the same: to give the reader a survey of a variety of measurement techniques and to develop a careful attitude toward experimental uncertainty and accuracy. Some new topics have been added. A more systematized approach to experiment planning has been presented in Chapter 2. The role of uncertainty analysis in experiment design and planning has been given additional emphasis in Chapter 3 through illustrative examples. Additional material on photoconductive transducers is given in Chapter 4. The dynamic response of pressure transducers is given a somewhat expanded discussion in Chapter 6, and the laser-anemometer flow-measurement technique is added to Chapter 7. The section on dynamic compensation of thermocouples has been expanded in Chapter 8, and sections on humidity and heat-flux measurements have been added to Chapter 9. The section on acoustic measurements has been expanded considerably in Chapter 11 with appropriate mention of psychoacoustic factors. A section on reflectance and transmittance measurements has been added to Chapter 12 because of their importance in modern thermal-radiation studies. Additional problems have been provided for all chapters. A new feature is the set of brief Review Questions which appears at the end of each chapter.

The author is particularly grateful to Professor S. J. Kline of Stanford University for reading the complete manuscript of the first edition and making many helpful suggestions.   The suggestions of W. S. Kut, W. L. Rogers, J. H. Dittfach, J. E. Blair, and others regarding content of the second edition are appreciated.   Grateful acknowledgement is also extended to Dr. C. A. Albritton and the Science Information Institute of Southern Methodist University for partial sponsorship of the first edition.   The encouragement of Dean H. J. Henry of the School of Engineering in the early stages of writing is also appreciated.   The manuscript for the first edition was typed by my wife, Katherine.   Her careful attention to detail helped correct many of my mistakes.   Mrs. Gretna Anglen typed the new material for the second edition.

It is fairly obvious that a book of modest length cannot cover all details of experimental methods which will be of interest to the reader.   For this reason, the following brief list of selected books and periodicals is provided for further study and reference.   This list covers mainly those sources of *general* information on instruments and measurements.   Techniques for measurement in any particular field will appear in those journals which regularly present research results in the field.   Thus, the reader should consult such specialized references for information on new measurement techniques which may be employed from time to time.

J. P. Holman

List of Selected Books
and Periodicals

Books

1. Ambrosius, E. E., R. D. Fellows, and A. D. Brickman: "Mechanical Measurement and Instrumentation," Ronald Press, New York, 1966.
2. Bartholomew, D.: "Electrical Measurements and Instrumentation," Allyn and Bacon, Inc., Boston, Mass., 1963.
3. Beckwith, T. G. and W. L. Buck: "Mechanical Measurements," 2d ed., Addison-Wesley Publishing Company, Inc., Reading, Mass., 1969.
4. Cady, W. M.: "Physical Measurements in Gas Dynamics and Combustion," Princeton University Press, Princeton, N.J., 1954.
5. Cerni, R. H., and L. E. Foster: "Instrumentation for Engineering Measurement," John Wiley & Sons, Inc., New York, 1962.
6. Considine, D. M. (ed.): "Process Instruments and Controls Handbook," McGraw-Hill Book Company, New York, 1957.
7. Cook, N. H. and E. Rabinowicz: "Physical Measurement and Analysis," Addison-Wesley Publishing Company, Inc., Reading, Mass., 1963.
8. Dally, J. W., and W. F. Riley: "Experimental Stress Analysis," McGraw-Hill Book Company, New York, 1965.
9. Doebelin, E. O.: "Measurement Systems: Application and Design," McGraw-Hill Book Company, New York, 1966.
10. Dove, R. C., and P. H. Adams: "Experimental Stress Analysis and Motion Measurement," Charles E. Merrill Books, Inc., Columbus, Ohio, 1964.
11. Eckman, D. P.: "Industrial Instrumentation," John Wiley & Sons, Inc., New York, 1950.
12. Ewing, G. W.: "Instrumental Methods of Chemical Analysis," 3rd Edition, McGraw-Hill Book Company, New York, 1969.
13. Frank E.: "Electrical Measurement Analysis," McGraw-Hill Book Company, New York, 1959.
14. Keast, D. N.: "Measurements in Mechnical Dynamics," McGraw-Hill Book Company, New York, 1967.
15. Lion, K. S.: "Instrumentation in Scientific Research," McGraw-Hill Book Company, New York, 1959.
16. Schenck, H.: "Theories of Engineering Experimentation," 2d ed., McGraw-Hill Book Company, New York, 1968.
17. Tuve, G. L., and L. C. Domholdt: "Engineering Experimentation," McGraw-Hill Book Company, New York, 1966.

18. Wilson, E. B., Jr.: "An Introduction to Scientific Research," McGraw-Hill Book Company, New York, 1952.

Periodicals

1. *Control Engineering*
2. *Industrial Laboratory* (USSR; English translation)
3. *Instruments and Control Systems*
4. *Instruments and Experimental Techniques* (USSR; English translation)
5. *Journal of the Instrument Society of America*
6. *Journal of Research of the National Bureau of Standards*
7. *Journal of Scientific Instruments* (Great Britain)
8. *Measurement Techniques* (USSR; English translation)
9. *The Review of Scientific Instruments*
10. *Transactions of the Instrument Society of America*
11. *Transactions of the Society of Instrument Technology* (Great Britain)

Experimental Methods for Engineers

1
Introduction

There is no such thing as an easy experiment, nor is there any substitute for careful experimentation in many areas of basic research and applied product development. Because experimentation is so important in all phases of engineering, there is a very definite need for the engineer to be familiar with methods of measurement as well as analysis techniques for interpreting experimental data.

Experimental techniques have changed quite rapidly with the development of electronic devices for sensing primary physical parameters and for controlling process variables. In many instances more precision is now possible in the measurement of basic physical quantities through the use of these new devices. Further development in instrumentation techniques is certain to be very rapid because of the increasing demand for measurement and control of physical variables in a wide variety of applications. To meet this demand the engineer must be familiar with the basic principles of instrumentation and the ideas which govern instrumentation development and usage.

Obviously, a sound knowledge of many engineering principles is necessary to perform successful experiments; it is for this reason that experimentation is so difficult.   To design the experiment the engineer must be able to specify the physical variables which he needs to investigate and the role they will play in later analytical work.   Then, to design or procure the instrumentation for the experiment he must have a knowledge of the governing principles of a broad range of instruments.   Finally, to analyze his data the experimental engineer must have a combination of keen insight into the physical principles of the processes he is investigating and a knowledge of the limitations of his data.

Research involves a combination of analytical and experimental work. The theoretician strives to explain or predict the results of experiments on the basis of analytical models which are in accordance with fundamental physical principles that have been well established over the years.   When experimental data are encountered which do not fit into the scheme of existing physical theories, a skeptical eye is cast first at the experimental data and then at appropriate theories.   In some cases the theories are modified or revised to take the results of the new experimental data into account, after being sure that the validity of the data has been ascertained.   In any event, all physical theories must eventually rely upon experiment for verification.

Whether the research is of a basic or developmental character, the dominant role of experimentation is still present.   The nuclear physicist must always test his theories in the laboratory if he is to be sure of their validity, just as the engineer who conducts research on a new electronic circuit or a new type of hydraulic flow system certainly must perform a significant number of experiments in order to establish the usefulness of the device. Physical experimentation is the ultimate test of most theories.

In many engineering applications certain basic physical phenomena are well known, and experience with devices using these phenomena is already available.   Examples are vacuum tubes, flowmeters, and certain mechanisms. There will always arise, however, new uses for such devices in combination with other devices, e.g., a new type of amplifier, a new flow-control system, etc.   In these cases the engineer must utilize all his experience with the previous devices in order to design the apparatus for a new application.   No matter how reliable the information on which the design is based, the engineer must insist on thorough experimental tests of the new device before the design is finalized and production is initiated.

There exists a whole spectrum of tests and experiments which an engineer may be called upon to perform.   These might vary from a very crude test to determine the weight of a device to some exceedingly precise electronic measurements of nuclear radioactivity.   Since the range of experiments is so broad, the experimental background of the engineer must be correspondingly diverse in order to function effectively in many experimental situations.

Obviously, it is quite hopeless to expect any one person to operate at a high level of effectiveness in all areas of experimental work. The primary capability of a particular individual will necessarily be developed in an area of experimentation closely connected with his analytical and theoretical capability and related interests. The broader the interests of the individual, the more likely that he will develop broad experimental interests and capabilities.

In the past, there have been some engineers who were primarily experimentalists—those individuals who designed devices by trial and error with very little analytical work as a preliminary to the experimentation. There are some older areas of engineering where this technique will still prevail, primarily where years of experience have built up a background knowledge to rely upon. But, in new fields, more emphasis must be placed on a combination of theory and experimentation. To create a rather absurd example, we might cite the development of a rocket engine. It would be possible to build different sizes of rockets and test them until a lucky combination of design parameters was found; however, the cost would be prohibitive. The proper approach is one of test and theoretical study where experimental data are constantly evaluated and compared with theoretical estimates. New theories are formulated on the basis of the experimental measurements, and these theories help to guide further tests and the final design.

The engineer should know what he is looking for before beginning his experiments. The objective of the experiments will dictate the accuracy required, expense justified, and level of human effort necessary. A simple calibration check of a mercury-in-glass thermometer would be a relatively simple matter requiring a limited amount of equipment and time; however, the accurate measurement of the temperature of a high-speed gas stream at 3000°F would involve more thought and care. A test of an amplifier for a home music system might be less exacting than a test of an amplifier to be used as part of the electronic equipment in a satellite, and so on.

The engineer is not only interested in the measurement of physical variables but is also concerned with their control. The two functions are closely related, however, because one must be able to measure a variable such as temperature or flow in order to control it. The accuracy of control is necessarily dependent on the accuracy of measurement. Hence, we see that a good knowledge of measurement techniques is necessary for the design of control systems. A detailed consideration of control systems is beyond the scope of our discussion, but the applicability of specific instruments and sensing devices to control systems will be indicated from time to time.

It is not enough for the engineer to be able to skillfully measure certain physical variables. For the data to have maximum significance he must be able to specify the precise degree of accuracy with which he has measured a certain variable. To specify this accuracy the limitations of the apparatus must be understood and full account must be taken of certain random and/or regular errors which may occur in the experimental data. Statistical techniques are available for analyzing data to determine expected errors and deviations from the true measurements. The engineer must be familiar with these techniques in order to analyze his data effectively.

All too frequently the engineer embarks on an experimental program in a stumbling blind-faith manner. Data are collected at random, many of which are not needed for later analysis. Certain ranges of operation are not investigated thoroughly enough, resulting in the collection of data which may have limited correlative value. The engineer must be sure to take enough data, but he should not waste time and money by taking too many. The obvious point is that experiments should be carefully planned. Most experimentalists do indeed plan tests in respect to the range of certain variables that they will want to investigate. But they often neglect the fact that in certain ranges of operation more data points may be necessary than in others in order to ensure the same degree of accuracy in the final data evaluation. In other words, the anticipated methods of data analysis, statistical or otherwise, should be taken into account in planning the experiment, just as one would take into account certain variables in designing the physical size of the experimental apparatus. The engineer should always ask the question: How many data do I need to ensure that my data are not just the result of luck? We will have more to say about experiment planning throughout the book, and the reader should consider these opening remarks as only the initial motivation.

A few remarks concerning experimental research are in order at this point. It is very difficult to describe the atmosphere and technique of performing research. For, unlike standard performance testing where experiments are conducted according to some well-established procedure, in research there is seldom a clear-cut way of proceeding. Each problem is a different one, and if the research is worthwhile, it has not been attacked extensively before. This means that the engineer engaged in research must be prepared to face numerous experimental difficulties of varying complexities. Some desirable objectives of the research may have to be relaxed because of the unavailability of instrumentation to measure the variables involved. Many seemingly trivial details become significant problems before a new experimental apparatus is functioning properly. One of the most basic problems is that the engineer seldom gets to measure in a direct manner the variable he really wants. There are always corrections to apply to the measurements, and seldom do they fall in the category of "standard" corrections. One trivial detail piles on another until the whole experimental problem takes on a complexity which is usually not anticipated at the start of the research. Again we state the truism: there is no such thing as an *easy* experiment.

The neophyte experimentalist frequently assumes that a certain experiment will be easy to perform. All he need do is hook up the apparatus, flip the switch, and out will come reams of significant data which will startle his colleagues (or supervisor). He does not realize that one simple instrument may not work and thus spoil the experiment. Once this instrument is functioning properly, another may go bad, and so on. When the apparatus is functioning, the neophyte is then tempted to take data at random without giving much consideration to the results that he will want to derive from the data. He tries to solve all problems at once, vary many parameters at the same time, so that little control is exerted on the data and it eventually becomes necessary to go back and do some of the work over. The important point, once again, is that careful planning is called for. In experimental research great care and patience will usually produce the best results in the *quickest* possible way.

All the above remarks may appear discouraging to the beginner for whom this book is written. On the contrary, they are intended to advise the beginner so that he can avoid some of the more obvious pitfalls. And, even more important, they are intended to let the beginner know that some troubles are to be expected and that perseverance in the face of these troubles combined with intelligent planning will almost always lead to the desired results—accurate and meaningful data.

The objective of the presentation in this book is to impart a broad knowledge of experimental methods and measurement techniques. To accomplish this objective a rather large number of instruments will be discussed from the standpoint of both theory of operation and specific functional characteristics. Emphasis will be placed on analytical calculations to familiarize the reader with important points in the theoretical development as well as in the descriptive information pertaining to operating characteristics. As a further means of emphasizing the discussions, the uncertainties which may arise in the various instruments are given particular attention.

The study of experimental methods is a necessary extension of all analytical subjects. A knowledge of the methods of verifying analytical work injects new life and vitality into the theories, and a clear understanding of the difficulties of experimental measurements creates a careful attitude in the theoretician which cannot be generated in any other way.

2
Basic Concepts

2-1 INTRODUCTION

In this chapter we seek to explain some of the terminology used in experimental methods and to show the generalized arrangement of an experimental system. We shall also discuss briefly the standards which are available and the importance of calibration in any experimental measurement. A major portion of the discussion on experimental errors is deferred until Chap. 3, and only the definition of certain terms is given here.

2-2 DEFINITION OF TERMS

We are frequently concerned with the *readability* of an instrument. This term indicates the closeness with which the scale of the instrument may be read; an instrument with a 12-in. scale would have a higher readability than an instrument with a 6-in. scale and the same range. The *least count* is the smallest difference between two indications that can be detected on the instrument scale. Both readability and least count are dependent on scale length, spacing of graduations, size of pointer (or pen if a recorder is used), and parallax effects.

The *sensitivity* of an instrument is the ratio of the linear movement of the pointer on the instrument to the change in the measured variable causing this motion. For example, a 1-mv recorder might have a 10-in. scale length. Its sensitivity would be 10 in./mv, assuming that the measurement was linear all across the scale.

An instrument is said to exhibit *hysteresis* when there is a difference in readings depending on whether the value of the measured quantity is approached from above or below. Hysteresis may be the result of mechanical friction, magnetic effects, elastic deformation, or thermal effects.

The *accuracy* of an instrument indicates the deviation of the reading from a known input. Accuracy is usually expressed as a percentage of full-scale reading, so that a 100-psi pressure gage having an accuracy of 1 percent would be accurate within ± 1 psi over the entire range of the gage.

The *precision* of an instrument indicates its ability to reproduce a certain reading with a given accuracy. As an example of the distinction between precision and accuracy, consider the measurement of a known voltage of 100 volts with a certain meter. Five readings are taken, and the indicated values are 104, 103, 105, 103, and 105 volts. From these values it is seen that the instrument could not be depended on for an accuracy of better than 5 percent (5 volts), while a precision of ± 1 percent is indicated since the maximum deviation from the mean reading of 104 volts is only 1 volt. It may be noted that the instrument could be calibrated so that it could be used to dependably measure voltages within ± 1 volt. This simple example illustrates an important point. Accuracy can be improved up to but not beyond the precision of the instrument by calibration. The precision of an instrument is usually subject to many complicated factors and requires special techniques of analysis, which will be discussed in Chap. 3.

2-3 CALIBRATION

The calibration of all instruments is important, for it affords the opportunity to check the instrument against a known standard and subsequently to reduce errors in accuracy. Calibration procedures involve a comparison of the particular instrument with either (1) a primary standard, (2) a secondary standard with a higher accuracy than the instrument to be calibrated, or (3) a known input source. For example, a flowmeter might be calibrated by (1) comparing it with a standard flow-measurement facility of the National Bureau of Standards, (2) comparing it with another flowmeter of known accuracy, or (3) direct calibration with a primary measurement such as weighing a certain amount of water in a tank and recording the time elapsed for this quantity to flow through the meter. In item 2, the key words are "known accuracy." The meaning here is that the accuracy of the meter must be specified by a reputable source.

The importance of calibration cannot be overemphasized because it is calibration which firmly establishes the accuracy of the instruments. Rather than accept the reading of an instrument, it is usually best to make at least a simple calibration check to be sure of the validity of the measurements. Not even manufacturers' specifications or calibrations can always be taken at face value. Most instrument manufacturers are reliable; some, alas, are not. We shall be able to give more information on calibration methods throughout the book as various instruments and their accuracies are discussed.

2-4 STANDARDS

In order that investigators in different parts of the country and different parts of the world may compare the results of their experiments on a consistent basis, it is necessary to establish certain standard units of length, weight, time, temperature, and electrical quantities. The National Bureau of Standards has the primary responsibility for maintaining these standards in the United States.

The meter and the kilogram are considered fundamental units upon which, through appropriate conversion factors, the English system of length and mass is based. The standard meter is defined as the length of a platinum-iridium bar maintained at very accurate conditions at the International Bureau of Weights and Measures in Sèvres, France. Similarly, the kilogram is defined in terms of a platinum-iridium mass maintained at this same bureau. The conversion factors for the English and metric systems in the United States are fixed by law as

1 meter = 39.37 inches
1 pound mass = 453.59237 grams

Secondary standards of length and mass are maintained at the National Bureau of Standards for calibration purposes. In 1960 the General Conference on Weights and Measures defined the standard meter in terms of the wavelength of the orange-red light of a krypton-86 lamp. The standard meter is thus

1 meter = 1,650,763.73 wavelengths

The inch is exactly defined as

1 inch = 2.54 centimeters

Standard units of time are established in terms of known frequencies of oscillation of certain devices. One of the simplest devices is a pendulum. A torsional vibrational system may also be used as a standard of frequency.

The torsional system is widely used in clocks and watches. Ordinary 60-cycle line voltage may be used as a frequency standard under certain circumstances. An electric clock uses this frequency as a standard because it operates from a synchronous electric motor whose speed depends on line frequency. A tuning fork is a suitable frequency source, as are piezoelectric crystals. Electronic oscillators may also be designed to serve as very precise frequency sources.

The fundamental unit of time, the second, has been defined in the past as $\frac{1}{86400}$ of a mean solar day. The solar day is measured as the time interval between two successive transits of the sun across a meridian of the earth. The time interval varies with location on the earth and time of year; however, the *mean solar day* for one year is constant. The solar year is the time required for the earth to make one revolution around the sun. The mean solar year is 365 days 5 hr 48 min 48 sec.

The above definition of the second is quite exact but is dependent on astronomical observations in order to establish the standard. In October, 1967, the Thirteenth General Conference on Weights and Measures adopted a definition of the second as the duration of 9,192,631,770 periods of the radiation corresponding to the transition between the two hyperfine levels of the fundamental state of the atom of Cesium-133 [7]. This standard can be readily duplicated in standards laboratories throughout the world. The estimated accuracy of this standard is 2 parts in 10^9.

Very precise frequency and time standards are broadcast by the National Bureau of Standards' radio station, WWV. This station furnishes standard time intervals, time signals, standard musical pitch, standard radio frequencies, and radio-propagation forecasts. The accuracy of the radio and audio frequencies as transmitted is better than 1 part in 100,000,000, and the time interval is accurate to 1 part in 100,000,000 $\pm$ 1 μsec.

Standard units of electrical quantities are derivable from the mechanical units of force, mass, length, and time. These units represent the absolute electrical units and differ slightly from the international system of electrical units established in 1948. A detailed description of the international system is given in Refs. [1] and [2]. The main advantage of this system is that it affords the establishment of a standard cell, the output of which may be directly related to the absolute electrical units. The conversion from the international system is established by the following relations:

1 international ohm = 1.00049 absolute ohms
1 international volt = 1.000330 absolute volts
1 international ampere = 0.99835 absolute ampere

Laboratory calibration is usually made with the aid of secondary standards such as standard cells for voltage sources and standard resistors as standards of comparison for measurement of electrical resistance.

Table 2-1 Primary points for the International Temperature Scale of 1948

	Temperature		
Point at standard pressure = 14.6959 psia	°F	°C	°K
Oxygen point: Temperature of equilibrium between oxygen and its vapor	−297.346	−182.970	90.19
Ice point: Temperature at equilibrium between ice and air-saturated water	32.0	0	273.16
Steam point: Temperature at equilibrium between liquid water and its vapor	212.0	100.0	373.16
Sulfur point: Temperature at equilibrium between liquid sulfur and its vapor	832.28	444.6	717.76
Silver point: Temperature of equilibrium between solid and liquid silver	1761.4	960.8	1233.96
Gold point: Temperature of equilibrium between solid and liquid gold	1945.4	1063.0	1336.16

An absolute temperature scale was proposed by Lord Kelvin in 1854. It forms the basis for thermodynamic calculations. This absolute scale is so defined that particular meaning is given to the second law of thermodynamics when this temperature scale is used. The International Temperature Scale of 1948 [6] furnishes an experimental basis for a temperature scale which approximates as closely as possible the absolute thermodynamic temperature

Table 2-2 Secondary fixed points for the International Temperature Scale of 1948

Point	*Temperature, °C*
Equilibrium between solid carbon dioxide and its vapor	−78.5
Freezing mercury	−38.87
Equilibrium between ice, water, and its vapor (triple point)	+0.0100
Freezing tin	231.9
Freezing cadmium	320.9
Freezing lead	327.3
Equilibrium between mercury and its vapor	356.58
Freezing zinc	419.5
Freezing antimony	630.5
Freezing aluminum	660.1
Freezing copper in a reducing atmosphere	1083
Freezing nickel	1453
Freezing cobalt	1492
Freezing platinum	1769
Freezing rhodium	1960
Freezing iridium	2443
Melting tungsten	3880

scale.   In the international scale, six primary points are established as shown in Table 2-1.   Secondary fixed points are also established as given in Table 2-2.   In addition to the fixed points, precise procedures are established for interpolating between these points.   These interpolation procedures are given in Table 2-3.

Table 2-3   Interpolation procedures for the International Temperature Scale of 1948

Range definition	Temperature, °C	Procedure
Between ice point and freezing point of antimony	0–630.5	Based on resistance of platinum resistance thermometer: $R_t = R_0(1 + AT + BT^2)$ R_t = resistance at temperature T, ohms R_0 = resistance at 0°C, ohms A, B = constants determined from measured values of R_t at steam and sulfur points T = temperature
Between oxygen and ice points	−182.97–0	Based on resistance of platinum resistance thermometer: $R_T = R_0[1 + AT + BT^2 + C(T - 100)T^3]$ C = a constant determined from resistance at the oxygen point A, B, R_0 = same constants as above
Between freezing point of antimony and gold point	630.5–1063.0	Standard thermocouple of platinum and platinum-10% rhodium: $E = a + bT + cT^2$ E = emf of thermocouple when one junction is at 0°C and the other at temperature T T = temperature in specified range a, b, c = constants determined from emf data taken at antimony, silver, and gold points
Above gold point	Above 1063.0	Temperature defined by: $$\frac{J_t}{J_{Au}} = \frac{e^{C_2/\lambda(T_{Au} + T_0)} - 1}{e^{C_2/\lambda(T + T_0)} - 1}$$ J_t, J_{Au} = radiant energy emitted per unit time, per unit area, and per unit wavelength at wavelength λ, at temperature T, and gold-point temperature T_{Au}, respectively C_2 = 1.438 cm − °K T_0 = 273.16°K λ = wavelength

Eshbach [3] gives a detailed discussion of various electrical and thermal units. The absolute thermodynamic temperature scale is discussed by Obert [4], and Constant [5] gives an excellent discussion of electromagnetic units and concepts.

2-5 THE GENERALIZED MEASUREMENT SYSTEM

Most measurement systems may be divided into three parts:

1. A detector-transducer stage, which detects the physical variable and performs either a mechanical or an electrical transformation to convert the signal into a more usable form. In the general sense, a transducer is a device that transforms one physical effect into another. In most cases, however, the physical variable is transformed into an electric signal because this is the form of signal that is most easily measured.
2. Some intermediate stage, which modifies the direct signal by amplification, filtering, or other means so that a desirable output is available.
3. A final or terminating stage, which acts to indicate, record, or control the variable being measured.

As an example of a measurement system, consider the measurement of a low-voltage signal at a low frequency. The detector in this case may be just two wires and possibly a resistance arrangement, which are attached to appropriate terminals. Since we want to indicate or record the voltage, it may be necessary to perform some amplification. The amplification stage is then stage 2 designated above. The final stage of the measurement system may be either a voltmeter or a recorder that operates in the range of the output voltage of the amplifier. In actuality, a vacuum-tube voltmeter is a measurement system like the one described here. The amplifier and the readout voltmeter are contained in one cabinet, and various switches enable the user to change the range of the instrument by varying the input conditions to the amplifier.

Consider the simple bourdon-tube pressure gage shown in Fig. 2-1. This gage offers another example of the generalized measurement system. In this case the bourdon tube is the detector-transducer stage because it converts the pressure signal into a mechanical displacement of the tube. The intermediate stage consists of the gearing arrangement, which amplifies the displacement of the end of the tube so that a relatively small displacement at that point produces as much as three-quarters of a revolution of the center gear. The final indicator stage consists of the pointer and the dial arrangement, which, when calibrated with known pressure inputs, gives an indication of the pressure signal impressed on the bourdon tube. A schematic diagram of the generalized measurement system is shown in Fig. 2-2.

When a control device is used for the final measurement stage, it is necessary to apply some feedback signal to the input signal to accomplish the control objectives. The control stage compares the signal representing the measured variable with some other signal in the same form representing the assigned value the measured variable *should have*. The assigned value is given by a predetermined setting of the controller. If the measured signal agrees with the predetermined setting, then the controller does nothing. If the signals do not agree, the controller issues a signal to a device, which acts to alter the value of the measured variable. This device can be many things, depending on the variable which is to be controlled. If the measured variable is the flow rate of a fluid, the control device might be a motorized valve placed in the flow system. If the measured flow rate is too high, then the controller would cause the motorized valve to close, thereby reducing the flow rate. If the flow rate were too low, the valve would be opened. Eventually the operation would cease when the desired flow rate was achieved. The control feedback function is indicated in Fig. 2-2.

It is very important to realize that the accuracy of control cannot be any better than the accuracy of the measurement of the control variable. Therefore, one must be able to measure a physical variable accurately before

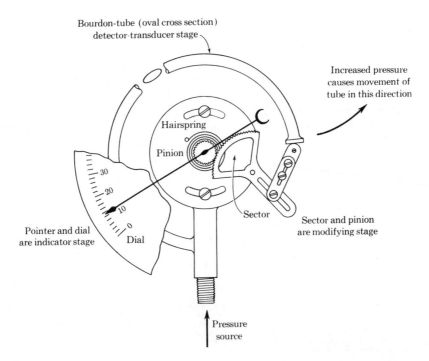

Fig. 2-1 Bourdon-tube pressure gage as the generalized measurement system.

he can hope to control the variable. In the flow system mentioned above, the most elaborate controller could not control the flow rate any more closely than the accuracy with which the primary sensing element measures the flow. We shall have more to say about some simple control systems in a later chapter. For the present we want to emphasize the importance of the measurement system in any control setup.

The overall schematic of the generalized measurement system is quite simple, and, as one might suspect, the difficult problems are encountered when suitable devices are sought to fill the requirements for each of the "boxes" on the schematic diagram. Most of the remaining chapters of the book are concerned with the types of detectors, transducers, modifying stages, etc., that may be used to fill these boxes.

2-6 BASIC CONCEPTS IN DYNAMIC MEASUREMENTS

Many experimental measurements are taken under such circumstances that ample time is available for the measurement system to reach steady state, and hence one need not be concerned with the behavior under non-steady-state conditions. In many other situations, however, it may be desirable to determine the behavior of a physical variable over a period of time. Sometimes the time interval is short, and sometimes it may be rather extended. In any event, the measurement problem usually becomes more complicated when the transient characteristics of a system need to be considered. In this section we wish to discuss some of the more important characteristics and parameters applicable to a measurement system under dynamic conditions.

To initiate the discussion let us consider a simple spring-mass damper system as shown in Fig. 2-3. We might consider this as a simple mechanical-measurement system where $x_1(t)$ is the input displacement variable which acts

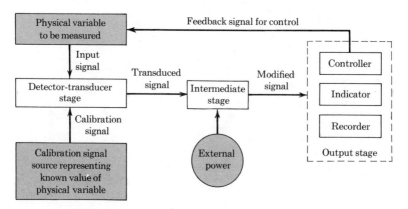

Fig. 2-2 Schematic of the generalized measurement system.

through the spring-mass damper arrangement to produce an output displacement $x_2(t)$. Both x_1 and x_2 vary with time. Suppose we wish to find $x_2(t)$, knowing $x_1(t)$, m, k, and the damping constant c. We assume that the damping force is proportional to velocity so that the differential equation governing the system is obtained from Newton's second law of motion as

$$k(x_1 - x_2) + c\left(\frac{dx_1}{dt} - \frac{dx_2}{dt}\right) = m\frac{d^2x_2}{dt^2} \tag{2-1}$$

Written in another form,

$$m\frac{d^2x_2}{dt^2} + c\frac{dx_2}{dt} + kx_2 = c\frac{dx_1}{dt} + kx_1 \tag{2-2}$$

Now suppose that $x_1(t)$ is the harmonic function

$$x_1(t) = x_0 \cos \omega_1 t \tag{2-3}$$

where x_0 is the amplitude of the displacement and ω_1 is the frequency.

We might imagine this simple vibrational system as being similar to a simple spring scale. The mass of the scale is m, the spring inside the scale is represented by the spring constant k, and whatever mechanical friction may be present is represented by c. We are subjecting the scale to an oscillating-displacement function and wish to know how the body of the scale will respond; i.e., we want to know $x_2(t)$. We might imagine that the spring scale is shaken by hand. When the oscillation $x_1(t)$ is very slow, we would note that the scale body very nearly follows the applied oscillation. When the frequency of the oscillation is increased, the scale body will react more violently until, at a certain frequency called the *natural frequency*, the amplitude of the displacement of the spring will take on its maximum value and could be *greater* than the amplitude of the impressed oscillation $x_1(t)$. The smaller the value of the damping constant, the larger the maximum amplitude at the natural frequency. If the impressed frequency is increased further, the

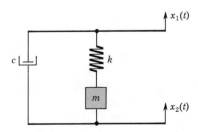

Fig. 2-3 Simple spring-mass damper system.

amplitude of the displacement of the spring body will decrease rather rapidly. The reader may conduct his own experiment to verify this behavior using a simple spring-mass system shaken by hand.

Clearly, the displacement function $x_2(t)$ depends on the frequency of the impressed function $x_1(t)$. We say that the system *responds* differently depending on the input frequency, and the overall behavior is designated as the *frequency response* of the system.

A simple experiment with the spring-mass system will show that the displacement of the mass is not in *phase* with the impressed displacement; i.e., the maximum displacement of the mass does not occur at the same time as the maximum displacement of the impressed function. This phenomenon is described as *phase shift*. We could solve Eq. (2-2) and determine the detailed characteristics of $x_2(t)$ including the frequency response and phase-shift behavior. However, we shall defer the solution for this system until Chap. 11, where vibration measurements are discussed. At this point in our discussion we have used this example because it is easy to visualize in a physical sense. Now, we shall consider a practical application, which involves a transformation of a *force-input function* into a displacement function. We shall present the solution to this problem because it shows very clearly the nature of frequency response and phase shift while emphasizing a practical system that might be used for a transient force or pressure measurement.

This system is shown in Fig. 2-4. The forcing function

$$F(t) = F_0 \cos \omega_1 t \tag{2-4}$$

is impressed on the spring-mass system, and we wish to determine the displacement of the mass $x(t)$ as a function of time. The differential equation for the system is

$$m \frac{d^2x}{dt^2} + c \frac{dx}{dt} + kx = F_0 \cos \omega_1 t \tag{2-5}$$

Equation (2-5) has the solution

$$x = \frac{(F_0/k) \cos (\omega_1 t - \phi)}{\{[1 - (\omega_1/\omega_n)^2]^2 + [2(c/c_c)(\omega_1/\omega_n)]^2\}^{\frac{1}{2}}} \tag{2-6}$$

where $\phi = \tan^{-1} \dfrac{2(c/c_c)(\omega_1/\omega_n)}{1 - (\omega_1/\omega_n)^2}$ (2-7)

$$\omega_n = \sqrt{\frac{k}{m}} \tag{2-8}$$

$$c_c = 2\sqrt{mk} \tag{2-9}$$

ϕ is called the phase angle, ω_n is the natural frequency, and c_c is called the critical damping coefficient.

The amplitude ratio $x_0/(F_0/k)$, where x_0 is the amplitude of the motion given by

$$x_0 = \frac{F_0/k}{\{[1 - (\omega_1/\omega_n)^2]^2 + [2(c/c_c)(\omega_1/\omega_n)]^2\}^{\frac{1}{2}}} \qquad (2\text{-}10)$$

is plotted in Fig. 2-5 to show the frequency response of the system, and the phase angle ϕ is plotted in Fig. 2-6 to illustrate the phase-shift characteristics. From these graphs we make several observations:

1. For low values of c/c_c the amplitude is very nearly constant up to a frequency ratio of about 0.3.
2. For large values of c/c_c (overdamped systems) the amplitude is reduced substantially.
3. The phase-shift characteristics are a strong function of damping ratio for all frequencies.

We might say that the system has good *linearity* for low damping ratios and up to a frequency ratio of 0.3 since the amplitude is essentially constant in this range.

While this brief discussion has been concerned with a simple mechanical system, we may remark that similar frequency and phase-shift characteristics are exhibited by electrical and thermal systems as well, and whenever time-varying measurements are made, due consideration must be given to these characteristics. Ideally, we should like to have a system with a linear frequency response over all ranges and with zero phase shift, but this is never completely attainable, although a certain instrument may be linear over a range of operation in which we are interested so that the behavior is good enough for the purposes intended. There are methods of providing compensation for the adverse frequency-response characteristics of an instrument, but these methods represent an extensive subject in themselves and cannot be discussed here. We shall have something to say about the dynamic characteristics of specific instruments in subsequent chapters.

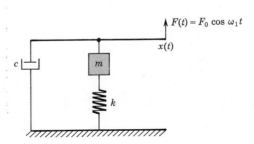

Fig. 2-4 Spring-mass damper system subjected to a force input.

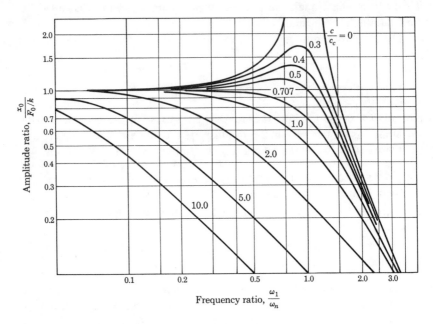

Fig. 2-5 Frequency response of the system in Fig. 2-4.

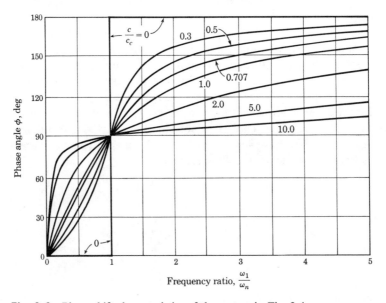

Fig. 2-6 Phase-shift characteristics of the system in Fig. 2-4.

2-7 SYSTEM RESPONSE

We have already discussed the meaning of frequency response and observed that in order for a system to have good response it must treat all frequencies the same within the range of application so that the ratio of output-to-input amplitude remains the same over the frequency range desired. We say that the system has linear frequency response if it follows this behavior.

Amplitude response pertains to the ability of the system to react in a linear way to various input amplitudes. In order for the system to have linear amplitude response, the ratio of output-to-input amplitude should remain constant over some specified range of input amplitudes. When this linear range is exceeded, the system is said to be overdriven, as in the case of a voltage amplifier where too high an input voltage is used.

We have already noted the significance of phase-shift response and its relation to frequency response. Phase shift is particularly important where complex waveforms are concerned because severe distortion may result if the system has poor phase-shift response.

Suppose a *step* or instantaneous input signal is applied to a system. In general there will be a slight delay in the output response, and this delay is called the *rise time* or *delay* of the system. This phenomenon is illustrated in Fig. 2-7.

Many systems exhibit an exponential decay type of behavior. When a capacitor is discharged through a resistance as shown in Fig. 2-8, the voltage varies with time according to

$$\frac{E(t)}{E_0} = e^{-(1/RC)t} \qquad\qquad (2\text{-}11)$$

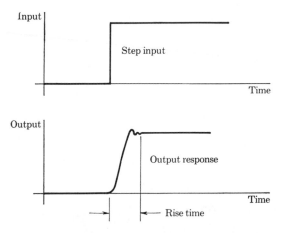

Fig. 2-7 Effect of rise time on output response to a step input.

where R is the value of the external resistance and C is the capacitance. The voltage across the capacitor as a function of time is $E(t)$, and the initial voltage is E_0. Some types of thermal systems also display this same kind of response. The temperature of a hot block of metal allowed to cool in a room varies with approximately the same kind of relation as shown in Eq. (2-11). For systems which have this kind of behavior we may note that when

$$t = RC \tag{2-12}$$

the ratio E/E_0 has a value of 36.8 percent of its initial value. Since the terms of Eq. (2-12) have the units of time, they are described as the *time constant* of the system. This meaning of time constant is carried into all types of systems—even those which do not have a simple behavior like Eq. (2-11). Thus, the time constant is usually taken as the time for the system to attain a state of 63.2 percent of its steady-state value. In the capacitance system we speak of voltage, in a thermal system we speak of temperature, and in some mechanical systems we might speak of velocity or displacement as the physical variables that change with time. The time constant would then be related to the steady-state values of these variables.

2-8 DISTORTION

Suppose a harmonic function of a complicated nature, i.e., composed of many frequencies, is transmitted through the mechanical system of Figs. 2-3 and 2-4. If the frequency spectrum of the incoming waveform is sufficiently broad, there would be different amplitude and phase-shift characteristics for each of the input-frequency components and the output waveform might bear little resemblance to the input. Thus, as a result of the frequency-response characteristics of the system, distortion in the waveform would be experienced. Distortion is a very general term that may be used to describe the variation of a signal from its true form. Depending on the system, the distortion may result from either poor frequency response or poor phase-shift response. In electronic devices various circuits are employed to reduce distortion to very small values. For mechanical systems the dynamic response characteristics are not as easily controlled and remain a subject for further

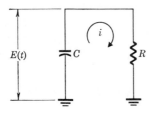

Fig. 2-8 Capacitor discharging through a resistance.

development. The effects of poor frequency and phase-shift response on a complex waveform are illustrated in Fig. 2-9.

2-9 IMPEDANCE MATCHING

In many experimental setups it is necessary to connect various items of electrical equipment in order to perform the overall measurement objective. When connections are made between electrical devices, proper care must be taken to avoid impedance mismatching. The input impedance of a two-terminal device may be illustrated as in Fig. 2-10. The device behaves as if the internal resistance R_i were connected in series with the internal voltage source E. The connecting terminals for the instrument are designated as A and B, and the open-circuit voltage presented at these terminals is the internal voltage E. Now, if an external load R is connected to the device and the internal voltage E remains constant, the voltage presented at the output

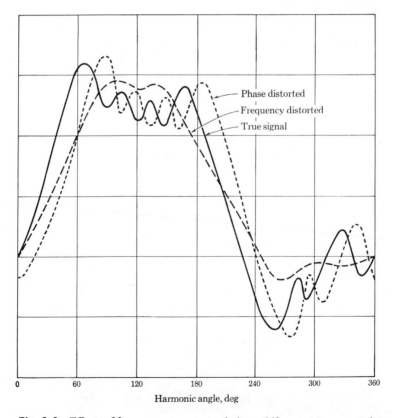

Fig. 2-9 Effects of frequency response and phase-shift response on complex waveform.

terminals A and B will be dependent on the value of R. The potential presented at the output terminals is

$$E_{AB} = E \frac{R}{R + R_i} \qquad (2\text{-}13)$$

The larger the value of R, the more closely the terminal voltage approaches the internal voltage E. Thus, if the device is used as a voltage source with some internal impedance, the external impedance (or load) should be large enough that the voltage is essentially preserved at the terminals. Or, if we wish to measure the internal voltage E, the impedance of the measuring device connected to the terminals should be large compared with the internal impedance.

Now suppose that we wish to deliver power from the device to the external load R. The power is given by

$$P = \frac{E_{AB}{}^2}{R} \qquad (2\text{-}14)$$

We ask for the value of the external load that will give the maximum power for a constant internal voltage E and internal impedance R_i. Equation (2-14) is rewritten

$$P = \frac{E^2}{R} \left(\frac{R}{R + R_i} \right)^2 \qquad (2\text{-}15)$$

and the maximizing condition

$$\frac{dP}{dR} = 0 \qquad (2\text{-}16)$$

is applied.   There results

$$R = R_i \qquad (2\text{-}17)$$

That is, the maximum amount of power may be drawn from the device when the impedance of the external load just matches the internal impedance. This is the essential principle of impedance matching in electric circuits.

Clearly, the internal impedance and external load of a complicated

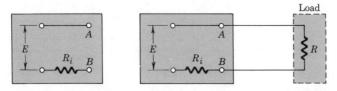

Fig. 2-10   Two-terminal device with internal impedance R_i.

electronic device may contain inductive and capacitative components that will be important in ac transmission and dissipation.   Nevertheless, the basic idea is the same.   The general principles of matching, then, are that the external impedance should match the internal impedance for maximum energy transmission (minimum attenuation) and the external impedance should be large compared with the internal impedance when a measurement of internal voltage of the device is desired.   It is this latter principle that makes a vacuum-tube voltmeter essential for measurement of voltages in electronic circuits.   The vacuum-tube voltmeter has a very high internal impedance (greater than 10 megohms) so that little current is drawn and the voltage presented to the terminals of the instrument is not altered appreciably by the measurement process.

Impedance-matching problems are usually encountered in electrical systems but can be important in mechanical systems as well.   We might, for example, imagine the simple spring-mass system of the previous section as a mechanical transmission system.   From the curves describing the system behavior it is seen that frequencies below a certain value are transmitted through the system; i.e., the force is converted to displacement with little attenuation.   Near the natural frequency undesirable amplification of the signal is performed, and above this frequency severe attenuation is present. We might say that this system exhibits a behavior characteristic of a variable impedance that is frequency-dependent.   When it is desired to transmit mechanical motion through a system, the natural-frequency and damping characteristics must be taken into account so that good "matching" is present. The problem is an impedance-matching situation, although it is usually treated as a subject in mechanical vibrations.

2-10 EXPERIMENT PLANNING

The key to success in experimental work is to ask continually: What am I looking for?   Why am I measuring this—does the measurement really answer any of my questions?   What does the measurement tell me?   These questions may seem rather elementary, but they should be asked frequently throughout the progress of any experimental program.   Some particular questions which should be asked in the initial phases of experiment planning are:

1. What primary variables shall be investigated?
2. What *control* must be exerted on the experiment?
3. What ranges of the primary variables will be necessary to describe the phenomena under study?
4. How many data points should be taken in the various ranges of operation to ensure good sampling of data considering instrument accuracy and other factors?   (See Chap. 3.)

5. What instrument accuracy is required for each measurement?
6. If a dynamic measurement is involved, what frequency response must the instruments have?
7. Are the instruments available commercially, or must they be constructed especially for the particular experiment?
8. What safety precautions are necessary if some kind of hazardous operation is involved in the experiment?
9. What financial resources are available to perform the experiment, and how do the various instrument requirements fit into the proposed budget?
10. What provisions have been made for recording the data?

The importance of control in any experiment should always be recognized. The physical principle, apparatus, or device under investigation will dictate the variables which must be controlled carefully. For example, a heat-transfer test of a particular apparatus might involve some heat loss to the surrounding air in the laboratory where the test equipment is located. Consequently, it would be wise to maintain (control) the surrounding temperature at a reasonably constant value. If one run is made with the room temperature at 90°F and another at 50°F, large unwanted effects may occur in the measurements. Suppose a test is to be made of the effect of cigarette smoke on the eating habits of mice. Clearly, we would want to control the concentration of smoke inhaled by the mice and also observe another group of mice which were not exposed to cigarette smoke at all. All other environmental variables should be the same if we are to establish the effect of the cigarette smoke on eating habits.

In the case of the heat-transfer test we make a series of measurements of the characteristics of a device under certain specified operating conditions— no comparison with other devices is made. For the smoke test with mice it is necessary to measure the performance of the mice under specified conditions and *also* to compare this performance with the performance of another group under different controlled conditions. For the heat-transfer test we establish an absolute measurement of performance, but for the mice a *relative performance* is all that can be ascertained. We have chosen two diverse examples of absolute and relative experiments, but the lesson is clear. Whenever a comparison test is performed to establish relative performance, control must be exerted over more than one experimental setup in order for the comparison to be significant.

It would seem obvious that very careful provisions should be made to record the data and all ideas and observations concerned with the experiment. Yet, many experimenters record data and important sketches on pieces of scratch paper or in such a disorganized manner that they may be lost or thrown away. In some experiments the readout instrument is a recording type so that a record is automatically obtained and there is little chance for loss. For many experiments, however, visual observations must be made and values recorded on an appropriate data sheet. This data sheet should be very carefully planned so that it may subsequently be used, if desired, for data reduction. Frequently, much time may be saved in the reduction process by eliminating unnecessary transferal of data from one sheet to another. If a computer is to be used for data reduction, then the primary data sheet should be so designed that the data may be easily transferred to the input device of the computer. *A bound notebook should be maintained to record sketches and significant observations of an unusual character which may occur during both the planning and the execution stages of the experiment.* The notebook is also used for recording thoughts and observations of a theoretical nature as the experiment progresses. Upon the completion of the experimental program the well-kept notebook forms a clear and sequential record of the experiment planning, observations during the experiment, and, where applicable, correspondence of important observations with theoretical predictions. Every experimenter should get into the habit of keeping a good notebook.

As a summary of our remarks on experimental planning we present the generalized experimental procedure given in Table 2-4. This procedure is, of course, a flexible one, and the reader should consider the importance of each item in relation to the entire experimental program. Notice particularly item 1*a*. The engineer should give careful thought to the *need* for the experiment. Perhaps, after some sober thinking, he will decide that a previously planned experiment is really not necessary at all and that he could get the desired information from an analytical study or from the results of experiments which have already been conducted. ***Do not take this item lightly.*** A great amount of money is wasted by individuals who rush into a program only to discover later that the experiments were unnecessary for their own particular purposes.

THE ROLE OF UNCERTAINTY ANALYSIS IN EXPERIMENT PLANNING

Items 3*b* and *d* in Table 2-4 note the need to perform preliminary analyses of experimental uncertainties in order to effect a proper selection of instruments and to design the apparatus to meet the overall goals of the experiment. These items are worthy of further amplification.

In Chap. 3 we shall see how one goes about estimating uncertainty in an experimental measurement. For now, let us consider how these estimates can aid our experiment planning. It is clear that certain variables we wish to measure are set by the particular experimental objectives, but there may be several choices open in the method we use to measure these variables. An electric-power measurement could be performed, for example, by measuring current and voltage and taking the product of these variables. The power might also be calculated by measuring the voltage drop across a known

Table 2-4   Generalized experimental procedure

1. *a.* Establish the need for the experiment.
   b. Establish the optimum budgetary, manpower, and time requirements, including time sequencing of the project.
   c. Modify scope of the experiment to *actual* budget, manpower, and time schedule which is allowable.

2. Begin detail planning for the experiment, clearly establish objectives of experiment (verify performance of production model, verify theoretical analysis of particular physical phenomenon, etc.). If experiments are similar to those of previous investigators, be sure to make use of experience of the previous workers. *Never* overlook the possibility that the work may have been done before and reported in the literature.

3. Continue planning by performing the following steps:
   a. Establish the primary variables which must be measured (force, strain, flow, pressure, temperature, etc.).
   b. Determine as nearly as possible the accuracy which may be required in the primary measurements and the number of such measurements which will be required for proper data analysis.
   c. Set up data reduction calculations *before* conducting the experiments to be sure that adequate data are being collected to meet the objectives of the experiment.
   d. Analyze the possible errors in the anticipated results *before* the experiments are conducted so that modifications in accuracy requirements on the various measurements may be changed if necessary.

4. Select instrumentation for the various measurements to match the anticipated accuracy requirements. Modify the instrumentation to match budgetary limitations if necessary.

5. Collect a few data points and conduct a preliminary analysis of these data to be sure that the experiment is going as planned.

6. Modify the experimental apparatus and/or procedure in accordance with the findings in item 5.

7. Collect the bulk of experimental data and analyze the results.

8. Organize, discuss, and publish the findings and results of the experiments, being sure to include information pertaining to all items 1 to 7, above.

resistor or possibly through some calorimetric determination of the heat dissipated from a resistor. The choice of the method used can be made on the basis of an uncertainty analysis, which indicates the relative accuracy of each method. A flow measurement might be performed by sensing the pressure drop across an obstruction meter or possibly by counting the number of revolutions of a turbine placed in the flow (see Chap. 7). In the first case the overall uncertainty depends on the accuracy of a measurement of pressure differential and other variables, such as flow area, while in the second case the overall uncertainty depends on the accuracy of counting and a time determination.

The point is that a careful uncertainty analysis during the experiment planning period may enable the investigator to make a better selection of instruments for his program. Briefly, then, an uncertainty analysis enters into the planning phase with the following approximate steps:

1. Once the variables to be measured have been established, several alternate measurement techniques are selected.
2. An uncertainty analysis is performed on each measurement technique, taking into account the estimated accuracies of the instruments that will actually be used.
3. The different measurement techniques are then compared on the basis of cost, availability of instrumentation, ease of data collection, and calculated uncertainty. The technique with the least uncertainty is clearly the most desirable from an experimental-accuracy standpoint, but it may be too expensive. Frequently, however, the investigator will find that cost is not a strong factor and the technique with the smallest uncertainty (within reason) is as easy to perform as some other less accurate method.

Figure 2-11 divides the procedure of Table 2-4 into a graphical pattern of preliminary, intermediate, and final stages of an experimental program. The feedback blocks in these diagrams are very important because they illustrate the need to continuously retrace one's steps and modify the program in accordance with the most current information that is available.

REPORT WRITING

Many books have been written on the subject of report writing, and the author has no intention of competing with these works. The importance of good report writing and data presentation cannot be overemphasized. No matter how good an experiment or how brilliant a discovery, it is worthless unless the information is communicated to other people. Only a few general remarks will be made concerning item 8 in Table 2-4.

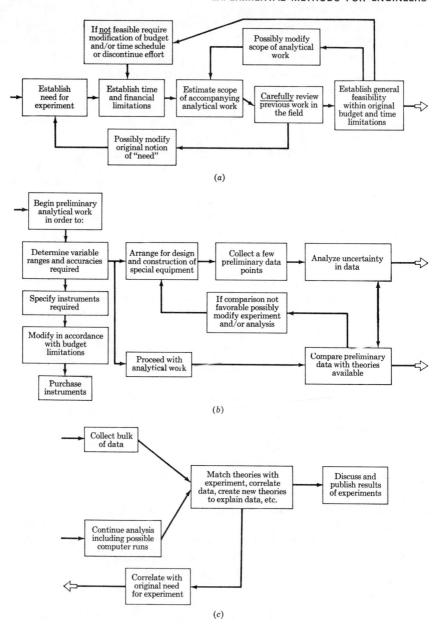

Fig. 2-11 (*a*) Preliminary stages of experiment planning; (*b*) intermediate stages of experiment planning; (*c*) final stages of experimental program.

Very good advice for those engaged in writing or speaking activities is contained in the aphorism:

> A man who uses a great many words to express his meaning is like a bad marksman who instead of aiming a single stone at an object takes up a handful and throws at it in hopes he may hit.
>
> *Samuel Johnson*

In other words, say what you have to say and then shut up! Be succinct but not laconic. When graphs or tables will present the idea clearly, use them, but do not also include a wordy explanation which tells the reader what he could plainly see for himself by careful inspection of the graph.

Third-person past tense is generally accepted as the most formal grammatical style for technical reports, and it is seldom incorrect to use such a style. In some instances, first person may be employed in order to emphasize a point or to stress the fact that a statement is primarily the opinion of the writer. The usual scientific writing style is also in the passive mode. Examples of the two styles are:

Third person Equation (5) is recommended for the final correlation in accordance with the limitations of the data as discussed above.

First person We (I) recommend Eq. (5) for the final correlation in accordance with the limitations of the data presented in our discussion above.

In the first-person statement, the writer is making the recommendation on a much more personal basis than he is in the third-person statement. The selection of the proper statement is a matter of idiom, which depends on many factors, including consideration of the person(s) who will read the report. For a formal paper in a scientific journal, the third-person statement might be preferable, whereas the first-person usage might be desirable in an engineering report to an individual.

Engineers could learn a few lessons from politicians and businessmen. Have you ever noticed the difference in writing style in the annual reports of corporations? If the earnings situation is good, the point is brought forth very quickly and very clearly. Wordiness is avoided in the presentation of good points so that the reader will not miss them. When the profit statement is poor, the discussion becomes more circumlocutious and clear graphical or tabular presentations are seldom employed. The point is that a writer should make sure that the strong points of his report are not buried in excess words. The degree to which weak points are submerged depends upon the audience for which the report is intended and the circumstances under which the report is written.

The above comments should not be misconstrued. The engineer should not bury bad data in a presentation in hopes that no one will notice!

On the other hand, one should recognize that some engineering reports are written for the layman, and it is extremely important for the presentation not to give conflicting or confusing results which may be misinterpreted. Of course, a paper for publication in the technical or scientific literature should be completely objective. The purpose of these brief paragraphs has been to illustrate the fact that an engineer may, in *some circumstances,* find it necessary to do a bit of selling in his writing. Such writing requires a great amount of skill and experience in order to be effective.

Be specific if you have something to be specific about. Consider the following two statements:

1. An analysis of the experimental data showed that the average deviation from the theoretical values was less than 1 percent. Uncertainties in the primary data were shown to account for a deviation of 0.5 percent. In view of this excellent agreement between Eq. (42) and the experimental data, this relation is believed to be an adequate representation of the physical phenomena and is recommended for calculation purposes.
2. The experimental data are in good agreement with the theoretical development. In view of this favorable comparison the assumptions pertaining to the derivation are verified.

Note the difference between the two statements. The first statement is quite specific and leaves the reader with a feeling of confidence in the experimental data and theoretical analysis. Upon examining the second statement the reader will immediately ask: How good is "good"; how favorable is "favorable"? The author has seen such statements applied to experimental data that differed from theoretical values by such a large factor that the veracity of the writer might be questioned by a careful reader.

A brief, general procedure for report writing might take the following form:

1. Make a written outline of the report with as much detail as possible.
2. Let the outline "cool" for a period of time while you direct your thoughts to other matters.
3. Go back to the outline and make whatever changes you feel are necessary.
4. Write the report in rough draft form as quickly as possible.
5. Let the report cool for a period of time, preferably a week or so.
6. Go back and make corrections on your draft. You will probably find that you did not say things quite the way you would like to, did not include as much information as you wanted to, or made several stupid mistakes in some of the data analysis.

7. If possible, have a colleague scrutinize the report carefully. This person should be one whose competence you respect.

8. Consider your colleague's comments very carefully, and rewrite the report in its final form.

REVIEW QUESTIONS

1. What is meant by sensitivity; accuracy; precision?

2. Why is instrument calibration necessary?

3. Why are standards necessary?

4. What is meant by frequency response?

5. Describe the meaning of phase shift.

6. Define time constant.

7. What kind of impedance matching is desired for (a) maximum power transmission and (b) minimum influence on the output of the system?

8. Why is a literature survey important in the preliminary stages of experiment planning?

9. Why is an uncertainty analysis important in the preliminary stages of experiment planning?

10. How can an uncertainty analysis help to reduce overall experimental uncertainty?

PROBLEMS

2-1. Consider an ordinary mercury-in-glass thermometer as a measurement system, and indicate which parts of the thermometer correspond to the boxes in the diagram of Fig. 2-2.

2-2. A thermometer is used for the range of 200 to 400°F, and it is stated that its accuracy is one-quarter of 1 percent. What does this mean in terms of temperature?

2-3. A sinusoidal forcing function is impressed on the system in Fig. 2-4. The natural frequency is 100 Hz, and the damping ratio c/c_c is 0.7. Calculate the amplitude ratio and time lag of the system for an input frequency of 40 Hz. (The time lag is the time interval between the maximum force input and maximum displacement output.)

2-4. For a natural frequency of 100 Hz and a damping ratio of 0.7, compute the input-frequency range for which the system in Fig. 2-4 will have an amplitude ratio of 1.00 ± 0.01.

2-5. A thermometer is initially at a temperature of 70°F and is suddenly placed in a liquid which is maintained at 300°F. The thermometer indicates 200 and 270°F after time intervals of 3 and 5 sec, respectively. Estimate the time constant for the thermometer.

2-6. Plot the power output of the circuit in Fig. 2-10 as a function of R/R_i. Assume R_i and E as constants and show the plot in dimensionless form; i.e., use PR_i/E^2 as the ordinate for the curve.

2-7. A steel scale is graduated in increments of $\frac{1}{32}$ in. What is the readability and least count of such a scale?

2-8. A 10-μf capacitor is charged to a potential of 100 volts. At time zero it is discharged through a 1-megohm resistor. What is the time constant for this system?

2-9. The two-terminal device shown in Fig. 2-10 has an internal resistance of 5,000 ohms. A meter with an impedance of 20,000 ohms is connected to the output to perform a voltage measurement. What is the percent error in determination of the internal voltage?

2-10. The device in Prob. 2-9 has an internal voltage of 100 volts. Calculate the power output for the loading conditions indicated. What would be the maximum power output? What power output would result for a load resistance of 1,000 ohms?

2-11. A 2-g mass is suspended from a simple spring. The deflection caused by this mass is 0.5 cm. What is the natural frequency of the system?

2-12. A 10-lb_m turntable is placed on rubber supports such that a deflection of 0.25 in. results from the weight. Calculate the natural frequency of such a system.

REFERENCES

1. Silsbee, F. B.: Extension and Dissemination of the Electrical and Magnetic Units by the National Bureau of Standards, *Natl. Bur. Std. (U.S.), Circ.* 531, 1952.
2. Silsbee, F. B.: Fundamental Units and Standards, *Instr. Control Systems*, vol. 26, p. 1,520, October, 1953.
3. Eshbach, O. N.: "Handbook of Engineering Fundamentals," 2d ed., John Wiley & Sons, Inc., New York, 1952.
4. Obert, E. F.: "Concepts of Thermodynamics," McGraw-Hill Book Company, New York, 1960.
5. Constant, F. W.: "Theoretical Physics," vol. 2, Addison-Wesley Publishing Company, Inc., Reading, Mass., 1958.
6. Stimson, H. F.: The International Temperature Scale of 1948, *J. Res. Natl. Bur. Std.* (paper 1962), vol. 42, p. 211, March, 1949.
7. *NBS Tech. News Bull.*, vol. 52, no. 1, p. 10, January, 1968.
8. Beckwith, T. G., and N. L. Buck: "Mechanical Measurements," 2d ed., Addison-Wesley Publishing Company, Inc., Reading, Mass., 1969.
9. Schenck, H., Jr.: "An Introduction to the Engineering Research Project," McGraw-Hill Book Company, New York, 1969.
10. Doebelin, E. O.: "Measurement Systems: Analysis and Design," McGraw-Hill Book Company, New York, 1966.
11. Time Standards, *Instr. Control Systems*, p. 87, October 1965.

3
Analysis of Experimental Data

3-1 INTRODUCTION

Some form of analysis must be performed on all experimental data. The analysis may be a simple verbal appraisal of the test results, or it may take the form of a complex theoretical analysis of the errors involved in the experiment and matching of the data with fundamental physical principles. Even new principles may be developed in order to explain some unusual phenomenon. Our discussion in this chapter will consider the analysis of data to determine errors, precision, and general validity of experimental measurements. The correspondence of the measurements with physical principles is another matter, quite beyond the scope of our discussion. Some methods of graphical data presentation will also be discussed. The interested reader should consult the monograph by Wilson [4] for many interesting observations concerning correspondence of physical theory and experiment.

The experimentalist should always know the validity of his data. The automobile test engineer must know the accuracy of his speedometer and gas gage if he is to express the miles-per-gallon performance with confidence. A nuclear engineer must know the accuracy and precision of many instruments just to make some simple radioactivity measurements with confidence. In order for an electrical engineer to specify the performance of an amplifier, he must know the accuracy with which he has conducted the appropriate measurements of voltage, distortion, etc. Many considerations enter into a final determination of the validity of the results of experimental data, and we wish to present some of these considerations in this chapter.

Errors will creep into all experiments regardless of the care which is exerted. Some of these errors are of a random nature, and some will be due to gross blunders on the part of the experimenter. Bad data due to obvious blunders may be discarded immediately. But what of the data points that just "look" bad? We cannot throw out data because they do not conform with our hopes and expectations unless we see something obviously wrong. If such "bad" points fall outside the range of normally expected random deviations, they may be discarded on the basis of some consistent statistical data analysis. The key word here is *consistent*. The elimination of data points must be consistent and should not be dependent on human whims and bias based on what "ought to be." In many instances it is very difficult for the individual to be consistent and unbiased. The pressure of a deadline, disgust with previous experimental failures, and normal impatience all can influence his rational thinking processes. However, the competent experimentalist will strive to maintain consistency in the primary data analysis. Our objective in this chapter is to show how one may go about maintaining this consistency.

3-2 CAUSES AND TYPES OF EXPERIMENTAL ERRORS

In this section we present a discussion of some of the types of errors that may be present in experimental data and begin to indicate the way these data may be handled. First, let us distinguish between single-sample and multisample data.

Single-sample data are those in which some uncertainties may not be discovered by repetition. Multisample data are obtained in those instances where enough experiments are performed so that the reliability of the results can be assured by statistics. Frequently, cost will prohibit the collection of multisample data, and the experimenter must be content with single-sample data and prepared to extract as much information as possible from such experiments. The reader should consult Refs. [1] and [4] for further discussions on this subject, but we state a simple example at this time. If one measures pressure with a pressure gage and a single instrument is the only one used for the entire set of observations, then some of the error that is present in the measurement will be sampled only once no matter how many times the reading is repeated. Consequently, such an experiment is a single-sample experiment.   On the other hand, if more than one pressure gage is used for the same total set of observations, then we might say that a multisample experiment has been performed.   The *number* of observations will then determine the success of this multisample experiment in accordance with accepted statistical principles.

An experimental error is an experimental error.   If the experimenter knew what the error was, he would correct it and it would no longer be an error.   In other words, the real errors in experimental data are those factors that are always vague to some extent and carry some amount of uncertainty. Our task is to determine just how uncertain a particular observation may be and to devise a consistent way of specifying the uncertainty in analytical form. A reasonable definition of experimental uncertainty may be taken as the *possible value* the error may have.   This uncertainty may vary a great deal depending upon the circumstances of the experiment.   Perhaps it is better to speak of experimental uncertainty instead of experimental error because the magnitude of an error is always uncertain.   Both terms are used in practice, however, so the reader should be familiar with the meaning attached to the terms and the ways that they relate to each other.

At this point we may mention some of the types of errors that may cause uncertainty in an experimental measurement.   First, there can always be those gross blunders in apparatus or instrument construction, which may invalidate the data.   Hopefully, the careful experimenter will be able to eliminate most of these errors.   Second, there may be certain *fixed errors*, which will cause repeated readings to be in error by roughly the same amount but for some unknown reason.   These fixed errors are sometimes called *systematic errors*.   Third, there are the *random errors*, which may be caused by personal fluctuations, random electronic fluctuations in the apparatus or instruments, various influences of friction, etc.   These random errors usually follow a certain statistical distribution, *but not always*.   In many instances it is very difficult to distinguish between fixed errors and random errors.

The experimentalist may sometimes use theoretical methods to estimate the magnitude of a fixed error.   For example, consider the measurement of the temperature of a hot gas stream flowing in a duct with a mercury-in-glass thermometer.   It is well known that heat may be conducted from the stem of the thermometer, out of the glass body, and into the surroundings. In other words, the fact that part of the thermometer is exposed to the surroundings at a temperature different from the gas temperature to be measured may influence the temperature of the stem of the thermometer. There is a heat flow from the gas to the stem of the thermometer, and, consequently, the temperature of the stem must be lower than that of the hot gas.   Therefore, the temperature we read on the thermometer is not the true temperature of the gas, and it will not make any difference how many readings are taken—we shall always have an error resulting from the heat-transfer condition of the stem of the thermometer.   This is a fixed error, and its magnitude may be estimated with theoretical calculations based upon known thermal properties of the gas and the glass thermometer.

3-3 ERROR ANALYSIS ON A COMMONSENSE BASIS

We have already noted that it is somewhat more explicit to speak of experimental uncertainty rather than experimental error.   Suppose that we have satisfied ourselves with the uncertainty in some basic experimental measurements, taking into consideration such factors as instrument accuracy, competence of the people using the instruments, etc.   Eventually, the primary measurements must be combined to calculate a particular result that is desired.   We shall be interested in knowing the uncertainty in the final result due to the uncertainties in the primary measurements.   This may be done by a commonsense analysis of the data, which may take many forms.   One rule of thumb that could be used is that the error in the result is equal to the maximum error in any parameter used to calculate the result.   Another commonsense analysis would combine all the errors in the most detrimental way in order to determine the maximum error in the final result.   Consider the calculation of electric power from

$$P = EI$$

where E and I are measured as

$$E = 100 \text{ volts} \pm 2 \text{ volts}$$
$$I = 10 \text{ amp} \pm 0.2 \text{ amp}$$

The nominal value of the power is $100 \times 10 = 1,000$ watts.   By taking the worst possible variations in voltage and current, we could calculate

$$P_{max} = (100 + 2)(10 + 0.2) = 1040.4 \text{ watts}$$
$$P_{min} = (100 - 2)(10 - 0.2) = 960.4 \text{ watts}$$

Thus, using this method of calculation, the uncertainty in the power is $+4.04$ percent, -3.96 percent.   It is quite unlikely that the power would be in error by these amounts because the voltmeter variations would probably not correspond with the ammeter variations.   When the voltmeter reads an extreme "high," there is no reason why the ammeter must also read an extreme "high" at that particular instant; indeed, this combination is most unlikely.

The simple calculation applied to the electric-power equation above is a useful way of inspecting experimental data to determine what errors *could* result in a final calculation; however, the test is too severe and should be used only for rough inspections of data.   It is significant to note, however, that if the results of the experiments appear to be in error by *more* than the amounts indicated by the above calculation, then the experimenter had better examine his data more closely. In particular, he should look for certain fixed errors in the instrumentation, which may be eliminated by applying either theoretical or empirical corrections.

The word "commonsense" has many connotations and means different things to different people. In the brief example given above, it is intended as a quick and expedient vehicle, which may be used to examine experimental data and results for gross errors and variations. In subsequent sections we shall present methods for determining experimental uncertainties in a more precise manner.

3-4 UNCERTAINTY ANALYSIS

A more precise method of estimating uncertainty in experimental results has been presented by Kline and McClintock [1]. The method is based on a careful specification of the uncertainties in the various primary experimental measurements. For example, a certain pressure reading might be expressed as

$$p = 100 \text{ psia} \pm 1 \text{ psia}$$

When the plus,or minus notation is used to designate the uncertainty, the person making this designation is stating in very precise terms the degree of accuracy with which he believes the measurement has been made. We may note that this specification is in itself uncertain because the experimenter is naturally uncertain about the accuracy of his measurements. To add a further specification of the uncertainty of a particular measurement, Kline and McClintock propose that the experimenter specify certain odds for the uncertainty. The above equation for pressure might thus be written

$$p = 100 \text{ psia} \pm 1 \text{ psia} \text{ (20 to 1)}$$

In other words, the experimenter is willing to bet with 20 to 1 odds that his pressure measurement is within ± 1 psia. It is important to note that the specification of such odds can *only* be made by the experimenter himself based on his total laboratory experience.

Suppose a set of measurements is made and the uncertainty in each measurement may be expressed with the same odds. These measurements are then used to calculate some desired result of the experiments. We wish to estimate the uncertainty in the calculated result on the basis of the uncertainties in the primary measurements. The result R is a given function of the independent variables $x_1, x_2, x_3, \ldots, x_n$. Thus,

$$R = R(x_1, x_2, x_3, \ldots, x_n) \tag{3-1}$$

Let w_R be the uncertainty in the result and $w_1, w_2, \ldots, w_n$ be the uncertainties in the independent variables.   If the uncertainties in the independent variables are all given with the same odds, then the uncertainty in the result having these odds is given in Ref. [1] as

$$w_R = \left[\left(\frac{\partial R}{\partial x_1} w_1 \right)^2 + \left(\frac{\partial R}{\partial x_2} w_2 \right)^2 + \cdots + \left(\frac{\partial R}{\partial x_n} w_n \right)^2 \right]^{\frac{1}{2}} \qquad (3\text{-}2)$$

If this relation is applied to the electric-power relation of the previous section, the expected uncertainty is 2.83 percent instead of 4.04 percent.

Example 3-1   The resistance of a certain size of copper wire is given as

$$R = R_0[1 + \alpha(T - 20)]$$

where $R_0 = 6$ ohms $\pm$ 0.3 percent is the resistance at 20°C, $\alpha = 0.004°C^{-1} \pm 1$ percent is the temperature coefficient of resistance, and the temperature of the wire is $T = 30 \pm 1°C$.   Calculate the resistance of the wire and its uncertainty.

Solution   The nominal resistance is

$$R = (6)[1 + (0.004)(30 - 20)] = 6.24 \text{ ohms}$$

The uncertainty in this value is calculated by applying Eq. (3-2).   The various terms are:

$$\frac{\partial R}{\partial R_0} = 1 + \alpha(T - 20) = 1 + (0.004)(30 - 20) = 1.04$$

$$\frac{\partial R}{\partial \alpha} = R_0(T - 20) = (6)(30 - 20) = 60$$

$$\frac{\partial R}{\partial T} = R_0\alpha = (6)(0.004) = 0.024$$

$$w_{R_0} = (6)(0.003) = 0.018 \text{ ohm}$$
$$w_\alpha = (0.004)(0.01) = 4 \times 10^{-5} \text{ °C}^{-1}$$
$$w_T = 1°C$$

Thus, the uncertainty in the resistance is

$$w_R = [(1.04)^2(0.018)^2 + (60)^2(4 \times 10^{-5})^2 + (0.024)^2(1)^2]^{\frac{1}{2}}$$
$$= 0.0305 \text{ ohm or } 0.49\%$$

Particular notice should be given to the fact that the uncertainty propagation in the result w_R predicted by Eq. (3-2) depends on the squares of the uncertainties in the independent variables w_n.   This means that if the uncertainty in one variable is significantly larger than the uncertainties in the other variables, say, by a factor of 5 or 10, then it is the largest uncertainty that predominates and the others may probably be neglected.   To illustrate, suppose there are three variables with an uncertainty of magnitude 1 and one variable with an uncertainty of magnitude 5.   The uncertainty in the result would be

$$(5^2 + 1^2 + 1^2 + 1^2)^{\frac{1}{2}} = \sqrt{28} = 5.29$$

The importance of this brief remark concerning the relative magnitude of uncertainties is evident when one considers the design of an experiment, procurement of instrumentation, etc. Very little is gained by trying to reduce the "small" uncertainties. Because of the square propagation it is the "large" ones that predominate, and any improvement in the overall experimental result must be achieved by improving the instrumentation or experimental technique connected with these relatively large uncertainties. In the examples and problems that follow, both in this chapter and throughout the book, the reader should always note the relative effect of uncertainties in primary measurements on the final result.

In Sec. 2-10 (Table 2-4) the reader was cautioned to examine possible experimental errors *before* the experiment is conducted. Equation (3-2) may be used very effectively for such analyses, as we shall see in the sections and chapters that follow. A further word of caution may be added here. It is equally as unfortunate to overestimate uncertainty as to underestimate it. An underestimate gives false security, while an overestimate may make one discard important results, miss a real effect, or buy much too expensive instruments. The purpose of this chapter is to indicate some of the methods for obtaining reasonable estimates of experimental uncertainty.

In the previous discussion of experiment planning we noted that an uncertainty analysis may aid the investigator in selecting alternate methods to measure a particular experimental variable. It may also indicate how one may improve the overall accuracy of a measurement by attacking certain critical variables in the measurement process. The next three examples illustrate these points.

Example 3-2 A resistor has a nominal stated value of 10 ohms $\pm$ 1 percent. A voltage is impressed on the resistor, and the power dissipation is to be calculated in two different ways: (1) from $P = E^2/R$ and (2) from $P = EI$. In (1) only a voltage measurement will be made, while both current and voltage will be measured in (2). Calculate the uncertainty in the power determination in each case when the measured values of E and I are:

$E = 100$ volts $\pm$ 1%   (for both cases)

$I = 10$ amp $\pm$ 1%

Solution The schematic is shown in the accompanying figure. For the first case we have

Fig. Example 3-2 Power measurement across a resistor.

$$\frac{\partial P}{\partial E} = \frac{2E}{R} \qquad \frac{\partial P}{\partial R} = -\frac{E^2}{R^2}$$

and we apply Eq. (3-2) to give

$$w_P = \left[\left(\frac{2E}{R}\right)^2 w_E{}^2 + \left(-\frac{E^2}{R^2}\right)^2 w_R{}^2 \right]^{\frac{1}{2}} \tag{a}$$

Dividing by $P = E^2/R$ gives

$$\frac{w_P}{P} = \left[4\left(\frac{w_E}{E}\right)^2 + \left(\frac{w_R}{R}\right)^2 \right]^{\frac{1}{2}} \tag{b}$$

Inserting the numerical values for uncertainty,

$$\frac{w_P}{P} = [4(0.01)^2 + (0.01)^2]^{\frac{1}{2}} = 2.236\%$$

For the second case we have

$$\frac{\partial P}{\partial E} = I \qquad \frac{\partial P}{\partial I} = E$$

and after similar algebraic manipulation, we obtain

$$\frac{w_P}{P} = \left[\left(\frac{w_E}{E}\right)^2 + \left(\frac{w_I}{I}\right)^2 \right]^{\frac{1}{2}} \tag{c}$$

Inserting the numerical values of uncertainty,

$$\frac{w_P}{P} = [(0.01)^2 + (0.01)^2]^{\frac{1}{2}} = 1.414\%$$

Thus, the second method of power determination provides considerably less uncertainty than the first method, even though the primary uncertainties in each quantity are the same.   In this example the utility of the uncertainty analysis is that it affords the individual a basis for *selection* of a measurement method to produce a result with less uncertainty.

Example 3-3   The power measurement in Example 3-2 is to be conducted by measuring voltage and current across the resistor with the circuit shown in the accompanying figure.   The voltmeter has an internal resistance R_m, and the value of R is known only approximately.   Calculate the nominal value of the power dissipated in R and the uncertainty for the following conditions:

$R = 100$ ohms   (not known exactly)

$R_m = 1,000$ ohms, $\pm 5\%$

$I = 5$ amp $\pm 1\%$

$E = 500$ volts $\pm 1\%$

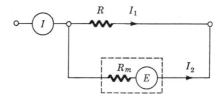

Fig. Example 3-3 Effect of meter impedance on measurement.

Solution A current balance on the circuit yields

$$I_1 + I_2 = I$$

$$\frac{E}{R} + \frac{E}{R_m} = I$$

and

$$I_1 = I - \frac{E}{R_m} \qquad\qquad (a)$$

The power dissipated in the resistor is

$$P = EI_1 = EI - \frac{E^2}{R_m} \qquad\qquad (b)$$

The nominal value of the power is thus calculated as

$$P = (500)(5) - \frac{500^2}{1,000} = 2,250 \text{ watts}$$

In terms of known quantities the power has the functional form $P = f(E, I, R_m)$, and so we form the derivatives

$$\frac{\partial P}{\partial E} = I - \frac{2E}{R_m} \qquad \frac{\partial P}{\partial I} = E$$

$$\frac{\partial P}{\partial R_m} = \frac{E^2}{R_m{}^2}$$

The uncertainty for the power is now written as

$$w_P = \left[\left(I - \frac{2E}{R_m} \right)^2 w_E{}^2 + E^2 w_I{}^2 + \left(\frac{E^2}{R_m{}^2} \right)^2 w_{R_m}{}^2 \right]^{\frac{1}{2}} \qquad\qquad (c)$$

Inserting the appropriate numerical values gives

$$w_P = \left[\left(5 - \frac{1,000}{1,000} \right)^2 5^2 + (25 \times 10^4)(25 \times 10^{-4}) + \left(25 \times \frac{10^4}{10^6} \right)^2 (2,500) \right]^{\frac{1}{2}}$$

$$= [16 + 25 + 6.25]^{\frac{1}{2}}(5)$$

$$= 34.4 \text{ watts}$$

or

$$\frac{w_P}{P} = \frac{34.4}{2,250} = 1.53\%$$

In order of influence on the final uncertainty in the power we have

1. Uncertainty of current determination
2. Uncertainty of voltage measurement
3. Uncertainty of knowledge of internal resistance of voltmeter

There are other conclusions we can draw from this example. The relative influence of the experimental quantities on the overall power determination are noted above. But this listing may be a bit misleading in that it implies that the uncertainty of the meter impedance does not have a large effect on the final uncertainty in the power determination. This results from the fact that $R_m \gg R$ ($R_m = 10R$). If the meter impedance were lower, say, 200 ohms, we would find that it was a dominant factor in the overall uncertainty. For a very *high* meter impedance there would be little influence, even with a very inaccurate knowledge of the exact value of R_m. Thus, we are led to the simple conclusion that we need not worry too much about the precise value of the internal impedance of the meter as long as it is very large compared with the resistance we are measuring the voltage across. This fact should influence instrument *selection* for a particular application.

Example 3-4   A certain obstruction-type flowmeter (orifice, venturi, nozzle), shown in the accompanying figure, is used to measure the flow of air at low velocities. The relation describing the flow rate is

$$\dot{m} = CA\left[\frac{2g_c p_1}{RT_1}(p_1 - p_2)\right]^{\frac{1}{2}} \tag{a}$$

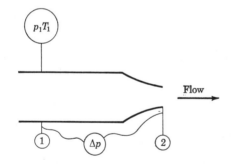

Fig. Example 3-4   Uncertainty in a flowmeter.

where C is an empirical-discharge coefficient, A is the flow area, p_1 and p_2 are the upstream and downstream pressures, T_1 is the upstream temperature, and R is the gas constant for air. Calculate the percent uncertainty in the mass flow rate for the following conditions:

$C = 0.92 \pm 0.005$ (from calibration data)

$p_1 = 25$ psia ± 0.5 psia

$T_1 = 70°F \pm 2°F$ $T_1 = 530°R$

$\Delta p = p_1 - p_2 = 1.4$ psia ± 0.005 psia   (measured directly)

$A = 1.0$ in.^{2} ± 0.001 in.^{2}

Solution   In this example the flow rate is a function of several variables, each subject to an uncertainty.

$$\dot{m} = f(C, A, p_1, \Delta p, T_1) \tag{b}$$

Thus, we form the derivatives:

$$\frac{\partial \dot{m}}{\partial C} = A\left(\frac{2g_c p_1}{RT_1} \Delta p\right)^{\frac{1}{2}}$$

$$\frac{\partial \dot{m}}{\partial A} = C\left(\frac{2g_c p_1}{RT_1} \Delta p\right)^{\frac{1}{2}}$$

$$\frac{\partial \dot{m}}{\partial p_1} = 0.5CA\left(\frac{2g_c}{RT_1} \Delta p\right)^{\frac{1}{2}} p_1^{-\frac{1}{2}} \tag{c}$$

$$\frac{\partial \dot{m}}{\partial \Delta p} = 0.5CA\left(\frac{2g_c p_1}{RT_1}\right)^{\frac{1}{2}} \Delta p^{-\frac{1}{2}}$$

$$\frac{\partial \dot{m}}{\partial T_1} = -0.5CA\left(\frac{2g_c p_1}{R} \Delta p\right)^{\frac{1}{2}} T_1^{-\frac{3}{2}}$$

The uncertainty in the mass flow rate may now be calculated by assembling these derivatives in accordance with Eq. (3-2). Designating this assembly as Eq. (c) and then dividing by Eq. (a) gives

$$\frac{w_{\dot{m}}}{\dot{m}} = \left[\left(\frac{w_C}{C}\right)^2 + \left(\frac{w_A}{A}\right)^2 + \frac{1}{4}\left(\frac{w_{p_1}}{p_1}\right)^2 + \frac{1}{4}\left(\frac{w_{\Delta p}}{\Delta p}\right)^2 + \frac{1}{4}\left(\frac{w_{T_1}}{T_1}\right)^2\right]^{\frac{1}{2}} \tag{d}$$

We may now insert the numerical values for the quantities to obtain the percent uncertainty in the mass flow rate.

$$\begin{aligned}
\frac{w_{\dot{m}}}{\dot{m}} &= \left[\left(\frac{0.005}{0.92}\right)^2 + \left(\frac{0.001}{1.0}\right)^2 + \frac{1}{4}\left(\frac{0.5}{25}\right)^2 + \frac{1}{4}\left(\frac{0.005}{1.4}\right)^2 + \frac{1}{4}\left(\frac{2}{530}\right)^2\right]^{\frac{1}{2}} \\
&= [29.5 \times 10^{-6} + 1.0 \times 10^{-6} + 1.0 \times 10^{-4} \\
&\quad + 3.19 \times 10^{-6} + 3.57 \times 10^{-6}]^{\frac{1}{2}} \\
&= [1.373 \times 10^{-4}]^{\frac{1}{2}} = 1.172\%
\end{aligned} \tag{e}$$

The main contribution to uncertainty is the p_1 measurement with its basic uncertainty of 2 percent. Thus, to improve the overall situation the accuracy of this measurement should be attacked first.

In order of influence on the flow-rate uncertainty, we have

1. Uncertainty in p_1 measurement (± 2 percent)
2. Uncertainty in value of C
3. Uncertainty in determination of T_1
4. Uncertainty in determination of Δp
5. Uncertainty in determination of A

By inspecting Eq. (e) we see that the first two items make practically the whole contribution to uncertainty. The value of the uncertainty analysis in this example is that it shows the investigator how to improve the overall measurement accuracy of this technique.   First, he should obtain a more precise measurement of p_1.   Then he should try to obtain a better calibration of the device, i.e., a better value of C.   In Chap. 7 we shall see how values of the discharge coefficient C are obtained.

3-5   STATISTICAL ANALYSIS OF EXPERIMENTAL DATA

We shall not be able to give an extensive presentation of the methods of statistical analysis of experimental data; we may only indicate some of the more important methods currently employed.   First, it is important to define some pertinent terms.

When a set of readings of an instrument is taken, the individual readings will vary somewhat from each other and the experimenter is usually concerned with the *mean* of all the readings.   If each reading is denoted by x_i and there are n readings, the *arithmetic mean* is given by

$$x_m = \frac{1}{n} \sum_{i=1}^{n} x_i \tag{3-3}$$

The *deviation* d_i for each reading is defined by

$$d_i = x_i - x_m \tag{3-4}$$

We may note that the average of the deviations of all the readings is zero since

$$\bar{d_i} = \frac{1}{n} \sum_{i=1}^{n} d_i = \frac{1}{n} \sum_{i=1}^{n} (x_i - x_m) \tag{3-5}$$

$$= x_m - \frac{1}{n}(nx_m) = 0$$

The average of the absolute values of the deviations is given by

$$|\bar{d_i}| = \frac{1}{n} \sum_{i=1}^{n} |d_i| = \frac{1}{n} \sum_{i=1}^{n} |x_i - x_m| \tag{3-6}$$

Note that this quantity is not necessarily zero.

The *standard deviation* or *root-mean-square deviation* is defined by

$$\sigma = \left[\frac{1}{n} \sum_{i=1}^{n} (x_i - x_m)^2 \right]^{\frac{1}{2}} \tag{3-7}$$

and the square of the standard deviation σ^2 is called the *variance*. We shall be interested in the determination of the standard deviation because it is important in all aspects of statistical data analysis.

Example 3-5 The following readings are taken of a certain physical length. Compute the mean reading, standard deviation, variance, and average of the absolute value of the deviation.

Reading	x, ft
1	5.30
2	5.73
3	6.77
4	5.26
5	4.33
6	5.45
7	6.09
8	5.64
9	5.81
10	5.75

Solution The mean value is given by

$$x_m = \frac{1}{n}\sum_{i=1}^{n} x_i = \tfrac{1}{10}(56.13) = 5.613 \text{ ft}$$

The other quantities are computed with the aid of the following table:

Reading	$d_i = x_i - x_m$	$(x_i - x_m)^2 \times 10^2$
1	−0.313	9.797
2	0.117	1.369
3	1.157	133.864
4	−0.353	12.461
5	−1.283	164.866
6	−0.163	2.657
7	0.477	22.753
8	0.027	0.0729
9	0.197	3.881
10	0.137	1.877

$$\sigma = \left[\frac{1}{n}\sum_{i=1}^{n}(x_i - x_m)^2\right]^{\frac{1}{2}} = [\tfrac{1}{10}(3.536)]^{\frac{1}{2}} = 0.595 \text{ ft}$$

$$\sigma^2 = 0.3536 \text{ ft}^2$$

$$|\bar{d_i}| = \frac{1}{n}\sum_{i=1}^{n}|d_i| = \frac{1}{n}\sum_{i=1}^{n}|x_i - x_m|$$

$$= \tfrac{1}{10}(4.224) = 0.4224 \text{ ft}$$

Suppose an "honest" coin is flipped a large number of times. It will be noted that after a large number of tosses heads will be observed about the same number of times as tails. If one were to consistently bet on either heads or tails, the best he could hope for would be a break-even proposition over a long period of time. In other words, the *frequency of occurrence* of either heads or tails is the same for a very large number of tosses. It is common knowledge that a few tosses of a coin, say, 5 or 10, may not be a break-even proposition, as in the case of a large number of tosses. This observation illustrates the fact that frequency of occurrence of an event may be dependent on the total number of events which are observed.

The probability that one will get a head when flipping an unweighted coin is $\frac{1}{2}$, regardless of the number of times the coin is tossed. The probability that a tail will occur is also $\frac{1}{2}$. The probability that either a head or a tail will occur is $\frac{1}{2} + \frac{1}{2}$ or unity. (We ignore the possibility that the coin will stand on edge.) *Probability* is a mathematical quantity that is linked to the frequency with which a certain phenomenon occurs after a large number of tries. In the case of the coin, it is the number of times heads would be expected to result in a large number of tosses divided by the total number of tosses. In a similar manner, the toss of an unloaded die would be expected to result in the occurrence of any one side one-sixth of the time in a large number of throws.

Probabilities are expressed in numerical values less than 1, and a probability of unity corresponds to certainty. In other words, if the probabilities for all possible events are added, the result must be unity. If we know the probability that separate events will occur, the probability that one of the events will occur is the sum of the individual probabilities for the events. For the tossing of a die, the probability that any one side will occur is $\frac{1}{6}$. The probability of getting *either* of two given numbers in a single throw of the die is $\frac{1}{6} + \frac{1}{6}$, or $\frac{1}{3}$. The probability for one of three given numbers is $\frac{1}{6} + \frac{1}{6} + \frac{1}{6}$, or $\frac{1}{2}$, and so on. As another example, the probability of drawing the ace of spades from a deck of cards is $\frac{1}{52}$, but the probability of drawing *any* ace is $(4)(\frac{1}{52})$, or $\frac{1}{13}$.

Suppose *two* dice are thrown and we wish to know the probability that both will display a 6. The probability for a 6 on a single die is $\frac{1}{6}$. By a short calculation or listing of the possible arrangements that the dice may have, it can be seen that there are 36 possibilities and that the desired result of two 6s represents only one of these possibilities. Thus, the probability is $\frac{1}{36}$. The reader who has a gambling heart will naturally want to know the probability of getting a 7 or 11 on the throw of the dice. There are 36 possible ways that the dice may be arranged and 6 possible ways of getting a 7; thus the probability of getting a 7 is $\frac{6}{36}$, or $\frac{1}{6}$. There are only 2 ways of getting an 11; thus the probability is $\frac{2}{36}$, or $\frac{1}{18}$. The probability of getting either a 7 or an 11 is $\frac{6}{36} + \frac{2}{36}$, or $\frac{1}{9}$.

If several independent events occur at the same time such that each event has a probability p_i, the probability that all events will occur is given as the product of the probabilities of the individual events. Thus, $p = \prod p_i$ where the $\prod$ sign designates a product. This rule could be applied to the problem of determining the probability of a double 6 in the throw of two dice. The probability of getting a 6 on each die is $\frac{1}{6}$, and the total probability is therefore $(\frac{1}{6})(\frac{1}{6})$, or $\frac{1}{36}$. This reasoning could not be applied to the problem of obtaining a 7 on the two dice because the number on each die is not independent of the number on the other die since a 7 can be obtained in more than one way.

As a final example we ask what the chances are of getting a royal flush in the first five cards drawn off the top of a deck. There are 20 suitable possibilities for the first draw (4 suits, 5 possible cards per suit) out of a total of 52 cards. On the second draw we have fixed the suit so that there are only 4 suitable cards out of the 51 remaining. There are three suitable cards on the third draw, two on the fourth, and only one on the fifth draw. The total probability of drawing the royal flush is thus the product of the probabilities of each draw, or

$$\frac{20}{52} \times \frac{4}{51} \times \frac{3}{50} \times \frac{2}{49} \times \frac{1}{48} = \frac{1}{649,740}$$

In the above discussion we have seen that the probability is related to the number of ways a certain event may occur. In a sense we are assuming that all events are equally likely, and hence the probability that an event will occur is the number of ways the event may occur divided by the number of possible events. For our purposes it is sufficient that we recognize the link between probability and the mathematics dealing with permutations and combinations without pursuing the matter further. Our primary concern is the application of probability and statistics to the analysis of experimental data. For this purpose we need to discuss next the meaning and use of *probability distributions*. We shall be concerned with a few particular distributions that are directly applicable to experimental data analysis.

3-6 PROBABILITY DISTRIBUTIONS

Suppose we toss a horseshoe some distance x. Even though we might be playing a game such that every effort was made to toss the horseshoe the same distance each time, we would not always meet with success. On the first toss the horseshoe might travel a distance x_1, on the second toss a distance of x_2, and so on. If one is a good player of the game, there would be more tosses which have an x distance equal to that of the objective. Also, we would expect fewer and fewer tosses for those x distances which are further and further away from the target. Suppose the horseshoe is tossed a large number of times. We might compute the probability that it will travel a distance x by dividing the number traveling this distance by the total number of tosses. Since each x distance will vary somewhat from other x distances, we might find it advantageous to calculate the probability of a toss landing in a certain increment of x between x and $x + \Delta x$. When this calculation is made, we might get something like the situation shown in Fig. 3-1. For a good player, the maximum probability is expected to surround the distance x_m designating the position of the target.

The curve shown in Fig. 3-1 is called a *probability distribution*. It shows how the probability of success in a certain event is distributed over the distance x. Each value of the ordinate $p(x)$ gives the probability that the horseshoe will land between x and $x + \Delta x$, where Δx is allowed to approach zero. We might consider the deviation from x_m as the error in the throw. If the horseshoes player has good aim, large errors are less likely than small errors. The area under the curve is unity since it is certain that the horseshoe will land somewhere.

A particular probability distribution is the *binomial distribution*. This distribution gives the number of successes n out of N possible independent

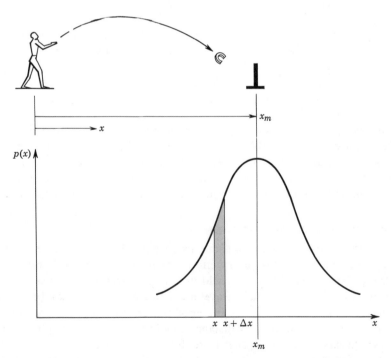

Fig. 3-1 Distribution of throws for a "good" horseshoes player.

events when each event has a probability of success p.   The probability that n events will succeed is given in Ref. [2] as

$$p(n) = \frac{N!}{(N-n)!n!} p^n (1-p)^{N-n} \tag{3-8}$$

It will be noted that the quantity $(1-p)$ is the probability of failure of each independent event.

Example 3-6   An unweighted coin is flipped three times.   Calculate the probability of getting zero, one, two, or three heads in these tosses.

Solution   The binomial distribution applies in this case.   The probability of getting a head on each throw is $p = \frac{1}{2}$ and $N = 3$, while n takes on the values 0, 1, 2, and 3. The probabilities are calculated as

$$p(0) = \frac{3!}{(3!)(0!)} \left(\frac{1}{2}\right)^0 \left(\frac{1}{2}\right)^3 = \frac{1}{8}$$

$$p(1) = \frac{3!}{(2!)(1!)} \left(\frac{1}{2}\right)^1 \left(\frac{1}{2}\right)^2 = \frac{3}{8}$$

$$p(2) = \frac{3!}{(1!)(2!)} \left(\frac{1}{2}\right)^2 \left(\frac{1}{2}\right)^1 = \frac{3}{8}$$

$$p(3) = \frac{3!}{(0!)(3!)} \left(\frac{1}{2}\right)^3 \left(\frac{1}{2}\right)^0 = \frac{1}{8}$$

Now suppose that the number of possible independent events N is very large and the probability of occurrence of each p is very small.   The calculation of the probability of n successes out of the N possible events using Eq. (3-8) would be most cumbersome because of the size of the numbers. The limit of the binomial distribution as $N \to \infty$ and $p \to 0$, such that

$$Np = a = \text{const}$$

is called the *Poisson distribution* and is given by

$$p_a(n) = \frac{a^n e^{-a}}{n!} \tag{3-9}$$

The Poisson distribution is applicable to the calculation of the decay of radioactive nuclei, as we shall see in a subsequent chapter.   It may be shown that the standard deviation of the Poisson distribution is

$$\sigma = \sqrt{a} \tag{3-10}$$

3-7   THE GAUSSIAN OR NORMAL ERROR DISTRIBUTION

Suppose an experimental observation is made and some particular result recorded. We know (or would strongly suspect) that the observation has been subjected to many random errors. These random errors may make the final reading either too large or too small, depending on many circumstances which are unknown to us. Assuming that there are many small errors that contribute to the final error and that each small error is of equal magnitude and equally likely to be positive or negative, the *gaussian* or *normal error distribution* may be derived. If the measurement is designated by x, the gaussian distribution gives the probability that the measurement will lie between x and $x + dx$ and is written

$$P(x) = \frac{1}{\sigma\sqrt{2\pi}}\, e^{-(x-x_m)^2/2\sigma^2} \tag{3-11}$$

In this expression, x_m is the mean reading and σ is the standard deviation. A plot of Eq. (3-11) is given in Fig. 3-2. Note that the most probable reading is x_m. The standard deviation is a measure of the width of the distribution curve; the larger the value of σ, the flatter the curve and hence the larger the expected error of all the measurements. Equation (3-11) is normalized so that the total area under the curve is unity. Thus,

$$\int_{-\infty}^{+\infty} P(x)\, dx = 1.0 \tag{3-12}$$

At this point we may note the similarity between the shape of the normal error curve and the expected experimental distribution for tossing horseshoes as shown in Fig. 3-1. This is what we would expect because the good horseshoes player will have his throws bunched around the target. The better he is at the game, the more closely they will be grouped around the mean and the more probable will be the mean distance x_m. Thus, in the case of the horseshoes player, a smaller standard deviation would mean a larger percentage of "ringers."

We may quickly anticipate the next step in the analysis as one of trying to determine the precision of a set of experimental measurements through an application of the normal error distribution. One may ask: But how do you know that the assumptions pertaining to the derivation of the normal error distribution apply to experimental data? The answer is that for sets of data where a large number of measurements are taken, experiments indicate that the measurements do indeed follow a distribution like that shown in Fig. 3-2 when the experiment is under control. If an important parameter is not controlled, one gets just scatter, i.e., no sensible distribution at all. Thus, as a matter of experimental verification the gaussian distribution is believed to represent the random errors in an adequate manner for a properly controlled experiment.

By inspection of the gaussian distribution function of Eq. (3-11) we see that the maximum probability occurs at $x = x_m$ and the value of this probability is

$$P(x_m) = \frac{1}{\sigma\sqrt{2\pi}} \tag{3-13}$$

It is seen from Eq. (3-13) that smaller values of the standard deviation produce larger values of the maximum probability, as would be expected in an intuitive sense. $P(x_m)$ is sometimes called a *measure of precision* of the data because it has a larger value for smaller values of the standard deviation.

We next wish to examine the gaussian distribution to determine the likelihood that certain data points will fall within a specified deviation from the mean of all the data points. The probability that a measurement will fall within a certain range x_1 of the mean reading is

$$P = \int_{x_m - x_1}^{x_m + x_1} \frac{1}{\sigma\sqrt{2\pi}}\, e^{-(x - x_m)^2/2\sigma^2}\, dx \tag{3-14}$$

Making the variable substitution,

$$\eta = \frac{x - x_m}{\sigma}$$

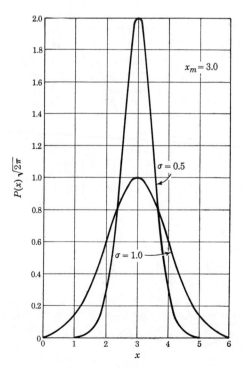

Fig. 3-2 The gaussian or normal error distribution for two values of the standard deviation.

EXPERIMENTAL METHODS FOR ENGINEERS

Table 3-1   Values of the gaussian normal error distribution

Values of the function $(1/\sqrt{2\pi})e^{-\eta^2/2}$ for different values of the argument η. (Each figure in the body of the table is preceded by a decimal point.)

η	0.00	0.01	0.02	0.03	0.04	0.05	0.06	0.07	0.08	0.09
0.0	39894	39892	39886	39876	39862	39844	39822	39797	39767	39733
0.1	39695	39654	39608	39559	39505	39448	39387	39322	39253	39181
0.2	39104	39024	38940	38853	38762	38667	38568	38466	38361	38251
0.3	38139	38023	37903	37780	37654	37524	37391	37255	37115	36973
0.4	36827	36678	36526	36371	36213	36053	35889	35723	35553	35381
0.5	35207	35029	34849	34667	34482	34294	34105	33912	33718	33521
0.6	33322	33121	32918	32713	32506	32297	32086	31874	31659	31443
0.7	31225	31006	30785	30563	30339	30114	29887	29658	29430	29200
0.8	28969	28737	28504	28269	28034	27798	27562	27324	27086	26848
0.9	26609	26369	26129	25888	25647	25406	25164	24923	24681	24439
1.0	24197	23955	23713	23471	23230	22988	22747	22506	22265	22025
1.1	21785	21546	21307	21069	20831	20594	20357	20121	19886	19652
1.2	19419	19186	18954	18724	18494	18265	18037	17810	17585	17360
1.3	17137	16915	16694	16474	16256	16038	15822	15608	15395	15183
1.4	14973	14764	14556	14350	14146	13943	13742	13542	13344	13147
1.5	12952	12758	12566	12376	12188	12001	11816	11632	11450	11270
1.6	11092	10915	10741	10567	10396	10226	10059	09893	09728	09566
1.7	09405	09246	09089	08933	08780	08628	08478	08329	08183	08038
1.8	07895	07754	07614	07477	07341	07206	07074	06943	06814	06687
1.9	06562	06438	06316	06195	06077	05959	05844	05730	05618	05508
2.0	05399	05292	05186	05082	04980	04879	04780	04682	04586	04491
2.1	04398	04307	04217	04128	04041	03955	03871	03788	03706	03626
2.2	03547	03470	03394	03319	03246	03174	03103	03034	02965	02898
2.3	02833	02768	02705	02643	02582	02522	02463	02406	02349	02294
2.4	02239	02186	02134	02083	02033	01984	01936	01888	01842	01797
2.5	01753	01709	01667	01625	01585	01545	01506	01468	01431	01394
2.6	01358	01323	01289	01256	01223	01191	01160	01130	01100	01071
2.7	01042	01014	00987	00961	00935	00909	00885	00861	00837	00814
2.8	00792	00770	00748	00727	00707	00687	00668	00649	00631	00613
2.9	00595	00578	00562	00545	00530	00514	00499	00485	00470	00457
3.0	00443									
3.5	008727									
4.0	0001338									
4.5	0000160									
5.0	000001487									

Table 3-2 Integrals of the gaussian normal error function

Values of the integral $(1/\sqrt{2\pi})\int_0^{\eta_1} e^{-\eta^2/2}\,d\eta$ are given for different values of the argument η_1. It may be observed that

$$\frac{1}{\sqrt{2\pi}}\int_{-\eta_1}^{+\eta_1} e^{-\eta^2/2}\,d\eta = 2\,\frac{1}{\sqrt{2\pi}}\int_0^{\eta_1} e^{-\eta^2/2}\,d\eta$$

The values are related to the error function since

$$\text{erf }\eta_1 = \frac{1}{\sqrt{\pi}}\int_{-\eta_1}^{+\eta_1} e^{-\eta^2}\,d\eta$$

so that the tabular values are equal to $\tfrac{1}{2}\text{erf }(\eta_1/\sqrt{2})$. (Each figure in the body of the table is preceded by a decimal point.)

η_1	0.00	0.01	0.02	0.03	0.04	0.05	0.06	0.07	0.08	0.09
0.0	00000	00399	00798	01197	01595	01994	02392	02790	03188	03586
0.1	03983	04380	04776	05172	05567	05962	06356	06749	07142	07355
0.2	07926	08317	08706	09095	09483	09871	10257	10642	11026	11409
0.3	11791	12172	12552	12930	13307	13683	14058	14431	14803	15173
0.4	15554	15910	16276	16640	17003	17364	17724	18082	18439	18793
0.5	19146	19497	19847	20194	20450	20884	21226	21566	21904	22240
0.6	22575	22907	23237	23565	23891	24215	24537	24857	25175	25490
0.7	25804	26115	26424	26730	27035	27337	27637	27935	28230	28524
0.8	28814	29103	29389	29673	29955	30234	30511	30785	31057	31327
0.9	31594	31859	32121	32381	32639	32894	33147	33398	33646	33891
1.0	34134	34375	34614	34850	35083	35313	35543	35769	35993	36214
1.1	36433	36650	36864	37076	37286	37493	37698	37900	38100	38298
1.2	38493	38686	38877	39065	39251	39435	39617	39796	39973	40147
1.3	40320	40490	40658	40824	40988	41198	41308	41466	41621	41774
1.4	41924	42073	42220	42364	42507	42647	42786	42922	43056	43189
1.5	43319	43448	43574	43699	43822	43943	44062	44179	44295	44408
1.6	44520	44630	44738	44845	44950	45053	45154	45254	45352	45449
1.7	45543	45637	45728	45818	45907	45994	46080	46164	46246	46327
1.8	46407	46485	46562	46638	46712	46784	46856	46926	46995	47062
1.9	47128	47193	47257	47320	47381	47441	47500	47558	47615	47670
2.0	47725	47778	47831	47882	47932	47962	48030	48077	48124	48169
2.1	48214	48257	48300	48341	48382	48422	48461	48500	48537	48574
2.2	48610	48645	48679	48713	48745	48778	48809	48840	48870	48899
2.3	48928	48956	48983	49010	49036	49061	49086	49111	49134	49158
2.4	49180	49202	49224	49245	49266	49286	49305	49324	49343	49361
2.5	49379	49396	49413	49430	49446	49461	49477	49492	49506	49520
2.6	49534	49547	49560	49573	49585	49598	49609	49621	49632	49643
2.7	49653	49664	49674	49683	49693	49702	49711	49720	49728	49736
2.8	49744	49752	49760	49767	49774	49781	49788	49795	49801	49807
2.9	49813	49819	49825	49831	49836	49841	49846	49851	49856	49861
3.0	49865									
3.5	4997674									
4.0	4999683									
4.5	4999966									
5.0	4999997133									

Equation (3-14) becomes

$$P = \frac{1}{\sqrt{2\pi}} \int_{-\eta_1}^{+\eta_1} e^{-\eta^2/2}\, d\eta \tag{3-15}$$

where $\eta_1 = \dfrac{x_1}{\sigma}$ \hfill (3-16)

Values of the gaussian normal error function

$$\frac{1}{\sqrt{2\pi}} e^{-\eta^2/2}$$

and integrals of the gaussian function corresponding to Eq. (3-15) are given in Tables 3-1 and 3-2.

If we have a sufficiently large number of data points, the error for each point should follow the gaussian distribution and we could determine the probability that certain data fall within a specified deviation from the mean value. Example 3-7 illustrates the method of computing the chances of finding data points within one or two standard deviations from the mean. (Table 3-3 gives the chances for certain deviations from the mean value of the normal-distribution curve.)

Example 3-7   Calculate the probabilities that a measurement will fall within one, two, and three standard deviations of the mean value, and compare them with the values in Table 3-3.

Solution   We perform the calculation using Eq. (3-15) with $\eta_1 = 1, 2,$ and 3.   The values of the integral may be obtained from Table 3-2.   We observe that

$$\int_{-\eta_1}^{+\eta_1} e^{-\eta^2/2}\, d\eta = 2\int_{0}^{\eta_1} e^{-\eta^2/2}\, d\eta$$

so that

$$P(1) = (2)(0.34134) = 0.683$$
$$P(2) = (2)(0.47725) = 0.954$$
$$P(3) = (2)(0.49865) = 0.997$$

Table 3-3 Chances for deviations from mean value of normal-distribution curve

Deviation	Chance of results falling within specified deviation
$\pm 0.6745\sigma$	1–1
σ	2.15–1
2σ	21–1
3σ	356–1

Using the odds given in Table 3-3, we would calculate the probabilities as

$$P(1) = \frac{2.15}{2.15 + 1} = 0.683$$

$$P(2) = \frac{21}{21 + 1} = 0.954$$

$$P(3) = \frac{356}{356 + 1} = 0.997$$

In many circumstances the engineer will not be able to collect as many data points as might be desired and only an approximation to the gaussian distribution may be obtained. Generally speaking, it is desirable to have about 20 measurements in order to obtain reliable estimates of standard deviation and general validity of the data. For smaller sets of data it is recommended that the following relation be used as the best estimate for the standard deviation:

$$\sigma = \left[\frac{\sum\limits_{i=1}^{n} (x_i - x_m)^2}{n - 1} \right]^{\frac{1}{2}} \tag{3-17}$$

Note that the factor $n - 1$ is used in Eq. (3-17) instead of n as in Eq. (3-7).

It is a rare circumstance indeed when an experimenter does not find that some of his data points look bad and out of place in comparison with the bulk of the data. He is therefore faced with the task of deciding if these points are the result of some gross experimental blunder and hence may be neglected or if they represent some new type of physical phenomenon that is peculiar to a certain operating condition. The engineer cannot just throw out those points that do not fit his expectations—he must have some consistent basis for elimination.

Example 3-8 Calculate the best estimate of standard deviation for the data of Example 3-5 based on Eq. (3-17).

Solution The calculation gives

$$\sigma = \left[\frac{1}{10 - 1} (3.536) \right]^{\frac{1}{2}} = (0.3929)^{\frac{1}{2}} = 0.627 \text{ ft}$$

Suppose n measurements of a quantity are taken and n is large enough that we may expect the results to follow the gaussian error distribution. This distribution may be used to compute the probability that a given reading will deviate a certain amount from the mean. We would not expect a probability much smaller than $1/n$ because this would be unlikely to occur in the set of n measurements. Thus, if the probability for the observed

Table 3-4   Chauvenet's criterion for rejecting a reading

Number of readings, n	Ratio of maximum acceptable deviation to standard deviation, d_{max}/σ
2	1.15
3	1.38
4	1.54
5	1.65
6	1.73
7	1.80
10	1.96
15	2.13
25	2.33
50	2.57
100	2.81
300	3.14
500	3.29
1,000	3.48

deviation of a certain point is less than $1/n$, a suspicious eye would be cast at that point with an idea toward eliminating it from the data. Actually, a more restrictive test is usually applied to eliminate data points. It is known as *Chauvenet's criterion* and specifies that a reading may be rejected if the probability of obtaining the particular deviation from the mean is less than $1/2n$. Table 3-4 lists values of the ratio of deviation to standard deviation for various values of n according to this criterion.

In applying Chauvenet's criterion to eliminate dubious data points one first calculates the mean value and standard deviation using all data points. The deviations of the individual points are then compared with the standard deviation in accordance with the information in Table 3-4 (or by a direct application of the criterion), and the dubious points are eliminated. For the final data presentation a new mean value and standard deviation are computed with the dubious points eliminated from the calculation. Note that Chauvenet's criterion might be applied a second or third time to eliminate additional points; but this practice is unacceptable, and only the first application may be used.

Example 3-9   Using Chauvenet's criterion, test the data points of Example 3-5 for possible inconsistency. Eliminate the questionable points and calculate a new standard deviation for the adjusted data.

Solution   The best estimate of the standard deviation is given in Example 3-8 as 0.627 ft. We first calculate the ratio d_i/σ and eliminate data points in accordance with Table 3-4.

Reading	d_i/σ
1	0.499
2	0.187
3	1.845
4	0.563
5	2.046
6	0.260
7	0.761
8	0.043
9	0.314
10	0.218

In accordance with Table 3-4, we may eliminate only point number 5. When this point is eliminated, the new mean value is

$$x_m = \tfrac{1}{9}(51.80) = 5.756 \text{ ft}$$

The new value of the standard deviation is now calculated with the following table:

Reading	$d_i = x_i - x_m$	$(x_i - x_m)^2 \times 10^2$
1	-0.436	19.010
2	-0.026	0.0676
3	1.034	106.916
4	-0.496	24.602
6	-0.306	9.364
7	0.334	11.156
8	-0.116	1.346
9	0.054	0.292
10	-0.006	0.0036

$$\sigma = \left[\frac{1}{n-1}\sum_{i=1}^{n}(x_i - x_m)^2\right]^{\frac{1}{2}} = [\tfrac{1}{8}(1.728)]^{\frac{1}{2}}$$

$$= (0.216)^{\frac{1}{2}} = 0.465 \text{ ft}$$

Thus, by the elimination of the one point, the standard deviation has been reduced from 0.627 to 0.465 ft. This is a 25.8 percent reduction.

3-8 PROBABILITY GRAPH PAPER

We have seen that the normal error distribution offers a means for examining experimental data for statistical consistency. In particular, it enables us to eliminate questionable readings with the Chauvenet criterion and thus obtain a better estimate of the standard deviation and mean reading. If the distribution of random errors is not *normal*, then this elimination technique will not apply. It is to our advantage, therefore, to determine if the data are following a normal distribution before making too many conclusions about the mean value, variances, etc. Specially constructed probability graph

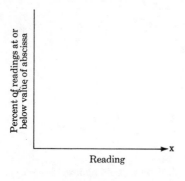

Fig. 3-3 Probability graph paper.

paper is available for this purpose and may be purchased from a technical drawing shop. The paper uses the coordinate system shown in Fig. 3-3. The ordinate has the percent of readings at or below the value of the abscissa, and the abscissa is the value of a particular reading. The ordinate spacings are arranged so that the gaussian-distribution curve will plot as a straight line on the graph. In addition, this straight line will intersect the 50 percent ordinate at an abscissa equal to the arithmetic mean of the data.

Thus, to determine if a set of data points are distributed normally we plot the data on probability paper and see how well they match with the theoretical straight line. It is to be noted that the largest reading cannot be plotted on the graph because the ordinate does not extend to 100 percent. In assessing the validity of the data we should not place as much reliance on the points near the upper and lower ends of the curve since they are closer to the "tails" of the probability distribution and are thus less likely to be valid.

Example 3-10 The following data are collected for a certain measurement. Plot the data on probability paper and comment on the normality of the distribution.

Reading	x_i, cm
1	4.62
2	4.69
3	4.86
4	4.53
5	4.60
6	4.65
7	4.59
8	4.70
9	4.58
10	4.63
	$\overline{\sum x_i = 46.45}$

From these data the mean value is calculated as

$$x_m = \tfrac{1}{10} \sum x_i = \tfrac{1}{10} (46.45) = 4.645 \text{ cm}$$

The data are plotted in the accompanying figure indicating a reasonably normal

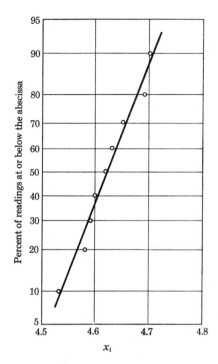

distribution. It should be noted that the straight line crosses the 50 percent ordinate at a value of approximately $x = 4.62$, which is not in agreement with the calculated value of x_m. Note that point 3, $x = 4.86$, does not appear on the plot since it would represent the 100 percent ordinate.

3-9 THE CHI-SQUARE TEST OF GOODNESS OF FIT

In the previous discussion we have noted that random experimental errors would be expected to follow the gaussian distribution, and the examples illustrated the method of calculating the probability of occurrence of a particular experimental determination. We might ask how it is known that the random errors or deviations do approximate a gaussian distribution. In general, we may ask how we can determine if experimental observations match some particular expected distribution for the data. As a simple example, consider the tossing of a coin. We would like to know if a certain coin is "honest," i.e., unweighted toward either heads or tails. If the coin is unweighted, then heads should occur half the time and tails should occur half the time. But suppose we do not want to take the time to make thousands of tosses to get the frequency distribution of heads and tails for a large number of tosses. Instead, we toss the coin a few times and wish to infer from these few tosses whether the coin is unweighted or weighted. Commonsense tells us not to expect exactly six heads and six tails out of,

EXPERIMENTAL METHODS FOR ENGINEERS

Table 3-5   Chi-squared. P is the probability that the value in the table will be exceeded for a given number of degrees of freedom F†

P \ F	0.995	0.990	0.975	0.950	0.900	0.750	0.500	0.250	0.100	0.050	0.025	0.010	0.005
1	0.0^4393	0.0^3157	0.0^3982	0.0^2393	0.0158	0.102	0.455	1.32	2.71	3.84	5.02	6.63	7.88
2	0.0100	0.0201	0.0506	0.103	0.211	0.575	1.39	2.77	4.61	5.99	7.38	9.21	10.6
3	0.0717	0.115	0.216	0.352	0.584	1.21	2.37	4.11	6.25	7.81	9.35	11.3	12.8
4	0.207	0.297	0.484	0.711	1.06	1.92	3.36	5.39	7.78	9.49	11.1	13.3	14.9
5	0.412	0.554	0.831	1.15	1.61	2.67	4.35	6.63	9.24	11.1	12.8	15.1	16.7
6	0.676	0.872	1.24	1.64	2.20	3.45	5.35	7.84	10.6	12.6	14.4	16.8	18.5
7	0.989	1.24	1.69	2.17	2.83	4.25	6.35	9.04	12.0	14.1	16.0	18.5	20.3
8	1.34	1.65	2.18	2.73	3.49	5.07	7.34	10.2	13.4	15.5	17.5	20.1	22.0
9	1.73	2.09	2.70	3.33	4.17	5.90	8.34	11.4	14.7	16.9	19.0	21.7	23.6
10	2.16	2.56	3.25	3.94	4.87	6.74	9.34	12.5	16.0	18.3	20.5	23.2	25.2
11	2.60	3.05	3.82	4.57	5.58	7.58	10.3	13.7	17.3	19.7	21.9	24.7	26.8
12	3.07	3.57	4.40	5.23	6.30	8.44	11.3	14.8	18.5	21.0	23.3	26.2	28.3
13	3.57	4.11	5.01	5.89	7.04	9.30	12.3	16.0	19.8	22.4	24.7	27.7	29.8
14	4.07	4.66	5.63	6.57	7.79	10.2	13.3	17.1	21.1	23.7	26.1	29.1	31.3
15	4.60	5.23	6.26	7.26	8.55	11.0	14.3	18.2	22.3	25.0	27.5	30.6	32.8
16	5.14	5.81	6.91	7.96	9.31	11.9	15.3	19.4	23.5	26.3	28.8	32.0	34.3
17	5.70	6.41	7.56	8.67	10.1	12.8	16.3	20.5	24.8	27.6	30.2	33.4	35.7
18	6.26	7.01	8.23	9.39	10.9	13.7	17.3	21.6	26.0	28.9	31.5	34.8	37.2
19	6.84	7.63	8.91	10.1	11.7	14.6	18.3	22.7	27.2	30.1	32.9	36.2	38.6
20	7.43	8.26	9.59	10.9	12.4	15.5	19.3	23.8	28.4	31.4	34.2	37.6	40.0
21	8.03	8.90	10.3	11.6	13.2	16.3	20.3	24.9	29.6	32.7	35.5	38.9	41.4
22	8.64	9.54	11.0	12.3	14.0	17.2	21.3	26.0	30.8	33.9	36.8	40.3	42.8
23	9.26	10.2	11.7	13.1	14.8	18.1	22.3	27.1	32.0	35.2	38.1	41.6	44.2
24	9.89	10.9	12.4	13.8	15.7	19.0	23.3	28.2	33.2	36.4	39.4	43.0	45.6
25	10.5	11.5	13.1	14.6	16.5	19.9	24.3	29.3	34.4	37.7	40.6	44.3	46.9
26	11.2	12.2	13.8	15.4	17.3	20.8	25.3	30.4	35.6	38.9	41.9	45.6	48.3
27	11.8	12.9	14.6	16.2	18.1	21.7	26.3	31.5	36.7	40.1	43.2	47.0	49.6
28	12.5	13.6	15.3	16.9	18.9	22.7	27.3	32.6	37.9	41.3	44.5	48.3	51.0
29	13.1	14.3	16.0	17.7	19.8	23.6	28.3	33.7	39.1	42.6	45.7	49.6	52.3
30	13.8	15.0	16.8	18.5	20.6	24.5	29.3	34.8	40.3	43.8	47.0	50.9	53.7

† From C. M. Thompson: *Biometrika*, vol. 32, 1941, as abridged by A. M. Mood and F. A. Graybill, "Introduction to the Theory of Statistics," 2d ed., McGraw-Hill Book Company, New York, 1963.

say, twelve tosses. But how much deviation from this arrangement could we tolerate and still expect the coin to be unweighted? The chi-square test of goodness of fit is a suitable way of answering this question. It is based on a calculation of the quantity chi squared, defined by

$$\chi^2 = \sum_{i=1}^{n} \frac{[(\text{observed value})_i - (\text{expected value})_i]^2}{(\text{expected value})_i} \tag{3-18}$$

where n is the number of observations. The expected value is the value which would be obtained if the measurements matched the expected distribution perfectly.

The chi-square test may be applied to check the validity of various distributions. Calculations have been made [2] of the probability that the actual measurements match the expected distribution, and these probabilities are given in Table 3-5. In this table, F represents the number of degrees of freedom in the measurements and is given by

$$F = n - k \tag{3-19}$$

where n is the number of observations and k is the number of imposed conditions on the expected distribution. A plot of the chi-square function is given in Fig. 3-4.

While we initiated the discussion on the chi-square test in terms of random errors following the gaussian distribution, the test is an important

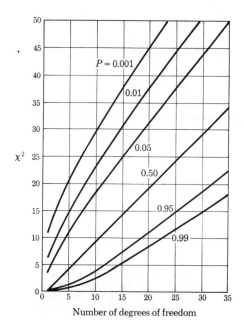

Fig. 3-4 The chi-square function.

tool for testing any expected experimental distribution. In other words, we might use the test to analyze random errors or to check the adherence of certain data to an expected distribution. We interpret the test by calculating the number of degrees of freedom and χ^2 from the experimental data. Then, consulting Table 3-5, we obtain the probability P that this value of χ^2 or higher values could occur by chance. If $\chi^2 = 0$, then the assumed or expected distribution and measured distribution match exactly. The larger the value of χ^2, the larger is the disagreement between the assumed distribution and the observed values, or the smaller the probability that the observed distribution matches the expected distribution. The reader should consult Refs. [2] and [4] for more specific information on the chi-square test and the derivation of the probabilities associated with it.

One may note that the heading of this section includes the term "goodness of fit." We see that the chi-square test may be used to determine how good a set of experimental observations fits an assumed distribution. In connection with this test we may remark that data may sometimes be "too good" or "too consistent." For example, we would be quite surprised if in the conduct of an experimental test, the results were found to check with theory *exactly* or to follow some well-defined relationship exactly. We might find, for instance, that a temperature controller maintained a set point temperature *exactly*, with no measurable deviation whatsoever. Experienced laboratory people know that controllers usually do not operate this way and would immediately suspect that the temperature recorder might be stuck or otherwise defective. The point of this brief remark is that one must be suspicious of high values of P as well as low values. A good rule of thumb is that if P lies between 0.1 and 0.9, the observed distribution may be considered to follow the assumed distribution. If P is either less than 0.02 or greater than 0.98, the assumed distribution may be considered unlikely. The use of the chi-square test is illustrated in the following examples.

Example 3-11 Two dice are rolled 100 times, and the following results are noted:

Number	Number of occurrences
2	2
3	3
4	9
5	12
6	14
7	19
8	15
9	13
10	8
11	4
12	1

Calculate the probability that the dice are unloaded.

Solution Eleven observations have been made with only one restriction: the number of rolls of the dice is fixed. Thus, $F = 11 - 1 = 10$. If the dice are unloaded, a short listing of the combinations of the dice will give the probability of occurrence for each number. The expected value of each number is then the probability multiplied by 100, the total number of throws. The values of interest are tabulated below.

Number	Observed	Probability	Expected
2	2	1/36	2.778
3	3	1/18	5.556
4	9	1/12	8.333
5	12	1/9	11.111
6	14	5/36	13.889
7	19	1/6	16.667
8	15	5/36	13.889
9	13	1/9	11.111
10	8	1/12	8.333
11	4	1/18	4.446
12	1	1/36	2.778

From these data the value of chi squared is calculated as 3.754. If Table 3-5 is consulted, the probability is calculated as $P = 0.957$.

Example 3-12 A coin is tossed 10 times, resulting in 3 heads and 7 tails. Using the chi-square test, estimate the probability that the coin is unweighted. Suppose another set of tosses of the same coin is made such that four heads and six tails are obtained. What is the probability of having an unweighted coin based on the information from both sets of data?

Solution For each set of data we may make only two observations: the number of heads and the number of tails. Thus, $n = 2$. Furthermore, we impose one restriction on the data: the number of tosses is fixed. Thus $k = 1$ and the number of degrees of freedom is

$$F = n - k = 2 - 1 = 1$$

The values of interest are

	Observed	Expected
Heads	3	5
Tails	7	5

For these values, χ^2 is calculated as

$$\chi^2 = \frac{(3 - 5)^2}{5} + \frac{(7 - 5)^2}{5} = 1.60$$

Consulting Table 3-5 we find $P = 0.22$; that is, there is a 22 percent chance that this distribution is just the result of random fluctuations and that the coin may be unweighted.

Now consider the additional information we gain about the coin from the second set of observations. We now have four observations: the number of heads and tails in each set. There are only two restrictions on the data: the total number of tosses is fixed in each set. Thus, the number of degrees of freedom is

$$F = n - k = 4 - 2 = 2$$

For the second set of data the values of interest are

	Observed	Expected
Heads	4	5
Tails	4	5

Chi squared is now calculated on the basis of all four observations.

$$\chi^2 = \frac{(3 - 5)^2}{5} + \frac{(7 - 5)^2}{5} + \frac{(4 - 5)^2}{5} + \frac{(6 - 5)^2}{5} = 2.0$$

Consulting Table 3-5 again we find $P = 0.39$. So, with the additional information we find a stronger likelihood that the tosses are following a random variation and that the coin is unweighted.

Example 3-13 A test is conducted to determine the effect of cigarette smoke on the eating habits and weight of mice. One group is fed a certain diet while being exposed to a controlled atmosphere containing cigarette smoke. A control group is fed the same diet but in the presence of clean air. The observations are given below. Does the presence of smoke cause a loss in weight?

	Gained weight	Lost weight	Total
Exposed to smoke	61	89	150
Exposed to clean air	65	77	142
Total	126	166	292

Solution Clearly, there are four observations in this experiment, but we are faced with the problem of deciding on the expected values. We cannot just take the "clean-air" data as the expected values because some of the behavior might be a result of the special diet that is fed to both groups of mice. Consequently, about the best estimate we can make is one based on the total sample of mice. Thus, the expected frequencies would be

Expected fraction to gain weight $= \frac{126}{292}$
Expected fraction to lose weight $= \frac{166}{292}$

The expected values for the groups would thus be

	Gained weight	Lost weight
Exposed to smoke	$\frac{126}{292}150 = 64.7$	$\frac{166}{292}150 = 85.3$
Exposed to clean air	$\frac{126}{292}142 = 61.3$	$\frac{166}{292}142 = 80.7$

We observe that there are three restrictions on the data: (1) the number exposed to smoke, (2) the number exposed to clean air, and (3) the additional restriction involved in the calculation of the expected fractions which gain and lose weight. The number of degrees of freedom is thus

$$F = 4 - 3 = 1$$

The value of chi squared is calculated from

$$\chi^2 = \frac{(61 - 64.7)^2}{64.7} + \frac{(89 - 85.3)^2}{85.3} + \frac{(65 - 61.3)^2}{61.3} + \frac{(77 - 80.7)^2}{80.7} = 0.767$$

From Table 3-5 we find $P = 0.41$, or there is a 41 percent chance that the difference in the observations for the two groups is just the result of random fluctuations. One may not conclude from this information that the presence of cigarette smoke causes a loss in weight for the mice.

3-10 METHOD OF LEAST SQUARES

Suppose we have a set of observations $x_1, x_2, \ldots, x_n$. The sum of the squares of their deviations from some mean value is

$$S = \sum_{i=1}^{n} (x_i - x_m)^2 \tag{3-20}$$

Now suppose we wish to minimize S with respect to the mean value x_m. We set

$$\frac{\partial S}{\partial x_m} = 0 = \sum_{i=1}^{n} -2(x_i - x_m) = -2\left(\sum_{i=1}^{n} x_i - nx_m\right) \tag{3-21}$$

where n is the number of observations. We find that

$$x_m = \frac{1}{n} \sum_{i=1}^{n} x_i \tag{3-22}$$

or the mean value which minimizes the sum of the squares of the deviations is the arithmetic mean. This example might be called the simplest application of the method of least squares. We shall be able to give only two other applications of the method, but it is of great utility in analyzing experimental data.

Suppose that the two variables x and y are measured over a range of values. Suppose further that we wish to obtain a simple analytical expression for y as a function of x. The simplest type of function is a linear one; hence we might try to establish y as a linear function of x. (Both x and y may be complicated functions of other parameters so arranged that x and y vary approximately in a linear manner. This matter will be discussed later.) The problem is one of finding the *best* linear function, for the data may scatter a considerable amount. We could solve the problem rather quickly by plotting the data points on graph paper and drawing a straight line through them by eye. Indeed this is common practice, but the method of least squares affords an opportunity to obtain a better functional relationship than by the guesswork of plotting. We seek an equation of the form

$$y = ax + b \tag{3-23}$$

We therefore wish to minimize the quantity

$$S = \sum_{i=1}^{n} [y_i - (ax_i + b)]^2 \tag{3-24}$$

This is accomplished by setting the derivatives with respect to a and b equal to zero. Performing these operations there results

$$nb + a \sum x_i = \sum y_i \tag{3-25}$$

$$b \sum x_i + a \sum x_i^2 = \sum x_i y_i \tag{3-26}$$

Solving Eqs. (3-25) and (3-26) simultaneously gives

$$a = \frac{n \sum x_i y_i - (\sum x_i)(\sum y_i)}{n \sum x_i^2 - (\sum x_i)^2} \tag{3-27}$$

$$b = \frac{(\sum y_i)(\sum x_i^2) - (\sum x_i y_i)(\sum x_i)}{n \sum x_i^2 - (\sum x_i)^2} \tag{3-28}$$

The method of least squares may also be used for determining higher-order polynomials for fitting data. One only needs to perform additional differentiations to determine additional constants. For example, if it were desired to obtain a least squares fit according to the quadratic function

$$y = ax^2 + bx + c$$

the quantity

$$S = \sum_{i=1}^{n} [y_i - (ax_i^2 + bx_i + c)]^2$$

would be minimized by setting the following derivatives equal to zero:

$$\frac{\partial S}{\partial a} = 0 = \sum 2[y_i - (ax_i^2 + bx_i + c)](-x_i^2)$$

$$\frac{\partial S}{\partial b} = 0 = \sum 2[y_i - (ax_i^2 + bx_i + c)](-x_i)$$

$$\frac{\partial S}{\partial c} = 0 = \sum 2[y_i - (ax_i^2 + bx_i + c)](-1)$$

Expanding and collecting terms,

$$a \sum x_i^4 + b \sum x_i^3 + c \sum x_i^2 = \sum x_i^2 y_i \qquad (3\text{-}29)$$
$$a \sum x_i^3 + b \sum x_i^2 + c \sum x_i = \sum x_i y_i \qquad (3\text{-}30)$$
$$a \sum x_i^2 + b \sum x_i + cn = \sum y_i \qquad (3\text{-}31)$$

These equations may then be solved for the constants a, b, and c.

In the above discussion of the method of least squares no mention has been made of the influence of experimental uncertainty on the calculation. We are considering the method primarily for its utility in fitting an algebraic relationship to a set of data points. Clearly, the various x_i and y_i could have different experimental uncertainties. To take all these into account requires a rather tedious calculation procedure which we shall not present here; however, we may state the following rules:

1. If the values of x_i and y_i are taken as the data value in y and the value of x *on the fitted curve for the same value of y*, then there is a presumption that the uncertainty in x is large compared with that in y.
2. If the values of x_i and y_i are taken as the data value in y and the value *on the fitted curve for the same value of x*, the presumption is that the uncertainty in y dominates.
3. If the uncertainties in x_i and y_i are believed to be of approximately equal magnitude, a special averaging technique must be used.

In Example 3-14, rule 2 is assumed to apply.

Example 3-14 From the following data obtain y as a linear function of x using the method of least squares:

y_i	x_i
1.2	1.0
2.0	1.6
2.4	3.4
3.5	4.0
3.5	5.2
$\sum y_i = 12.6$	$\sum x_i = 15.2$

Solution We seek an equation of the form

$$y = ax + b$$

We first calculate the quantities indicated in the following table:

$x_i y_i$	x_i^2
1.2	1.0
3.2	2.56
8.16	11.56
14.0	16.0
18.2	27.04
$\sum x_i y_i = 44.76$	$\sum x_i^2 = 58.16$

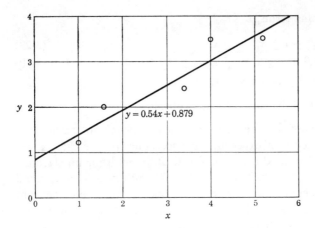

We calculate the value of a and b, using Eqs. (3-27) and (3-28) with $n = 5$:

$$a = \frac{(5)(44.76) - (15.2)(12.6)}{(5)(58.16) - (15.2)^2} = 0.540$$

$$b = \frac{(12.6)(58.16) - (44.76)(15.2)}{(5)(58.16) - (15.2)^2} = 0.879$$

Thus, the desired relation is

$$y = 0.540x + 0.879$$

A plot of this relation and the data points from which it was derived is shown in the accompanying figure.

3-11 STANDARD DEVIATION OF THE MEAN

We have taken the arithmetic mean value as the best estimate of the true value of a set of experimental measurements. Considerable discussion has been devoted to the gaussian normal error distribution and to an examination of the various types of errors and deviations that may occur in an experimental measurement. But one very important question has not yet been answered; i.e., how *good* (or precise) is this arithmetic mean value which is taken as the best estimate of the true value of a set of readings? To obtain an experimental answer to this question it would be necessary to repeat the set of measurements and find a new arithmetic mean. In general, we would find that this new arithmetic mean would differ from the previous value, and thus we would not be able to resolve the problem until a large number of *sets of data* were collected. We would then know how well the mean of a single set approximated the mean which would be obtained with a large number of sets. The mean value of a large number of sets is presumably the true value. Consequently, we wish to know the standard deviation of the mean of a single set of data from this true value.

It turns out that the problem may be resolved with a statistical analysis, which we shall not present here. The result is

$$\sigma_m = \frac{\sigma}{\sqrt{n}} \tag{3-32}$$

where σ_m = standard deviation of the mean value
σ = standard deviation of the set of measurements
n = number of measurements in the set

The following example illustrates the use of Eq. (3-32).

Example 3-15   For the data of Example 3-5, estimate the uncertainty in the calculated mean value of the readings.

Solution   We shall make this estimate for the original data and for the reduced data of Example 3-9.   For the original data the standard deviation of the mean is

$$\sigma_m = \frac{\sigma}{\sqrt{n}} = \frac{0.627}{\sqrt{10}} = 0.198 \text{ ft}$$

The arithmetic mean value calculated in Example 3-5 was $x_m = 5.613$ ft.   We could now specify the uncertainty of this value by using the odds of Table 3-3:

$x_m = 5.613 \pm 0.198$ ft   (2.15 to 1)
$\quad = 5.613 \pm 0.396$ ft   (21 to 1)
$\quad = 5.613 \pm 0.594$ ft   (356 to 1)

Using the data of Example 3-9, where one point has been eliminated by Chauvenet's criterion, we may make a better estimate of the mean value with less uncertainty.   The standard deviation of the mean is calculated as

$$\sigma_m = \frac{\sigma}{\sqrt{n}} = \frac{0.465}{\sqrt{9}} = 0.155 \text{ ft}$$

for the mean value of 5.756 ft.   Thus, we would estimate the uncertainty as

$x_m = 5.756 \pm 0.155$ ft   (2.15 to 1)
$\quad = 5.756 \pm 0.310$ ft   (21 to 1)
$\quad = 5.756 \pm 0.465$ ft   (356 to 1)

3-12   GRAPHICAL ANALYSIS AND CURVE FITTING

Engineers are well known for their ability to plot many curves of experimental data and to extract all sorts of significant facts from these curves.   The better one understands the physical phenomena involved in a certain experiment, the better is he able to extract a wide variety of information from graphical displays of experimental data.   Because these physical phenomena may encompass all engineering science, we cannot discuss them here except to emphasize that the person who is usually most successful in analyzing experimental data is the one who understands the physical processes behind the data.   Blind curve-plotting and cross-plotting usually generate an excess of displays, which are confusing not only to the management or

Table 3-6   Methods of plotting various functions to obtain straight lines

Functional relationship	Method of plot	Graphical determination of parameters
$y = ax + b$	y versus x on linear paper	
$y = ax^b$	log y versus log x on log-log paper	
$y = ae^{bx}$	log y versus x on semilog paper	
$y = \dfrac{x}{a + bx}$	$\dfrac{1}{y}$ versus $\dfrac{1}{x}$ on linear paper	

Table 3-6 Methods of plotting various functions to obtain straight lines *(Continued)*

Functional relationship	Method of plot	Graphical determination of parameters
$y = a + bx + cx^2$	$\dfrac{y - y_1}{x - x_1}$ versus x on linear paper	
$y = \dfrac{x}{a + bx} + c$	$\dfrac{x - x_1}{y - y_1}$ versus x on linear paper	
$y = ae^{bx + cx^2}$	$\log\left[\left(\dfrac{y}{y_1}\right)^{1/(x - x_1)}\right]$ versus x on semilog paper	

For the first row: Slope $= c$, intercept $b + cx_1$.

For the second row: Slope $= b + \dfrac{b^2}{a}x_1$, intercept $a + bx_1$.

For the third row: Slope $= c \log e$, intercept $b + cx_1 \log e$.

supervisory personnel who must pass on the experiments, but sometimes even to the experimenter himself. To be blunt, the engineer should give considerable thought to the kind of information that he is looking for before he even takes the graph paper out of the package.

Assuming that the engineer knows what he wants to examine with graphical presentations, the plots may be carefully prepared and checked against appropriate theories. Frequently, a *correlation* of the experimental data is desired in terms of an analytical expression between variables that were measured in the experiment. When the data may be approximated by a straight line, the analytical relation is easy to obtain; but when almost any other functional variation is present, difficulties are usually encountered. This fact is easy to understand since a straight line is easily recognizable on graph paper whereas the functional form of a curve is rather doubtful. The curve could be a polynomial, exponential, or complicated logarithmic function and still present roughly the same appearance to the eye. It is most convenient, then, to try to plot the data in such a form that a straight line will be obtained for certain types of functional relationships. If the experimenter has a good idea of the type of function that will represent the data, then the type of plot is easily selected. He frequently is able to estimate the functional form that the data will take on the basis of theoretical considerations and the results of previous experiments of a similar nature.

Table 3-6 summarizes several different types of functions and plotting methods that may be used to produce straight lines on graph paper. The graphical measurements, which may be made to determine the various constants, are also shown. It may be remarked that the method of least squares may be applied to all these relations to obtain the best straight line to fit the experimental data.

3-13 GENERAL CONSIDERATIONS IN DATA ANALYSIS

Our discussions in this chapter have considered a variety of topics: statistical analysis, uncertainty analysis, curve plotting, least squares, etc. With these tools the reader is equipped to handle a variety of circumstances that may occur in experimental investigations. As a summary to this chapter let us now give an approximate outline of the manner in which one would go about analyzing a set of experimental data.

1. *Examine the data for consistency.* No matter how hard one tries, there will always be some data points that appear to be grossly in error. If we add heat to a container of water, the temperature must rise, and so if a particular data point indicates a *drop* in temperature for a heat *input*, that point might be eliminated. In other words, the data should follow commonsense consistency and points eliminated that do not

appear proper. If very many data points fall in the category of "inconsistent," perhaps the entire experimental procedure should be investigated for gross mistakes or miscalculation.

2. *Perform a statistical analysis of data where appropriate.* A statistical analysis is only appropriate when measurements are repeated several times. If this is the case, make estimates of such parameters as standard deviation, etc.

3. *Estimate the uncertainties in the results.* We have discussed uncertainties at length. Hopefully, these calculations will have been performed in advance, and the investigator will already know the influence of different variables by the time he has obtained the final results.

4. *Anticipate the results from theory.* Before trying to obtain correlations of the experimental data, the investigator should carefully review the theory appropriate to the subject and try to glean some information that will indicate the trends his results may take. Important dimensionless groups, pertinent functional relations, and other information may lead him to a fruitful interpretation of his own data.

5. *Correlate the data.* The word "correlate" is subject to misinterpretation. In the context here we mean that the experimental investigator should make sense of the data in terms of physical theories or on the basis of previous experimental work in the field. Certainly, the results of the experiments should be analyzed to show how they conform or differ from previous investigations or from standards that may be employed for such measurements.

REVIEW QUESTIONS

1. How does an error differ from an uncertainty?

2. What is a fixed error; random error?

3. Define standard deviation and variance.

4. In the normal error distribution what does $P(x)$ represent?

5. What is meant by measure of precision?

6. How does one compensate for a relatively few number of measurements in the computation of standard deviation?

7. What is Chauvenet's criterion and how is it applied?

8. What are some purposes of uncertainty analyses?

9. Why is an uncertainty analysis important in the preliminary stages of experiment planning?

10. How can an uncertainty analysis help to reduce overall experimental uncertainty?

11. What is meant by standard deviation of the mean?

12. What is a least squares analysis?

PROBLEMS

3-1. The resistance of a resistor is measured 10 times, and the values determined are 100.0, 100.9, 99.3, 99.9, 100.1, 100.2, 99.9, 100.1, 100.0, and 100.5. Calculate the uncertainty in the resistance.

3-2. A certain resistor draws 110.2 volts and 5.3 amp. The uncertainties in the measurements are ± 0.2 volt and ± 0.06 amp, respectively. Calculate the power dissipated in the resistor and the uncertainty in the power.

3-3. A small plot of land has measured dimensions of 50.0 by 150.0 ft. The uncertainty in the 50-ft dimension is ± 0.01 ft. Calculate the uncertainty with which the 150-ft dimension must be measured to ensure that the total uncertainty in the area is not greater than 150 percent of that value it would have if the 150-ft dimension were exact.

3-4. Two resistors R_1 and R_2 are connected in series and parallel. The values of the resistances are

$R_1 = 100.0 \pm 0.1$ ohms
$R_2 = 50.0 \pm 0.03$ ohms

Calculate the uncertainty in the combined resistance for both the series and the parallel arrangements.

3-5. A resistance arrangement of 50 ohms is desired. Two resistances of 100.0 ± 0.1 ohms and two resistances of 25.0 ± 0.02 ohms are available. Which should be used, a series arrangement with the 25-ohm resistors or a parallel arrangement with the 100-ohm resistors? Calculate the uncertainty for each arrangement.

3-6. The following data are taken from a certain heat-transfer test. The expected correlation equation is $y = ax^b$. Plot the data in an appropriate manner, and use the method of least squares to obtain the best correlation.

x	2040	2580	2980	3220	3870	1690	2130	2420	2900	3310	1020	1240	1360	1710	2070
y	33.2	32.0	42.7	57.8	126.0	17.4	21.4	27.8	52.1	43.1	18.8	19.2	15.1	12.9	78.5

Calculate the mean deviation of these data from the best correlation.

3-7. A horseshoes player stands 30 ft from his target. The results of the tosses are

Toss	Deviation from target, ft	Toss	Deviation from target, ft
1	0	6	+2.4
2	+3	7	−2.6
3	−4.2	8	+3.5
4	0	9	+2.7
5	+1.5	10	0

On the basis of these data would you say that he is a good player or a poor player? What advice would you give in regard to improving his game?

3-8. Calculate the probability of drawing a full house (three of a kind and two of a kind) in the first 5 cards from a 52-card deck.

3-9. Calculate the probability of filling an inside straight with one draw from the remaining 48 cards of a 52-card deck.

3-10. A voltmeter is used to measure a known voltage of 100 volts. Forty percent of the readings are within 0.5 volt of the true value. Estimate the standard deviation for the meter. What is the probability of an error of 0.75 volt?

3-11. In a certain mathematics course the instructor informs the class that grades will be distributed according to the following scale provided that the average class score is 75:

Grade	A	B	C	D	F
Score	90–100	80–90	70–80	60–70	Below 60

Estimate the percentage distribution of grades for 5, 10, and 15 percent failing. Assume that there are just as many A's as F's.

3-12. For the following data points y is expected to be a quadratic function of x. Obtain this quadratic function by means of a graphical plot and also by the method of least squares.

x	1	2	3	4	5
y	1.9	9.3	21.5	42.0	115.7

3-13. It is suspected that the rejection rate for a plastic-cup-molding machine is dependent on the temperature at which the cups are molded. A series of short tests is conducted to examine this hypothesis with the following results:

Temperature	Total production	Number rejected
T_1	150	12
T_2	75	8
T_3	120	10
T_4	200	13

On the basis of these data do you agree with the hypothesis?

3-14. A capacitor discharges through a resistor according to the relation $\dfrac{E}{E_0} = e^{-t/RC}$ where $E_0 =$ voltage at time zero; $R =$ resistance; $C =$ capacitance. The value of the capacitance is to be measured by recording the time necessary for the voltage to drop to a value E_1. Assuming that the resistance is known accurately, derive an expression for the percent uncertainty in the capacitance as a function of the uncertainty in the measurements of E_1 and t.

3.15. In heat-exchanger applications, a log mean temperature difference is defined by

$$\Delta T_m = \frac{(T_{h1} - T_{c1}) - (T_{h2} - T_{c2})}{\ln\left[(T_{h1} - T_{c1})/(T_{h2} - T_{c2})\right]}$$

where the four temperatures are measured at appropriate inlet and outlet conditions for the heat-exchanger fluids. Assuming that all four temperatures are measured with the same absolute uncertainty w_T, derive an expression for the percentage uncertainty in ΔT_m in terms of the four temperatures and the value of w_T. Recall that the percentage uncertainty is

$$\frac{w_{\Delta T_m}}{\Delta T_m} \times 100$$

3.16. A certain length measurement is made with the following results:

Reading	1	2	3	4	5	6	7	8	9	10
x, in.	49.36	50.12	48.98	49.24	49.26	50.56	49.18	49.89	49.33	49.39

Calculate the standard deviation, the mean reading, and the uncertainty. Apply Chauvenet's criterion as needed.

3.17. Devise a method for plotting the Gaussian normal error distribution such that a straight line will result. (*Ans.* $(1/\eta) \ln [\sqrt{2\pi} P(\eta)]$ versus η.) Show how such a plot may be labeled so that it can be used to estimate the fraction of points which lie below a certain value of η. Subsequently show that this plot may be used to investigate the normality of a set of data points. Apply this reasoning to the data points of Example 3-5 and Probs. 3-6 and 3-7.

3-18. A citizens' traffic committee decides to conduct their own survey and analysis of the influence of drinking on car accidents. By some judicious estimates the committee determines that in their community 30 percent of the drivers on a Saturday evening between 10 P.M. and 2 A.M. have consumed some alcohol. During this same period there were 50 accidents, varying from minor scratched fenders to fatalities. In these 50 accidents 50 of the drivers had had something to drink (there are 100 drivers for 50 accidents). From these data what conclusions do you draw about the influence of drinking on car accidents? Can you devise a better way to perform this analysis?

3-19. The grades for a certain class fall in the following ranges:

Number	10	30	50	40	10	8
Score	90–100	80–90	70–80	60–70	50–60	Below 50

The arithmetic mean grade is 68. Devise your own grade distribution for this class. Be sure to establish the criteria for the distribution.

3-20. A certain length measurement is performed 100 times. The arithmetic mean reading is 6.823 ft, and the standard deviation is 0.01 ft. How many readings fall within (*a*) ± 0.005 ft, (*b*) ± 0.02 ft, (*c*) ± 0.05 ft, and (*d*) 0.001 ft of the mean value?

3-21. A series of calibration tests is conducted on a pressure gage. At a known pressure of 1,000 psia, it is found that 30 percent of the readings are within 1 psia of the true value. At a known pressure of 500 psia, 40 percent of the readings are within 1 psia. At a pressure of 200 psia, 45 percent of the readings are within 1 psia. What conclusions do you draw from these readings? Can you estimate a standard deviation for the pressure gage?

3-22. Two resistors are connected in series and have the following values:

$$R_1 = 10,000 \text{ ohms} \pm 5\%; \quad R_2 = 1 \text{ megohm} \pm 10\%.$$

Calculate the percent uncertainty for the series total resistance.

3-23. Apply Chauvenet's criterion to the data of Example 3-10 and then replot the data on probability paper, omitting any excluded points.

3-24. Plot the data of Example 3-5 on probability paper. Replot the data, taking into account the point eliminated in Example 3-9. Comment on the normality of these two sets of data.

3-25. Two groups of secretaries operate under the same manager. Both groups have the same number of people, use the same equipment, and turn out about the same amount of work.　During one maintenance period, group A had 10 service calls on the equipment while group B had only 6 calls.　From these data would you conclude that group A was harder on the equipment?

3-26.　A laboratory experiment is conducted to measure the viscosity of a certain oil. A series of tests gives the values as 0.040, 0.041, 0.041, 0.042, 0.039, 0.040, 0.043, 0.041, and 0.039 ft²/sec.　Calculate the mean reading, the variance, and the standard deviation. Eliminate any data points as necessary.

3-27.　The following data are expected to follow a linear relation of the form $y = ax + b$. Obtain the best linear relation in accordance with a least squares analysis.　Calculate the standard deviation of the data from the predicted straight-line relation.

x	0.9	2.3	3.3	4.5	5.7	6.7
y	1.1	1.6	2.6	3.2	4.0	5.0

3-28.　The following data points are expected to follow a functional variation of $y = ax^b$. Obtain the values of a and b from a graphical analysis.

x	1.21	1.35	2.40	2.75	4.50	5.1	7.1	8.1
y	1.20	1.82	5.0	8.80	19.5	32.5	55.0	80.0

3-29.　The following data points are expected to follow a functional variation of $y = ae^{bx}$. Obtain the values of a and b from a graphical analysis.

x	0	0.43	1.25	1.40	2.60	2.9	4.3
y	9.4	7.1	5.35	4.20	2.60	1.95	1.15

REFERENCES

1. Kline, S. J., and F. A. McClintock: Describing Uncertainties in Single-sample Experiments, *Mech. Eng.*, p. 3, January, 1953.
2. Mood, A. M., and F. A. Graybill: "Introduction to the Theory of Statistics," 2d ed., McGraw-Hill Book Company, New York, 1963.
3. Schenck, H.: "Theories of Engineering Experimentation," McGraw-Hill Book Company, New York, 1961.
4. Wilson, E. B.: "An Introduction to Scientific Research," McGraw-Hill Book Company, New York, 1952.
5. Young, H. D.: "Statistical Treatment of Experimental Data," McGraw-Hill Book Company, New York, 1962.
6. Hicks, C. R.: "Fundamental Concepts in the Design of Experiments," Holt, Rinehart, and Winston, Inc., New York, 1964.

4
Basic Electrical Measurements and Sensing Devices

4-1 INTRODUCTION

A large majority of measuring devices either use some basic electrical principle for their operation or rely on an electronic device for the intermediate modifying and final readout stages. As a consequence, it is to our advantage to discuss some of the more important electric devices currently employed and to emphasize their relationship to the measurement process. We shall first consider the measurement of the basic electrical quantities of current and voltage. Next, we shall examine some simple circuits that may be used for modification and measurement of input signals. The physical principles and operating characteristics of some of the more important electric transducers will then be studied and their application indicated.

In all the following discussions due consideration must be given to impedance matching. The principles of matching were discussed in Sec. 2-9, and the reader should keep this discussion in mind as the exposition of the present chapter progresses.

4-2 THE MEASUREMENT OF CURRENT

If a point charge of q coulombs moving with a velocity v is acted upon by an electric and magnetic field, the resultant force on the point charge is given by the basic relation

$$\mathbf{F} = q(\mathbf{E} + \mathbf{v} \times \mathbf{B}) \text{ newtons} \tag{4-1}$$

where $\mathbf{E}$ = electric-field intensity, volts/m
  $\mathbf{v}$ = velocity, m/sec
  $\mathbf{B}$ = magnetic flux density, webers/m²

Now suppose that a current-carrying conductor is placed in a magnetic field as shown in Fig. 4-1. Let us consider the case where $E = 0$. The electric current through the conductor is the time rate of motion of charge or

$$i = \frac{dq}{d\tau} \tag{4-2}$$

We could write

$$i \, ds = \frac{dq}{d\tau} \, ds = v \, dq \tag{4-3}$$

and the force exerted on the charge is given by Eq. (4-1) as

$$dF = dq(\mathbf{v} \times \mathbf{B}) = i \, d\mathbf{s} \times \mathbf{B} \tag{4-4}$$

for $E = 0$. Integrating along the length of the conductor,

$$F = \int_0^L i \, d\mathbf{s} \times \mathbf{B} \tag{4-5}$$

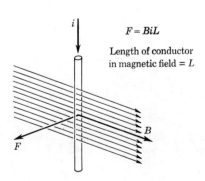

$F = BiL$

Length of conductor
in magnetic field $= L$

Fig. 4-1 Current-carrying conductor in a magnetic field.

If the magnetic flux vector and length vector are oriented at an angle of 90°, Eq. (4-5) becomes

$$F = BiL \qquad\qquad (4\text{-}6)$$

Equation (4-6) gives the magnitude of the force on a current-carrying conductor placed at a right angle with respect to a magnetic field. In order to integrate Eq. (4-5) and obtain Eq. (4-6) it has been assumed that end effects on the conductor are negligible and that the region L is sharply defined so that the magnetic field is zero outside this region and constant inside the region.

Clearly, the above principles may be used to construct an instrument to measure electric current. We construct a coil, as shown in Fig. 4-2, place it in a magnetic field, and measure the force exerted on the coil as the result of the electric current. If the coil has N turns and the length of each turn in the magnetic field is L, the force on the coil is

$$F = NBiL \qquad\qquad (4\text{-}7)$$

The force is measured by observing the deflection of a force-restraining device such as a spring. The above principles form the basis of the construction of the mirror galvanometer shown in Fig. 4-3. A permanent magnet is used to produce the magnetic field, while the telescope arrangement and expanded scale improve the readability of the instrument. The meter shown in Fig. 4-3 is designated as the D'Arsonval moving-coil type. The metal ribbon furnishes the torsional-spring restraining force in this case, while a filamentary suspension would be used for a more sensitive instrument.

Instead of the mirror and light-beam arrangement shown in Fig. 4-3, the D'Arsonval movement could be used as a pointer-type instrument, as shown in Fig. 4-4; however, such an instrument has a lower sensitivity than the mirror galvanometer because of the additional mass of the pointer and decreased readability resulting from the relatively shorter scale length. The

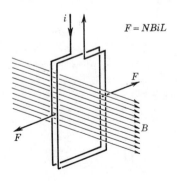

Fig. 4-2   Current-carrying coil in a magnetic field.

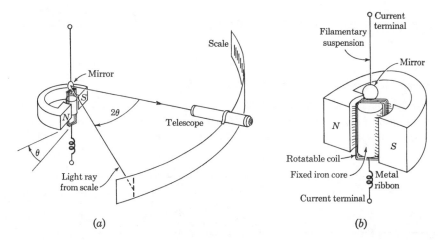

Fig. 4-3 A typical mirror galvanometer. (*a*) Optical system; (*b*) D'Arsonval movement.

sensitivities of various commercial galvanometers as compiled by Sweeney [5] are given in Table 4-1.   In this table the period is the undamped natural period of oscillation of the moving coil.   The sensitivity values are those obtained with an external circuit resistance necessary to give critical damping.

It is clear that the D'Arsonval movement, in one form or another, may be used for the measurement of dc current.   When this movement is connected to an ac current, the meter will either vibrate or, if the frequency is sufficiently high, indicate zero.   In either event, the D'Arsonval movement is not directly applicable to the measurement of ac current.   A detailed analysis of the response of galvanometer movement to various ac waveforms is given by Frank [2].

The two most common types of movements used for ac current measurement are the iron-vane, or moving-iron, and electrodynamometer arrangements.   In the iron-vane instrument, as shown in Fig. 4-5, the current

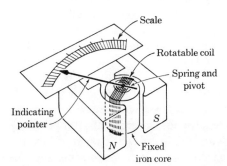

Fig. 4-4 D'Arsonval movement used as a pointer-type instrument.

Table 4-1   Typical galvanometer characteristics†

Construction type	Sensitivity per mm or scale division	Period, sec	Resistance, ohms	
			External CDR	Coil
Slack suspension, one-meter light beam	0.00001 μa	40	100,000	800
	0.00004 μa	20	70,000	800
	0.0005 μa	6	10,000	650
	0.008 μa	1.5	2,500	500
	0.05 μv	7	10	16
	0.2 μv	5	40	16
	0.5 μv	1.5	40	21
Taut suspension, self-contained optical system	0.0005 μa	3	25,000	550
	0.005 μa	3	6,000	1,000
	0.05 μa	2	2,500	350
	0.5 μv	3	50	17
	2.5 μv	2	12	13
	2.5 μv	2	50	13
Taut suspension, pointer type	0.030 μa	5.3	42,000	1,900
	0.125 μa	3.5	10,000	1,000
	1.0 μa	3	950	250
	46 μv	4.5	30	16
	280 μv	3	50	20

† According to Sweeney [5].

is applied to a *fixed coil*. The iron vane is movable and connected to a re-straining spring as shown. The displacement of the vane is then proportional to the inductive force exerted by the coil. The meter is subject to eddy-current losses in the iron vane and various hysteresis effects which limit its accuracy.

The features of the electrodynamometer movement are shown in Fig. 4-6. This movement is similar to the D'Arsonval movement except that the permanent magnet is replaced by an electromagnet, which may be actuated by an ac current. Consequently, the field in the electromagnet may be made to operate in synchronization with an ac current in the moving coil. In order to use the electrodynamometer movement for ac measurements, it is necessary to connect the electromagnet and moving coil in series as shown in Fig. 4-7.

Both the iron-vane and the electrodynamometer movements are nor-

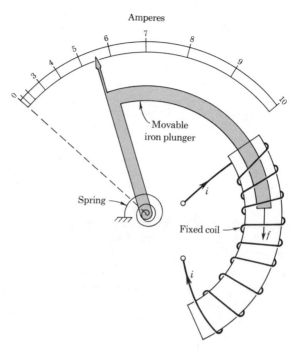

Fig. 4-5 Principle of operation of the iron-vane or moving-iron instrument.

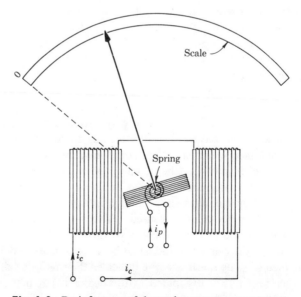

Fig. 4-6 Basic features of electrodynamometer movement.

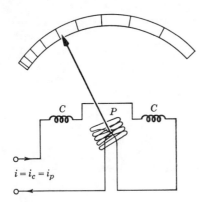

$i = i_c = i_p$

Fig. 4-7 Electrodynamometer movement used as an ammeter.

mally used for low-frequency applications with frequencies from 25 to 125 Hz. Special designs of the electrodynamometer movement may be used to extend its range to about 2,000 Hz. Various theoretical considerations for these movements are discussed by Frank [2]. Both the iron-vane and the electrodynamometer instruments indicate the rms value of the alternating current, and the meter deflection varies with I_{rms}^2 where

$$I_{rms} = \left(\frac{1}{T}\int_0^T i^2 \, d\tau\right)^{\frac{1}{2}} \tag{4-8}$$

The scale of the instrument is not necessarily based on a square law because the proportionality constant between I_{rms}^2 and the meter deflection changes somewhat with the current.

An important feature of the electrodynamometer instrument is that it may be calibrated with direct current and that the calibration will hold for ac applications within the frequency range of the instrument. The iron-vane instrument is not as versatile because of the residual magnetism in the iron when direct current is used.

A rectifier arrangement may also be used for ac measurements. In this device an ac waveform is modified by some type of rectifier such that current is obtained with a steady dc component. A D'Arsonval movement may then be used to measure the dc component and thereby indicate the value of the alternating current which was applied to the rectifier.

For measurements of high-frequency alternating currents, a thermocouple meter is usually used. This type of meter is indicated in Fig. 4-8. The alternating current is passed through a heater element, and the temperature of the element is indicated by a thermocouple connected to a D'Arsonval meter. The thermocouple indicates the rms value of the current because the average power dissipated in the heater is equal to $I_{rms}^2 R$. The instrument reading is independent of waveform because of this relationship between the

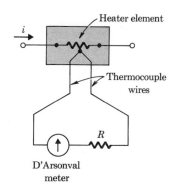

Fig. 4-8 Schematic of a thermocouple meter.

thermal emf generated in the thermocouple and the power dissipated in the heater. The thermal emf generated in the thermocouple varies approximately with the square of current, although slight deviations from the square law may be obtained because of change in heater resistance with temperature and other side effects. Alternating currents with frequencies up to 100 MHz may be measured with thermocouple meters.

4-3 VOLTMETERS

A dc voltmeter may be constructed very easily by modifying a D'Arsonval movement, as shown in Fig. 4-9. In this arrangement, a large resistor is placed in series with the movement; thus, when the instrument is connected to a voltage source, the current in the instrument is an indication of the voltage. The range of the voltmeter may be altered by changing the internal series resistor. The voltmeter is usually rated in terms of the input voltage for full-scale deflection or in terms of the ratio of internal resistance to the voltage for full-scale deflection. A series-resistor arrangement may also be used with the iron-vane and electrodynamometer instruments for measurements of rms values of ac voltages. For low frequencies up to about 125 Hz the electrodynamometer meter may be calibrated with dc voltage and the calibration used for ac measurements.

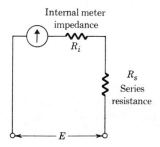

Fig. 4-9 D'Arsonval meter used as a voltmeter.

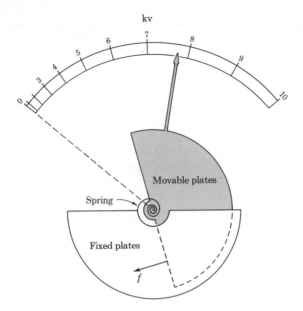

Fig. 4-10   Electrostatic-voltmeter movement.

Electrostatic forces may also be used to indicate electric potential difference.   For this purpose two plates are arranged as shown in Fig. 4-10. One plate is fixed, and the other is mounted in jeweled bearings so that it may move freely.   A spiral spring provides a restraining force on the movable plate.   The electric potential difference to be measured is impressed on the two plates.   If two complete disks were used instead of the sectoral plate arrangement, the net torque would be zero; however, with the arrangement shown, the fringe effects of the electric field produce a net force in the indicated direction which is proportional to the square of the rms voltage.   As the movable plate changes position, the capacitance changes and hence the proportionality between the stored energy and the voltage varies with the impressed voltage.

The electrostatic voltmeter may be used for either ac or dc voltage measurements, but potentials above 100 volts are required in order to produce a sufficiently strong torque in the system.   The meter may be calibrated with direct current and then used for measurement of rms values of ac voltages, regardless of the waveform.   Electrostatic voltmeters are generally applicable up to frequencies of 50 MHz.   It may be noted that the electrostatic voltmeter has an extremely high input impedance for dc applications but a much lower ac impedance as a result of the capacitance reactance.   The capacitance may be about 20 pf for a 5,000-volt meter.

In the preceding paragraphs we have discussed some of the more important devices that are used for measurement of electric current and voltage. We have not discussed electronic methods because they are generally considered as composite measurement systems and use one or more of the above devices for readout purposes.

4-4 BASIC INPUT CIRCUITRY

The output of an electric transducer is frequently of such a nature that it must be modified to produce a signal in a more usable form. At this point we wish to discuss the circuits which may be used as input-modifying circuits. The transducer may be of the *active* or *passive* type; an active element furnishes an energy input to the circuit, while a passive element merely modifies the electrical properties of the circuit through a change in some characteristic like resistance, capacitance, or inductance. A pressure transducer whose electrical resistance changes with pressure is a passive element, while a piezoelectric crystal or photoelectric cell is an example of an active element. The particular transducer must be connected to appropriate circuitry which will modify the signal so that it may eventually be displayed on an indicator or recorder. Clearly, there are many types of circuits that may be used, and we shall be able to discuss only a few representative cases.

We have already discussed the measurement of electric current with appropriate meter movements. A simple type of input circuit might use the current flow through a passive-resistance transducer as an indication of the value of the resistance. The schematic arrangement for this type of measurement is shown in Fig. 4-11. A change in the physical variable to be measured is represented in this case by a change in the resistance R, as indicated by the movable contact. The current is given by

$$i = \frac{E_i}{R + R_i} \tag{4-9}$$

Let the maximum resistance of the transducer be R_m. Then the current may be written in dimensionless form as

$$\frac{i}{E_i/R_i} = \frac{1}{(R/R_m)(R_m/R_i) + 1} \tag{4-10}$$

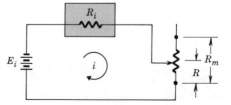

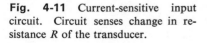

Fig. 4-11 Current-sensitive input circuit. Circuit senses change in resistance R of the transducer.

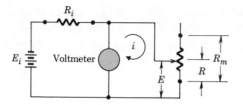

Fig. 4-12 Voltage-sensitive input circuit. Change in resistance R is indicated through change in voltage indication.

It would be desirable to have the current output vary in a linear manner with the resistance of the transducer. Unfortunately this is not the case as shown in Eq. (4-10), although the output may be approximately linear for some ranges of operation.

The circuit shown above may be modified slightly by using a voltmeter as in Fig. 4-12. Let us assume that the internal impedance of the voltmeter is very large compared with the resistance in the circuit so that we may neglect the current drawn by the meter. The current flow is still given by

$$i = \frac{E_i}{R + R_i} \tag{4-11}$$

Let E be the voltage across the transducer, as indicated in Fig. 4-12. Then

$$\frac{E}{E_i} = \frac{iR}{i(R + R_i)} = \frac{(R/R_m)(R_m/R_i)}{1 + (R/R_m)(R_m/R_i)} \tag{4-12}$$

Now we have obtained a voltage indication as a measurement of the resistance R, but a nonlinear output is still obtained. The advantage of the circuit in Fig. 4-12 over the one in Fig. 4-11 is that a voltage measurement is frequently easier to perform than a current measurement. The voltage-sensitive circuit is called a *ballast circuit*.

We might define the sensitivity of the ballast circuit as the rate of change of the voltage indication with respect to the resistance R. Thus,

$$S = \frac{dE}{dR} = \frac{E_i R_i}{(R_i + R)^2} \tag{4-13}$$

We would like to design the circuit so that the sensitivity S is a maximum. The circuit-design variable which is at our disposal is the fixed resistance R_i so that we wish to maximize the sensitivity with respect to this variable. The maximizing condition

$$\frac{dS}{dR_i} = 0 = \frac{E_i(R - R_i)}{(R_i + R)^3} \tag{4-14}$$

is applied. Thus, for maximum sensitivity we should take $R_i = R$. But, since R is a variable, we may select the value of R_i only for the range of R where the sensitivity is to be a maximum.

In both of the above circuits a current measurement has been used as an indication of the value of the variable resistance in the transducer. In some instances it is more convenient to use a voltage-divider circuit, as indicated in Fig. 4-13. In this arrangement a fixed voltage E_0 is impressed across the total transducer resistance R_m while the sliding contact is connected to the voltmeter with internal resistance R_i. If the impedance of the meter is sufficiently high, the indicated voltage E will be directly proportional to the variable resistance R; that is,

$$\frac{E}{E_0} = \frac{R}{R_m} \qquad \text{for } R_i \gg R \tag{4-15}$$

With a finite meter resistance a current is drawn which affects the voltage measurement. Considering the internal resistance of the meter, the current drawn from the voltage source is

$$i = \frac{E_0}{R_m - R + R_i R/(R + R_i)} \tag{4-16}$$

The indicated voltage is therefore

$$E = E_0 - i(R_m - R)$$

or

$$\frac{E}{E_0} = \frac{R/R_m}{(R/R_m)(1 - R/R_m)(R_m/R_i) + 1} \tag{4-17}$$

As a result of the loading action of the meter, the voltage does not vary in a linear manner with the resistance R. If Eq. (4-17) is taken as the true relationship between voltage and resistance, then an expression for the loading

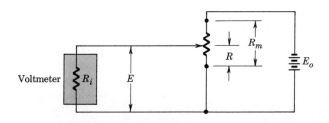

Fig. 4-13   Simple voltage-divider circuit.

error may be written as

$$\text{Loading error} = \frac{(E/E_0)_{\text{true}} - (E/E_0)_{\text{ind}}}{(E/E_0)_{\text{true}}}$$

The true value is taken from Eq. (4-17) and the indicated value from Eq. (4-15). The error based on full-scale reading is

$$\text{Loading error} = - \frac{(R/R_m)^2(1 - R/R_m)}{(R/R_m)(1 - R/R_m) + R_i/R_m} \tag{4-18}$$

The voltage-divider circuit shown in Fig. 4-13 has the disadvantage that the indicated voltage is affected by the loading of the meter. This difficulty may be alleviated by utilizing a voltage-balancing potentiometer circuit, as shown in Fig. 4-14. In this arrangement a known voltage E_0 is impressed on the resistor R_m while the unknown voltage is impressed on the same resistor through the galvanometer with internal resistance R_i and the movable contact on the resistor R_m. At some position of the movable contact the galvanometer will indicate zero current, and the unknown voltage may be calculated from

$$\frac{E}{E_0} = \frac{R}{R_m} \tag{4-19}$$

Notice that the internal resistance of the galvanometer does not affect the reading in this case; however, it does influence the sensitivity of the circuit. The voltage-balancing potentiometer circuit is widely used for precise measurements of small electric potentials, particularly those generated by thermocouples. In order to accurately determine the unknown voltage E, the supply voltage E_0 must be accurately known. A battery is ordinarily used for the supply voltage, but this represents an unreliable source because of aging characteristics. The aging problem is solved by using a standard-cell arrangement to standardize periodically the battery voltage, as shown

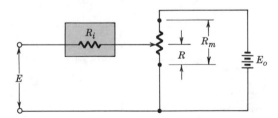

Fig. 4-14 Simple voltage-balancing potentiometer circuit.

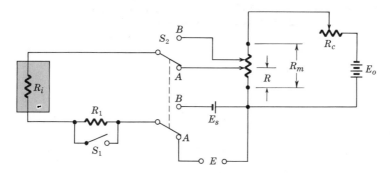

Fig. 4-15 Potentiometer circuit incorporating features for standardization of battery voltage.

schematically in Fig. 4-15.   When switch S_2 is in position A and switch S_1 is closed, the circuit is the same as Fig. 4-14.   Then, when switch S_1 is opened and switch S_2 is placed in position B, the standard cell E_s is connected in the circuit.   The variable compensator resistance R_c is adjusted until the gal- vanometer indicates balance conditions.   The protective resistor R_1 may then be bypassed by closing switch S_1 and a fine adjustment of the compensator resistor effected.   The protective resistor R_1 is placed in the circuit to avoid excessive current drain on the standard cell and also to protect the gal- vanometer.   Note that the standard-cell voltage is impressed on a fixed portion of the resistance R.   Once the battery has been standardized, switch S_1 may be placed in the A position and the unknown voltage E measured.   The battery may be standardized as often as necessary for the particular application.

Example 4-1   The output of a transducer with a total resistance of 150 ohms is to be measured with a voltage-sensitive circuit like that shown in Fig. 4-12.   The sensitivity is to be a maximum at the midpoint of the transducer.   Calculate the sensitivity at the 25 and 75 percent positions, assuming a voltage source E_i of 100 volts.

Solution   For maximum sensitivity at the midpoint of the range we take

$R_i = R = \frac{1}{2}R_m = 75$ ohms

At the 25 percent position, $R = (0.25)(150) = 37.5$ ohms and the sensitivity is calculated from Eq. (4-13):

$$S = \frac{dE}{dR} = \frac{E_i R_i}{(R_i + R)^2} = \frac{(100)(75)}{(75 + 37.5)^2} = 0.592 \text{ volt/ohm}$$

At the 75 percent position the corresponding sensitivity is

$$S = \frac{(100)(75)}{(75 + 112.5)^2} = 0.213 \text{ volt/ohm}$$

Example 4-2 The voltage-divider circuit is used to measure the output of the transducer in Example 4-1. A 100-volt source is used (E_0 = 100 volts), and the internal resistance of the meter R_i is 10,000 ohms. Calculate the loading error at the 25 and 75 percent positions.

Solution We use Eq. (4-18) for the calculation with

$$\frac{R_i}{R_m} = \frac{10,000}{150} = 66.7$$

At the 25 percent position R/R_m = 0.25 and

$$\text{Loading error} = -\frac{(0.25)^2(1 - 0.25)}{(0.25)(1 - 0.25) + 66.7}$$

$$= -0.000703 = -0.0703\%$$

At the 75 percent position R/R_m = 0.75 and

$$\text{Loading error} = -\frac{(0.75)^2(1 - 0.75)}{(0.75)(1 - 0.75) + 66.7}$$

$$= -0.00211 = -0.211\%$$

4-5 BRIDGE CIRCUITS

Bridge circuits are employed in a variety of applications for the measurement of resistance, inductance, and capacitance under both steady-state and transient conditions. We shall be concerned with the characteristics of some of the more prominent types of circuits and their application to various measurements and control.

The Wheatstone bridge is normally used for the comparison and measurement of resistances in the range of 1 ohm to 1 megohm. A schematic of the bridge is given in Fig. 4-16. It is composed of four resistances:

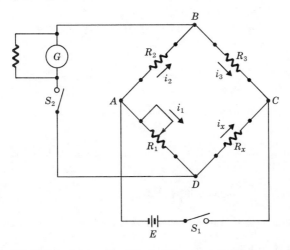

Fig. 4-16 Schematic of basic Wheatstone bridge.

R_1, R_2, R_3, R_x. R_1 is a variable resistance, whereas R_x is some unknown. A voltage is applied to the bridge by closing the switch S_1, and by adjusting the variable resistance R_1, the bridge may be balanced so that the potential at B equals the potential at D. This balance condition may be attained by connecting the galvanometer in the circuit through S_2 and adjusting the value of R_1 until the galvanometer indicates zero-current flow. At this condition the voltage drop through resistance R_2 must equal the voltage drop through R_1. Thus,

$$i_2 R_2 = i_1 R_1$$

Also,

$$i_2 = i_3 = \frac{E}{R_2 + R_3}$$

and

$$i_1 = i_x = \frac{E}{R_1 + R_x}$$

If the currents are eliminated from these relations, we obtain

$$\frac{R_2}{R_3} = \frac{R_1}{R_x} \tag{4-20}$$

or

$$R_x = \frac{R_1 R_3}{R_2} \tag{4-21}$$

If the resistances R_1, R_2, and R_3 are known, then the value of the unknown resistance R_x is easily determined.

The term "ratio arms" is frequently used in describing two known adjacent arms in a Wheatstone bridge. The galvanometer is usually connected to the junction of these two known resistors. In Fig. 4-16, R_2 and R_3 would normally be called the ratio arms.

If accurate measurements are to be made with the bridge circuit, the values of the resistors must be precisely known and the galvanometer must be sufficiently sensitive to detect small degrees of imbalance in the circuit. When the unknown resistance R_x is placed in the circuit, care must be taken to use leads which have a resistance which is small in comparison with the unknown.

Example 4-3   For the basic Wheatstone bridge in Fig. 4-16, determine the uncertainty in the measured resistance R_x as a result of an uncertainty of 1 percent in the known resistances.   Repeat for 0.05 percent.

Solution   We use Eq. (3-2) to estimate the uncertainty.   We have

$$\frac{\partial R_x}{\partial R_1} = \frac{R_3}{R_2} \qquad \frac{\partial R_x}{\partial R_2} = -\frac{R_1 R_3}{R_2{}^2} \qquad \frac{\partial R_x}{\partial R_3} = \frac{R_1}{R_2}$$

For a 1 percent uncertainty in the known resistances this gives

$$\frac{w_{R_x}}{R_x} = (0.01^2 + 0.01^2 + 0.01^2)^{\frac{1}{2}} = 0.01732$$

$$= 1.732\%$$

For a 0.05 percent uncertainty in the known resistances the corresponding uncertainty in R_x is 0.0866 percent.

The basic Wheatstone bridge circuit may also be employed for the measurement of ac impedances.   The main problem is that more than one balance condition must be made in order to obtain the null condition; one balance obtains the null on the real part of the waveform, while another obtains the null on the imaginary part.   There are some types of ac bridges that may be balanced with two *independent adjustments*.   Several of these types are shown in Table 4-2 along with the balance conditions which may be used to determine values of the unknown quantities.

Since an alternating current is involved, the null condition may not be sensed by a galvanometer as in the case of the dc Wheatstone bridge; some type of ac instrument must be used.   This might be a vacuum-tube voltmeter, an oscilloscope, or a rectifier type of meter.   A pair of headphones might also be used when audio-frequency signals are involved.   For radio-frequency signals a communication receiver could be used as the detector in the bridge circuit.

It may be noted from Table 4-2 that for some types of ac bridge circuits the balance condition is independent of frequency (types *a*, *c*, and *e*), while for others the frequency must be known in order to apply the balance conditions (types *b*, *d*, and *f*).   Thus, with the capacitances and resistances known for the Wien bridge we could use the bridge as a frequency-measuring device.

Bridge circuits are very useful for experimental measurements.   The dc bridge is widely used for measuring the resistance of various transducers such as resistance thermometers, strain gages, and other devices that register the change in a physical variable through a change in resistance. The ac bridge circuits are used for inductance and capacitance measurements. Numerous transducers operate on the principle of a change in capacitance or inductance with a change in the measured quantity, and the bridge circuits may be employed to detect these changes.   Bridge circuits are also useful in an unbalanced condition because a rather small change in one of the bridge arms can produce a very large change in the detector signal, which may be used to control other circuits.

Bridge circuits may operate on either a *null* or a *deflection principle*. The null condition has been described above as where the galvanometer or sensing device reads zero at balance conditions.   At any other condition the galvanometer reading will be *deflected* from the null condition by a certain amount, which depends on the degree of unbalance.   Thus, the signal at the galvanometer or detector may be used as an indication of the unbalance of the bridge and may indicate the deviation of one of the arms from some specified balance condition.   The use of the deflection bridge is particularly important for the measurements of dynamic signals where insufficient time is available for achieving balance conditions.

Consider the bridge circuit shown in Fig. 4-17. R_1, R_2, R_3, and R_4 are the four arms of the bridge; R_g is the galvanometer resistance; and i_b and i_g are the battery and galvanometer currents, respectively. R_b represents the resistance of the battery circuit.   When the bridge is only *slightly* unbalanced, it can be shown that the value of R_b does not appreciably influence the effective resistance of the bridge circuit as presented to the galvanometer (see Ref. [2]). As a result, the following relation for the galvanometer current may be derived:

$$i_g = \frac{E_g}{R + R_g} \tag{4-22}$$

where R is the effective resistance of the bridge circuit presented to the galvanometer and is given by

$$R = \frac{R_1 R_4}{R_1 + R_4} + \frac{R_2 R_3}{R_2 + R_3} \tag{4-23}$$

This effective resistance R is indicated in Fig. 4-18.   The voltage presented at the terminals of the galvanometer E_g is

$$E_g = E\left(\frac{R_1}{R_1 + R_4} - \frac{R_2}{R_2 + R_3}\right) \tag{4-24}$$

The voltage impressed on the bridge E depends on the battery or external circuit resistance R_b and the resistance of the total bridge circuit as presented to the battery circuit, which we shall designate as R_0.   It can be shown that for small unbalances the resistance R_0 may be calculated by assuming that

Table 4-2 Summary of bridge circuits

Circuit	Balance relations	Name of bridge and remarks
(a)	$$C_x = \frac{C_3 R_2}{R_1}$$ $$R_x = \frac{R_3 R_1}{R_2}$$	Basic Wheatstone bridge. Greatest sensitivity when bridge arms are equal.
(b)	$$\frac{C_x}{C_3} = \frac{R_2}{R_1} - \frac{R_3}{R_x}$$ $$C_x C_3 = \frac{1}{\omega^2 R_3 R_x}$$ If $\quad C_3 = C_x$ and $R_3 = R_x$ $$f = \frac{1}{2\pi R_3 C_3}$$	Wien bridge. May be used for frequency measurement with indicated relations.
(c)	$$L_x = C_1 R_1 R_2$$ $$R_x = \frac{C_1 R_1}{C_2} - R_3$$	Owen bridge.

Table 4-2 Summary of bridge circuits (*Continued*)

Circuit	Balance relations	Name of bridge and remarks
(d)	$\omega^2 LC = 1$ $R_x = \dfrac{R_3 R_1}{R_2}$	Resonance bridge. At balance conditions may be used for frequency measurement with $f = \dfrac{1}{2\pi\sqrt{LC}}$
(e)	$L_x = R_1 R_3 C$ $R_x = \dfrac{R_1 R_3}{R_2}$	Maxwell bridge.
(f)	$L_x = \dfrac{R_1 R_3 C}{1 + \omega^2 C^2 R_2{}^2}$ $R_x = \dfrac{\omega^2 C^2 R_1 R_2 R_3}{1 + \omega^2 C^2 R_2{}^2}$	Hay bridge.

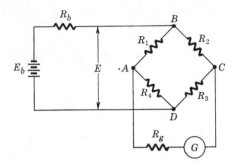

Fig. 4-17 Schematic for analysis of unbalanced bridge.

the galvanometer is not connected in the circuit.   Thus,

$$R_0 = \frac{(R_1 + R_4)(R_2 + R_3)}{R_1 + R_4 + R_2 + R_3} \tag{4-25}$$

The voltage impressed on the bridge is then

$$E = E_b \frac{R_0}{R_0 + R_b} \tag{4-26}$$

Because the deflection described above and indicated by Eq. (4-22) is based on the determination of the galvanometer current, the circuit is said to be current-sensitive. If the deflection measurement were made with a vacuum-tube voltmeter, oscilloscope, or other high-impedance device, the current flow in the detector circuit would be essentially zero since as $R_g \to \infty$, $i_g \to 0$ in Eq. (4-22).   Even so, there is still an unbalanced condition at the galvanometer or detector terminals of the bridge.   This condition is represented by the voltage between terminals A and C for the case of a high-impedance detector.   Such an arrangement is called a voltage-sensitive deflection-bridge circuit.   The voltage indication for such a bridge may be determined from Eq. (4-24).

For relatively large unbalances in a bridge circuit it may become necessary to shunt the galvanometer to prevent damage.   Such an arrangement is shown in Fig. 4-19a.   In this case the voltage across the galvanometer may still be given to a sufficiently close approximation by Eq. (4-24), but the current through the galvanometer must be computed by considering the

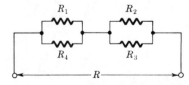

Fig. 4-18 Equivalent circuit of bridge as presented to the galvanometer.

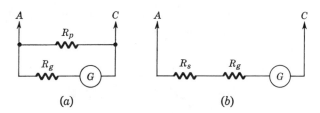

Fig. 4-19 (*a*) Galvanometer shunt arrangement for use with bridge of Fig. 4-18; (*b*) galvanometer with series resistance for use with bridge in Fig. 4-18.

shunt resistance R_p. The total resistance of the bridge and shunt combination *as seen by the galvanometer circuit* is given by

$$R_e = \frac{R_p R}{R_p + R} \qquad (4\text{-}27)$$

and the galvanometer current is now given by

$$i_g = \frac{E_g}{R_e + R_g} \qquad (4\text{-}28)$$

The galvanometer current may also be reduced by using a series resistor, as shown in Fig. 4-19*b*. For this case the galvanometer current becomes

$$i_g = \frac{E_g}{R + R_g + R_s} \qquad (4\text{-}29)$$

where R_s is the value of the series resistor.

Example 4-4 The Wheatstone bridge circuit of Fig. 4-16 has ratio arms (R_2 and R_3) of 6,000 and 600 ohms. A galvanometer with a resistance of 70 ohms and a sensitivity of 0.04 μa/mm is connected between B and D, and the adjustable resistance R_1 reads 340 ohms. The galvanometer deflection is 39 mm, and the battery voltage is 4 volts. Assuming no internal battery resistance, calculate the value of the unknown resistance R_x. Repeat for R_2 and R_3 having values of 600 and 60 ohms.

Solution In this instance the bridge is operated on the deflection principle. For purposes of analyzing the circuit we use Figs. 4-17 and 4-18. The galvanometer current is calculated from the deflection and sensitivity as

$$i_g = (39)(0.04 \times 10^{-6}) = 1.56\ \mu\text{a}$$

In the circuit of Fig. 4-17 the resistances are:

$$R_b = 0 \qquad R_1 = 340 \qquad R_2 = 6{,}000 \qquad R_3 = 600 \qquad R_4 = R_x \qquad R_g = 70$$

We also have $E = 4.0$ volts.

Combining Eqs. (4-22) to (4-24), we obtain

$$i_g = \frac{E[R_1/(R_1 + R_4) - R_2/(R_2 + R_3)]}{R_g + [R_1R_4/(R_1 + R_4) + R_2R_3/(R_2 + R_3)]}$$

Solving for R_4, we have

$$R_4 = \frac{ER_1R_3 - i_g[R_gR_1(R_2 + R_3) + R_2R_3R_1]}{i_g(1 + R_1 + R_g)(R_2 + R_3) + ER_2}$$

Using numerical values for the various quantities, we obtain

$R_4 = 33.93$ ohms

Taking $R_2 = 600$ and $R_3 = 60$, we obtain

$R_4 = 33.98$ ohms

4-6 AMPLIFIERS

Signals from experimental measurements are presented in a variety of forms: the voltage output of a bridge circuit, the frequency signal of a counting circuit, voltage signals representative of a change in capacitance, etc. In many cases these signals are rather faint and must be amplified before they can be represented adequately on a readout device. In other instances there is a serious mismatch between the measurement transducer and readout circuits so that an amplification device is called upon to provide a proper impedance conversion.

The simple *voltage amplifier* is shown in Fig. 4-20 for both the vacuum-tube and solid-state configurations. For the vacuum-tube amplifier the input voltage is applied to the grid to control the flow of electrons from cathode to plate. The output voltage is then a function of the input and the dc potential applied to the plate. The voltage gain of such devices can be quite substantial. The transistor voltage amplifier operates in a similar fashion with the input voltage applied to the base and the output present at the cathode. It is well to remark that the behavior of these amplifiers is frequency-dependent.

A fairly common application is the use of an amplifier stage to transform a signal from a high-impedance network to a low-impedance signal. The *impedance-transforming amplifiers* in Fig. 4-21 accomplish this objective. For the vacuum-tube case the input voltage is again presented at the grid, but now the output is taken across the cathode resistor R. The gain of this circuit is less than unity, but the output can draw a significant current without substantial alteration of the waveform of the signal. The circuit is called a *cathode follower*. Its transistor counterpart in Fig. 4-21b is called an *emitter follower* and provides essentially the same function. It may be noted, however, that the vacuum-tube circuit can be used for much higher impedance inputs; for example, capacitive transducers usually have too high an impedance for use with emitter-follower circuits. A condenser microphone is

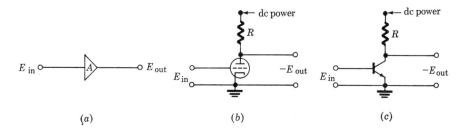

Fig. 4-20 Voltage amplifiers. (*a*) Measurement-system symbol for an amplifier; (*b*) single-stage vacuum-tube amplifier; (*c*) single-stage transistor amplifier.

basically a capacitive transducer and normally requires a cathode follower to transform the high-impedance signal to a low impedance for transmission along the microphone cable. In this case, the cathode-follower amplifier is located near the microphone in order to reduce loss in the cables.

Piezoelectric transducers produce voltage outputs at very high impedances of the order of several megohms. (See Sec. 4-17.) In order to handle such sources, the *charge amplifier* shown in Fig. 4-22 is employed. In this figure $Q(\tau)$ represents the fluctuating charge of a piezoelectric source, C_s is the capacitance of the source, R_L is the source impedance (which is very large and usually may be neglected), C_c is the capacitance of a connecting cable, A designates a voltage amplifier with high gain (1,000 or greater), and D_f is a capacitive feedback on the amplifier. For this circuit it can be shown that the overall gain (neglecting R_L) is

$$\frac{E_{\text{out}}}{Q(\tau)} = \frac{G_0}{C_s + C_c - C_f(1 + G_0)} \tag{4-30}$$

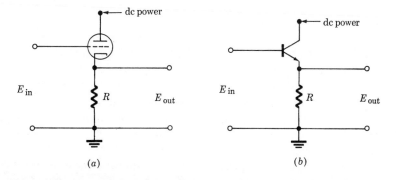

Fig. 4-21 Impedance-transforming amplifiers. (*a*) Cathode-follower circuit; (*b*) emitter-follower circuit.

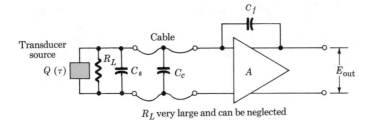

Fig. 4-22 Schematic of charge amplifier.

where G_0 is the open-circuit gain of the amplifier. Now, if the circuit is constructed so that C_f is very large and with a large gain G_0, we find that Eq. (4-30) reduces to

$$\frac{E_{\text{out}}}{Q(\tau)} \approx \frac{G_0}{-C_f G_0} = -\frac{1}{C_f} \qquad (C_f G_0 \gg C_s + C_s) \qquad (4\text{-}31)$$

Under these conditions we find that the amplifier-output voltage is directly proportional to the charge on the transducer and the transducer and cable capacitances do not influence the output. The charge amplifier is thus a very useful device for piezoelectric applications.

4-7 FILTER CIRCUITS

It is often desirable to use only a certain range of frequencies that are obtained from a transducer input. Perhaps these frequencies represent the signal, while other frequencies may be unwanted harmonics and noise resulting from some sort of distortion in the input arrangement. If a relatively narrow range of frequencies is used, it is usually possible to operate with much simpler electronic circuits for amplification purposes than when a broad frequency range is used.

Various arrangements may be used for filter circuits, but they all fall into three categories: (1) low-pass, (2) high-pass, and (3) bandpass circuits. The circuit may be constructed from passive elements or may involve electronic amplification to eliminate losses. The low-pass filter permits the transmission of frequencies below a certain cutoff value with little or no attenuation, while the high-pass filter permits the transmission of frequencies above a cutoff value. The bandpass filter permits the transmission of a certain range or band of frequencies while attenuating frequencies both above and below the limits of this range. The approximate performance curves for the three types of filters are shown in Fig. 4-23. The cutoff frequency is designated by f_c for the high- and low-pass filters, and the limits of the frequency transmission in the bandpass filter are defined by the frequencies f_1

and f_2. It must be noted that the various filter circuits do not provide a *sharp* cutoff frequency; i.e., there is a transmission of some frequencies above or below the cutoff value, although the signal attenuation becomes more pronounced as the frequency becomes further removed from cutoff conditions.

A summary of several types of passive filter circuits is given in Fig. 4-24, as recommended in Ref. [8]. The various filter sections may be used separately or in combination to produce more desirable cutoff performance. In the *m-derived sections* the quantity m is defined by

$$m = \sqrt{1 - \left(\frac{f_c}{f_\infty}\right)^2} \tag{4-32}$$

for a low-pass filter and

$$m = \sqrt{1 - \left(\frac{f_\infty}{f_c}\right)^2} \tag{4-33}$$

for a high-pass filter, where f_c is the desired cutoff frequency and f_∞ is a frequency having high attenuation. When only one m-derived section is used, a value of $m = 0.6$ is recommended [8]. When a filter is designed for a particular application, it is usually important that the impedance of the filter circuit be matched with the connecting circuitry. A consideration of the matching of impedances in filter circuits is beyond the scope of our discussion, and the interested reader should consult Ref. [9] for more information.

A measurement of the degree of amplification or attenuation of a circuit is its *gain* or *amplification ratio*. Gain is defined as

$$\text{Gain} = \frac{\text{output}}{\text{input}} \tag{4-34}$$

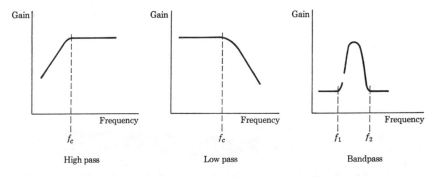

Fig. 4-23 Approximate performance curves for three types of filters.

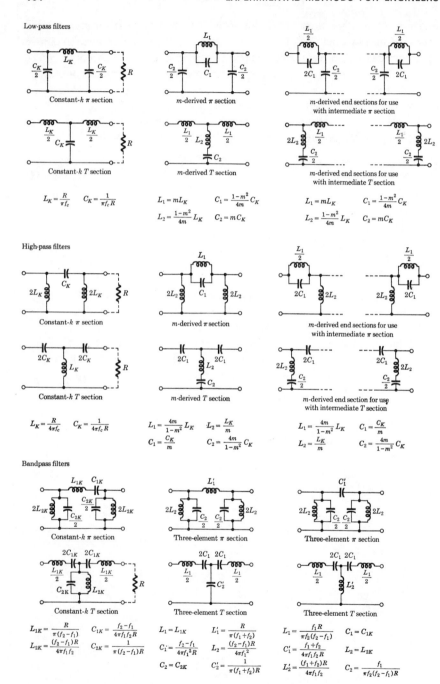

Fig. 4-24 Basic filter sections and design formulas according to Ref. [8]. R is in ohms, C is in farads, L is in henrys, and f is in cycles per second.

The output and input quantities may be voltage, current, or power, depending on the application. Gain usually is measured in *decibel units*, which are defined in terms of a logarithmic ratio.

$$\text{Decibels} = 10 \log \frac{P_2}{P_1} \qquad (4\text{-}35)$$

where P_1 and P_2 are the input and output powers, respectively. The voltage or current gain may be defined in a similar manner.

Since power may be expressed as E^2/R or I^2R, for a resistive load we would have

$$\frac{P_2}{P_1} = \frac{E_2{}^2}{E_1{}^2} = \frac{I_2{}^2}{I_1{}^2}$$

Table 4-3 Some simple *RL* and *RC* filter sections according to Ref. [9].

Diagram	Type	Time constant or resonant freq.	Formula and approximation
	A low-pass RC	$T = RC$	$\dfrac{E_{\text{out}}}{E_{\text{in}}} = \dfrac{1}{\sqrt{1 + \omega^2 T^2}} \approx \dfrac{1}{\omega T}$ $\phi_A = -\tan^{-1}(R\omega C)$
	B high-pass RC	$T = RC$	$\dfrac{E_{\text{out}}}{E_{\text{in}}} = \dfrac{1}{\sqrt{1 + \dfrac{1}{\omega^2 T^2}}} \approx \omega T$ $\phi_B = \tan^{-1}\dfrac{1}{R\omega C}$
	C low-pass RL	$T = \dfrac{L}{R}$	$\dfrac{E_{\text{out}}}{E_{\text{in}}} = \dfrac{1}{\sqrt{1 + \omega^2 T^2}} \approx \dfrac{1}{\omega T}$ $\phi_C = -\tan^{-1}\dfrac{\omega L}{R}$
	D high-pass RL	$T = \dfrac{L}{R}$	$\dfrac{E_{\text{out}}}{E_{\text{in}}} = \dfrac{1}{\sqrt{1 + \dfrac{1}{\omega^2 T^2}}} \approx \omega T$ $\phi_D = \tan^{-1}\dfrac{R}{\omega L}$

Expressed in terms of voltages or current, the decibel notation would then be

$$\text{Decibels} = 20 \log \frac{E_2}{E_1} \tag{4-36}$$

$$\text{Decibels} = 20 \log \frac{I_2}{I_1} \tag{4-37}$$

The filter circuits shown in Fig. 4-24 are the most commonly employed and are generally the most efficient passive-element types. It is possible to construct filters in RC and RL arrangements instead of the LC designs given in Fig. 4-24; however, the circuits which employ resistive elements are generally not as efficient because they remove energy from the system. Some simple RL and RC filter sections are shown in Table 4-3, along with the appropriate equations describing their input-output voltage characteristics. The phase-lag angle ϕ is also given.

Example 4-5   A 1.0-mv signal is applied to an amplifier such that an output of 1.0 volt is produced. The 1.0-volt signal is then applied to a second amplification stage to produce an output of 25 volts. For the three voltage points indicated, calculate the voltage in decibel notation referenced to (a) 1.0 mv and (b) 1.0 volt. Also calculate the overall voltage amplification in decibels.

Solution   The schematic for the amplifiers is shown in the accompanying figure.
   (a) For $E_0 = 1.0$ mv, we have

$$20 \log \frac{E_1}{E} = \text{db}_1$$

$$20 \log \frac{1.0 \times 10^{-3}}{1.0 \times 10^{-3}} = \text{db}_1 \qquad \text{db}_1 = 0$$

$$20 \log \frac{1.0}{1.0 \times 10^{-3}} = \text{db}_2 \qquad \text{db}_2 = 60$$

$$20 \log \frac{25}{1.0 \times 10^{-3}} = \text{db}_3 \qquad \text{db}_3 = 87.92$$

   (b) For $E_0 = 1.0$ volt, we have

$$20 \log \frac{1.0 \times 10^{-3}}{1.0} = \text{db}_1 \qquad \text{db}_1 = -60$$

$$20 \log \frac{1.0}{1.0} = \text{db}_2 \qquad \text{db}_2 = 0$$

$$20 \log \frac{25}{1.0} = \text{db}_3 \qquad \text{db}_3 = 27.92$$

The overall amplification is calculated from

$$20 \log \frac{E_3}{E_1} = 20 \log \frac{25}{1.0 \times 10^{-3}} = 87.92 \text{ db}$$

Notice that this result could have been obtained from the decibel calculations in (a) and (b) for either reference level. In other words

$$\text{Gain}_{13} \text{ (db)} = db_3 - db_1$$

when all are expressed on the same reference basis.

Example 4-6 A simple RC circuit is to be used as a low-pass filter. It is desired that the output voltage be attenuated 3 db at 100 Hz. Calculate the required value of the time constant $T = RC$.

Solution Network A of Table 4-3 is the desired arrangement. We wish to have

$$\frac{E_{\text{out}}}{E_{\text{in}}} = -3 \text{ db}$$

so that

$$-3 = 10 \log \frac{E_{\text{out}}}{E_{\text{in}}}$$

or

$$\frac{E_{\text{out}}}{E_{\text{in}}} = 0.501$$

Thus,

$$\frac{E_{\text{out}}}{E_{\text{in}}} = \frac{1}{(1 + \omega^2 T^2)^{\frac{1}{2}}} = 0.501$$

where $\omega^2 = (2\pi f)^2 = [2\pi(100)]^2$

We find that $T = 2.75 \times 10^{-3}$ sec. The circuit might be constructed using a 0.1-μf capacitor and a 2.75-kilohm resistor.

4-8 THE VACUUM-TUBE VOLTMETER (VTVM)

The vacuum-tube voltmeter is one of the most useful laboratory devices for the measurement of voltage. It may be used for both ac and dc measurements and is particularly valuable because of its high input-impedance characteristics, which make it applicable to the measurement of voltages in electronic circuits.

A block-diagram schematic of a simple VTVM is shown in Fig. 4-25. The input voltage is connected through appropriate terminals to the function switch. If a dc voltage is to be measured, the signal is fed directly to the range selector switch operating as a voltage-divider circuit where the signal is reduced to a suitable range for the succeeding amplifier circuit. The voltage signal from the voltage divider is then impressed on the grid circuit of the vacuum-tube–amplifier stage whose voltage output is used to drive a conventional D'Arsonval movement for readout purposes. The output from the amplifier stage could also be connected to a recorder or oscilloscope for the measurement of transient voltages.

For an ac voltage measurement the signal is fed through a voltage

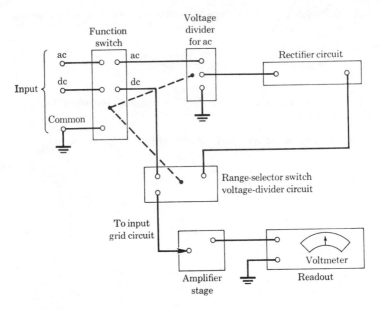

Fig. 4-25 Block-diagram schematic of a vacuum-tube voltmeter (VTVM).

divider and then to a rectifier circuit, which produces a dc voltage proportional to the input ac signal.   The dc voltage is then impressed on the voltage-divider network of the range-selector switch.   The remainder of the circuit is the same as in the case of the dc measurement.

Various modifications of the above arrangement may be used, depending on the range of voltages to be measured.   For very low-voltage ac signals an amplification stage may be added to the input instead of the voltage-divider arrangement.   For very low dc signals a chopper device might be used at the input to produce an ac signal, which is more easily amplified to the voltage levels that may be handled with conventional circuitry.

The important point is that the input impedance of the VTVM is very high, usually greater than 10 megohms, so that the measured circuit is not loaded appreciably and the indicated voltage more closely represents the true voltage to be measured.   Example 4-6 illustrates the influence that the meter can exert on a circuit and the reduction in error of measurement that can result when a VTVM is used.   It may be noted that a correction may be made for the meter impedance when the impedance of the measured circuit is known. A photograph of a typical commercial VTVM is shown in Fig. 4-26.

Although the VTVM carries the term *vacuum tube* in its name, it must be noted that many meters now employ transistors and solid-state devices for amplification and signal-modification stages.

Fig. 4-26 Photograph of a commercial vacuum-tube voltmeter. (*Courtesy Hewlett-Packard Company.*)

4-9 DIGITAL VOLTMETERS

A wide variety of voltmeters are now available that afford a digital output instead of the conventional pointer-scale arrangement. Such instruments have the advantage of reducing human reading error while at the same time increasing reading speed. Digital voltmeters operate on several different principles, and we shall not discuss all types here. We may, however, illustrate the operating characteristics of a fairly common digital-meter technique, as shown in Fig. 4-27. This meter (Hewlett-Packard HP 3440A) measures voltage by determining the time required for a linear ramp function to move from the input voltage to ground. This time period is measured with electronic-counting circuits and displayed on digital-readout tubes. In Fig. 4-27 we see the start of a voltage-measurement cycle when the voltage ramp starts. The input comparator illustrated in Fig. 4-28 continuously monitors both the voltage ramp and the input voltage. When they become equal, a gate starts the counting circuits. Another comparator monitors the

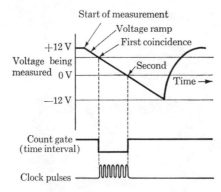

Fig. 4-27 Voltage-to-time conversion in digital voltmeter.

ramp and ground voltages, and when they become equal, a gate stops the counting circuits. The number of counts during this period is directly proportional to the input voltage. Readout circuits then convert the number of counts to a digital expression of voltage. The particular meter we have discussed here has an accuracy of about ± 0.05 percent and can perform readings at the rate of about 5/sec.

There are also digital voltmeters that are of the *integrating* type; i.e., they can determine the true average of the input voltage over a fixed measuring period. *Potentiometric*-type digital voltmeters are also available. A wide choice of input-signal modifiers, ac-dc converters, resistance-dc converters, amplifiers, etc., are available so that, in a real sense, the digital voltmeter now enables the experimentalist to make precision measurements over a broad range of variables. Not surprisingly, the cost of digital voltmeters is directly related to their accuracy and versatility.

Example 4-7 The voltage at points A and B is to be measured. A constant 100 volts is impressed on the circuit as shown. Two meters are available for the measure-

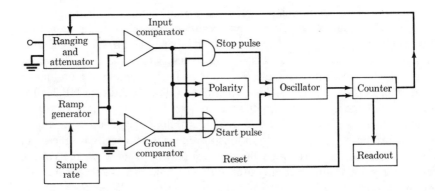

Fig. 4-28 Block diagram of HP 3440A digital voltmeter.

ment: a small volt-ohmmeter with an internal impedance of 100,000 ohms and a range of 100 volts and a VTVM with an internal impedance of 17 megohms.

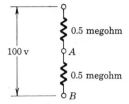

Compare the error in measurement with each of these devices.

Solution   The true voltage, by inspection, is 50 volts.   With the volt-ohmmeter connected in the circuit there results

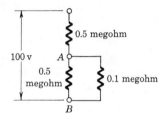

and the voltage at *A* and *B* is

$$E_{AB} = 100 \frac{1/[(1/0.5) + (1/0.1)]}{1/[(1/0.5) + (1/0.1)] + 0.5}$$

$$= 100 \left(\frac{0.0833}{0.0833 + 0.5} \right) = 14.3 \text{ volts}$$

or an error of −71 percent.
    With the VTVM connected in the circuit there results

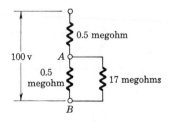

and the voltage at *A* and *B* is

$$E_{AB} = 100 \frac{1/[(1/0.5) + (1/17)]}{1/[(1/0.5 + 1/17)] + 0.5}$$

$$= 100 \left(\frac{0.4857}{0.4857 + 0.5} \right) = 49.27 \text{ volts}$$

or an error of −1.46 percent.

4-10 THE OSCILLOSCOPE

We have seen that the vacuum-tube voltmeter offers the advantage that it may be used to measure voltage without a substantial loading of the input circuit. The cathode-ray oscilloscope (CRO) is similar to the VTVM in that it has a high input impedance and is used as a voltage-measuring device. It offers the additional feature that it may be used for the measurement of rapidly varying transients with any type of waveform.

The heart of the oscilloscope is the cathode-ray tube (CRT), which is shown schematically in Fig. 4-29. Electrons are released from the hot cathode and accelerated toward the screen by the use of a positively charged anode. An appropriate grid arrangement then governs the focus of the electron beam on the screen. The exact position of the spot on the screen is controlled by the use of the horizontal and vertical deflection plates. A voltage applied on one set of plates produces the x deflection, while a voltage on the other set produces the y deflection. Thus, with appropriate voltages on the two sets of plates the electron beam may be made to fall on any particular spot on the screen of the tube. The screen is coated with a phosphorescent material, which emits light when struck by the electron beam. If the deflection of the beam against a known voltage input is calibrated, the oscilloscope may serve as a voltmeter. Since voltages of the order of several hundred volts are usually required to produce beam deflections across the entire diameter of the screen, the cathode-ray tube is not directly applicable for many low-level voltage measurements and amplification must be provided to bring the input signal up to the operating conditions for the CRT.

A schematic of the CRO is shown in Fig. 4-30. The main features are the CRT, as described above, the horizontal and vertical amplifiers, and the sweep and synchronization circuits. The sweep generator produces a sawtooth wave which may be used to provide a periodic horizontal deflection of the electron beam, in accordance with some desired frequency. This sweep then provides a time base for transient voltage measurements by means of the vertical deflection. Oscilloscopes provide internal circuits to vary the horizontal sweep frequency over a rather wide range as well as external connections for introducing other sweep frequencies. Internal switching is

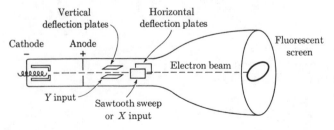

Fig. 4-29 Schematic diagram of cathode-ray tube (CRT).

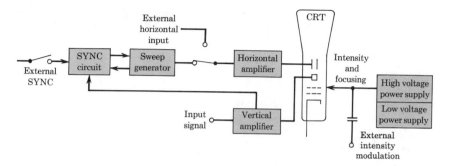

Fig. 4-30 Block-diagram schematic of an oscilloscope.

also provided which enables the operator of the scope to "lock" the sweep frequency onto the frequency impressed on the vertical input.   Provisions are also made for external modulation of the intensity of the electron beam. This is sometimes called the *z-axis input*.   This modulation may be used to cause the trace to appear on the screen during certain portions of a waveform and disappear during other portions.   It may also be used to produce traces of a specified time duration on the screen of the CRT so that a time base is obtained along with the waveform under study.   A photograph of a commercial single-beam oscilloscope is given in Fig. 4-31.

A *dual-beam oscilloscope* provides for amplification and display of two

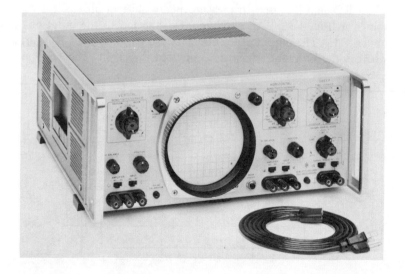

Fig. 4-31 Photograph of a commercial oscilloscope.   (*Courtesy Hewlett-Packard Company.*)

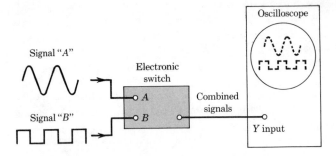

Fig. 4-32 Use of an electronic switch to achieve a dual-beam function with a single-beam oscilloscope.

signals at the same time, thereby permitting direct comparison of the signals on the CRT screen.   A single-beam oscilloscope may be given a dual-beam function through the use of an electronic switch.   This device switches between two inputs and alternately impresses the two signals at the input terminals of the single-beam oscilloscope.   The switching rate may be varied to accommodate different frequency input signals.   A schematic of the system is given in Fig. 4-32.   Notice the chopped nature of the signal displays on the CRT screen as a result of the switching process.   The width of the gaps in the displays is determined by the switching rate in the electronic switch.

PHASE-SHIFT MEASUREMENTS

The CRO may be used to measure phase shift in an electronic circuit, as shown in Fig. 4-33.   An oscillator is connected to the input of the circuit under test.   The output of the circuit is connected to the CRO vertical input whereas the oscillator signal is connected directly to the horizontal input.

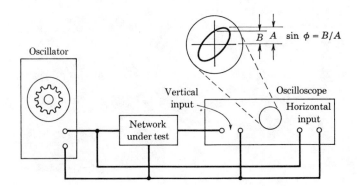

Fig. 4-33 Schematic illustrating use of oscilloscope for phase measurements.

The phase-shift angle ϕ may be determined from the relation

$$\phi = \sin^{-1} \frac{B}{A} \tag{4-38}$$

where B and A are measured as shown in Fig. 4-33.   For zero phase shift the ellipse will become a straight line with a slope of 45° to the right; for 90° phase shift it will become a circle; and for 180° phase shift it will become a straight line with a slope of 45° to the left.

LISSAJOUS DIAGRAMS

The CRO offers a convenient means of comparing signal frequencies through the use of Lissajous diagrams.   Two frequencies are impressed on the CRO inputs, one on the horizontal input and one on the vertical input.   One of these frequencies may be a known frequency as obtained from a variable frequency oscillator or signal generator.   If the two input frequencies are the same, an ellipse will be displayed on the CRT screen.   If the two input frequencies are not the same, the patterns that are displayed on the CRT screen are called Lissajous diagrams.   There is a distinct relationship that governs the shape of these diagrams in accordance with the input frequencies:

$$\frac{\text{Vertical input frequency}}{\text{Horizontal input frequency}}$$

$$= \frac{\text{number of vertical maxima on Lissajous diagram}}{\text{number of horizontal maxima on Lissajous diagram}} \tag{4-39}$$

Some typical shapes for the Lissajous diagrams are shown in Fig. 4-34.   It may be noted that these shapes can vary somewhat, depending on the phase relation between the input signals.

Oscilloscope traces may be recorded by various photographic methods, and several cameras are manufactured especially for oscilloscope applications.

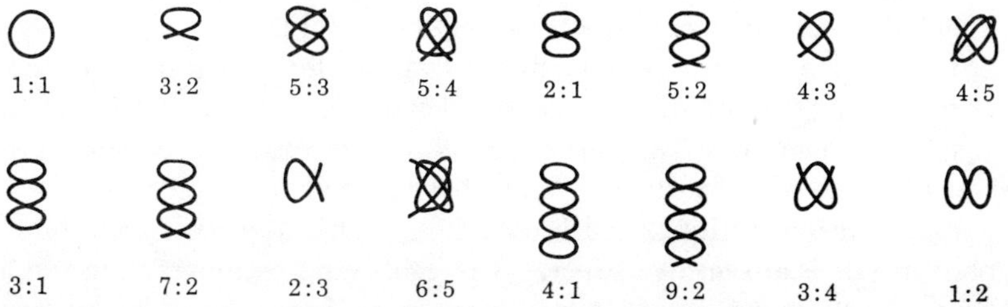

| 1:1 | 3:2 | 5:3 | 5:4 | 2:1 | 5:2 | 4:3 | 4:5 |
| 3:1 | 7:2 | 2:3 | 6:5 | 4:1 | 9:2 | 3:4 | 1:2 |

Fig. 4-34   Lissajous diagrams for various frequency ratios as indicated.

Example 4-8 An oscilloscope is used for a phase-shift measurement in accordance with the arrangement in Fig. 4-33. The measurements are

$A = 2.5 \pm 0.05$ cm

$B = 1.5 \pm 0.05$ cm

Calculate the nominal value of the phase angle, and estimate the uncertainty in the measurement.

Solution We calculate the nominal value with Eq. (4-38).

$$\phi = \sin^{-1}\frac{B}{A} = \sin^{-1}\frac{1.5}{2.5} = 36.9°$$

We may estimate the uncertainty with the use of Eq. (3-2). Observing that

$$\frac{\partial[\sin^{-1}(B/A)]}{\partial(B/A)} = \frac{1}{[1-(B/A)^2]^{\frac{1}{2}}}$$

we have

$$\frac{\partial\phi}{\partial B} = \frac{1}{[1-(B/A)^2]^{\frac{1}{2}}}\frac{1}{A} = \frac{1}{(1-0.36)^{\frac{1}{2}}}\frac{1}{1.5} = 0.833 \text{ rad/cm}$$

$$\frac{\partial\phi}{\partial A} = -\frac{B}{[1-(B/A)^2]^{\frac{1}{2}}}\frac{1}{A^2} = \frac{-1.5}{(1-0.36)^{\frac{1}{2}}}\left(\frac{1}{2.5}\right)^2 = -0.3 \text{ rad/cm}$$

Also,

$w_B = 0.05$ cm

$w_A = 0.05$ cm

so that the uncertainty in ϕ is

$$w_\phi = [(0.833)^2(0.05)^2 + (0.3)^2(0.05)^2]^{\frac{1}{2}}$$

$$= 1.4 \times 10^{-2} \text{ rad} = 0.80°$$

or

$$\frac{w_\phi}{\phi} = \frac{0.8}{36.9} = 0.0215 = 2.15 \text{ percent}$$

4-11 OSCILLOGRAPHS

The oscillograph is basically a recording D'Arsonval movement. Various input amplifiers are available for raising the signal to an acceptable level for driving the movement. Two types of galvanometer movements are used: the direct-writing, or stylus, movement and the light-beam movement. In the stylus movement the galvanometer deflection is recorded directly on recording paper by action of the stylus. Three types of stylus are in common use: ink-writing, pressure-writing, and heated styli. The latter two write on either pressure- or heat-sensitive paper. The light-beam oscillograph uses a light-beam galvanometer, a light source to produce a fine beam, and light-sensitive paper for recording. The movement of a stylus oscillograph is similar to that of the indicating D'Arsonval movement shown in Fig. 4-4 except that the pointer is replaced by a stylus, which writes on the recording paper. The construction of the light-beam oscillograph is somewhat similar to that of the light-beam galvanometer except that a lens system is used to focus the light beam to a small point in order to produce a fine trace on the recording paper. Since input amplification is provided, the galvanometer may be constructed in a more substantial fashion than in the system of Fig. 4-3. Typically, the mirror and coil are placed in an integral unit, which may be inserted in the oscillograph instrument. The ends of the unit allow it to pivot freely and reflect the light beam onto the recording paper. This unit is frequently not much larger than a matchstick. A photograph of a typical mirror or galvanometer for use in an oscillograph is shown in Fig. 4-35a, and the construction details are shown in Fig. 4-35b.

The frequency response of an oscillograph is primarily a function of the mechanical frequency response of the galvanometer, which, in turn, is primarily dependent on its mass moment of inertia and damping characteristics. The smaller the mass of the galvanometer movement, the higher the frequency response. For this reason, the light-beam oscillograph has a much higher frequency response than the stylus type because of the higher mass associated with the stylus movement. In general, stylus-type oscillographs exhibit flat frequency responses from 0 to 100 Hz, while light-beam oscillographs may be used up to 5,000 Hz.

A multitude of preamplifiers, chart speeds, and remote operation facilities are available on commercial oscillographs. Commercial models are available with 1 to 50 separate channels. A photograph of a typical eight-channel light-beam oscillograph is shown in Fig. 4-36. The input preamplifiers and controls for each channel are shown at the top of the cabinet, while the chart magazine is shown in the right-center portion of the cabinet.

The oscillograph has an accuracy of about ± 2 percent of full scale. For more precise recording of signals a chart recorder is used with an internal servo control, which acts as a self-balancing potentiometer circuit. Such recorders have accuracies of the order of 0.2 percent of full scale but are only adaptable to low-frequency signals below about 5 Hz. These recorders are widely used for thermocouple measurements and standard models are available with a full-scale range as low as 1 mv. Most of these recorders use a Zener-diode device to produce the reference voltage rather than a standard cell because such a source is quite stable and does not require frequent standardization.

4-12 COUNTERS, TIME, AND FREQUENCY MEASUREMENTS

The engineer is called upon to perform counting-rate and frequency measurements over an extremely broad range of time intervals. A determination of

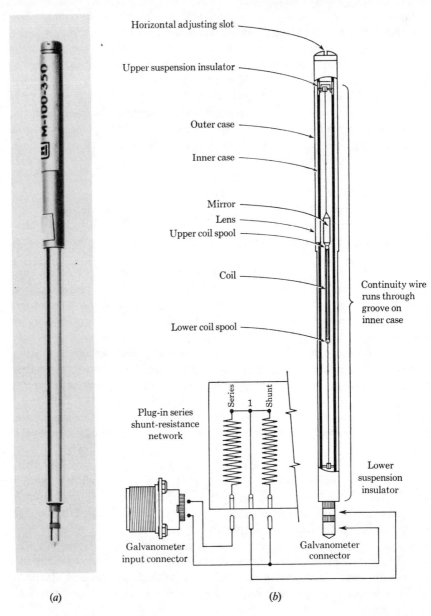

Fig. 4-35 (*a*) Photograph of a typical oscillograph galvanometer (the galvanometer is approximately the size of a matchstick); (*b*) schematic diagram of matchstick galvanometer and electric connections. (*Courtesy Honeywell, Inc.*)

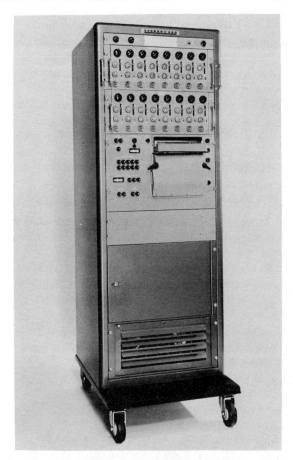

Fig. 4-36 Photograph of a commercial eight-channel oscillograph system. (*Courtesy Sanborn Company.*)

revolutions per minute on a slow-speed steam engine might involve a simple mechanical revolutions counter and a hand-operated stopwatch. In this case, the accuracy of the measurement would depend on the human-response time in starting and stopping the watch. The same measurement performed on an automobile engine might utilize an electric transducer, which generated a pulse for each revolution of the engine. The pulses could be fed to some type of counting device, which established the number of pulses produced for a given increment of time. The accuracy of measurement would again depend on the accuracy of the specification of the time interval. In both of these simple cases a circular frequency measurement is being performed through the combination of a counting measurement and time-interval measurement. Aside from the simple frequency measurements discussed with the CRO, most frequency measurements are performed by some type of counting operation.

A large variety of electronic counters are available commercially. These instruments have internal circuitry which enables them to be used for measurement of frequency, period, or time intervals over a very wide range. These instruments usually contain four sets of internal circuits:

1. Input-signal conditioning circuits, which transform the input signal into a series of pulses for counting
2. The time base, which provides precise time increments during which the pulses are counted
3. A signal "gate," which starts and stops the counting device
4. A decade counter and display, which counts the pulses and provides a digital readout

A photograph of a typical commercial counter for measurements of frequencies between 10 and 1.2 Hz is given in Fig. 4-37.

The functions of the electronic counter are shown schematically in Fig. 4-38. In Fig. 4-38a a frequency measurement is performed. The input signal is fed to the signal shaper, which produces an equivalent pulse train, which in turn is fed to the gate along with a predetermined time-base signal. The counters subsequently register the number of pulses in the given time,

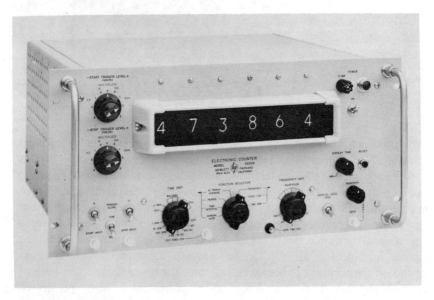

Fig. 4-37 Photograph of a commercial electronic counter. (*Courtesy Hewlett-Packard Company.*)

which is interpreted in terms of frequency. A period measurement is indicated in Fig. 4-38*b*. In this instance the time base is used to provide a known number of pulses per unit time, while the input signal is used to actuate the gate circuit. The measured count is thus an indication of the period of the input signal. For time-interval measurements the arrangement in Fig. 4-38*c* is used. Again, the time base provides a known number of pulses per unit time. Either external or internal trigger circuits provide the signals to the gate circuits to start and stop the count. These trigger circuits may be made to actuate on certain polarities or peak values of a waveform, depending on the measurement which is desired.

Phase-angle measurements may be made by measuring the time interval between similar points on the two waveforms. Calibration of the time-base circuits may be achieved by the use of special frequency standards based on tuning-fork or crystal oscillators, or with NBS standard broadcast frequencies. A very detailed discussion of frequency and time standards is given in Ref. [6].

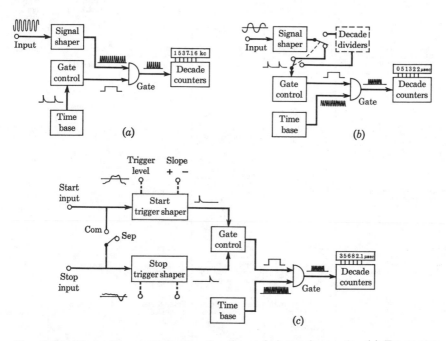

Fig. 4-38 Block diagrams indicating functions of electronic counter. (*a*) Frequency measurement: function selector is set to "frequency," and appropriate gate time is selected by time-base switch (combined with function selector on some counters); (*b*) period measurement: function selector is set to "period," and counted frequency is chosen by time-base switch; (*c*) time-interval measurements: start and stop triggers may be derived from two separate sources, as shown, or from different points of one waveform by setting COM-SEP switch to COM. (*Courtesy Hewlett-Packard Company.*)

Table 4-4 Summary of transducer characteristics

Type of transducer and principle of operation	Type of input	Input range or level	Input-impedance characteristics	Input sensitivity	Error and noise characteristics
Variable resistance: Movement of contact on slide-wire; also called resistance potentiometers.	Linear displacement or angular displacement	Minimum level as low as 0.1% of total resistance	Varies widely, depending on the total resistance characteristics and physical size	Commercial potentiometers can have sensitivity of less than 0.002 in., or 0.2° in an angular measurement	Deviation from nonlinearity of the order of 0.5% of total resistance. Noise is usually negligible of the order of 10 μv at the contact. Noise increases with "chatter" of contact
Differential transformer: See Fig. and text discussion. Linear variable differential transformer most widely used. Converts displacement to voltage.	Linear displacement	Total range from ± 0.005 to ± 3 in.	Depends on size. Forces from 0.1 to 0.3 g usually required	0.5% of total input range	Deviation from linearity about 0.5%; generally accurate to ± 1%
Capacitive transducer: Variable distance between plates registered as a change in capacitance.	Displacement or change in dielectric constant between plates. Also change in area of plates	Very broad. From 10^{-8} cm to several meters	The input force requirements are very small, of the order of a few dynes	Highly variable; can obtain sensitivities of the order of 1 pf/0.0001 in. for air-gap measurements of displacement	Errors may result from careless mechanical construction, humidity variations, noise, and stray capacitance in cable connections
Piezoelectric effect: Force impressed on crystals with asymmetrical charge distributions produces a potential difference at the surface of the crystal.	Force or stress	Varies widely with crystal material. See sensitivity	Input force requirements are relatively large compared with other transducers	Varies with material: Quartz 0.05 v-m/newton Rochelle salt 0.15 v-m/newton Barium titanate 0.007 v-m/newton	Subject to hysteresis and temperature effects
Photoelectric effect: Light striking metal cathode causes liberation of electrons which may be attracted to anode to produce electric current.	Light	Wavelength range depends on glass-tube enclosure. Photoemissive materials respond between 0.2 and 0.8 μ	Not applicable	Vacuum tubes: 0.002–0.1 μa/μwatt Gas-filled tubes: 0.01–0.15 μa/μwatt	Depends on plate voltage but of order of 10^{-8} amp at room temperature
Photoconductive transducer: Light striking a semiconductor material, such as selenium, metallic sulfides, or germanium, produces a decrease in resistance of the material.	Light	Very broad. From thermal radiation through the X-ray region	Not applicable	About 300 μa/μwatt at maximum sensitivity of the device	Very low noise, usually less than associated circuit
Photovoltaic cell: Light falling upon a semiconducting material in contact with a metal plate produces a potential.	Light	Depends on material. Selenium 0.2 to 0.7 μ, CuO 0.5 to 1.4 μ, germanium 1.0 to 1.7	Not applicable	1 ma/lumen or 10^{-7} watt/cm²-lumen	Low noise
Ionization transducer: Displacement converted to voltage through a capacitance change.	Displacement, 0.1 to 10 Hz excitation frequency	Less than 1 mm to several in.	Small force required	1–10 volts/mm	Can be accurate to microinches
Magnetometer search coil: Changing magnetic field impressed on coil generates an emf proportional to the time rate of change of the field.	Changing magnetic field	10^{-3} oersted to highest values obtainable	Not applicable	Depends on coil dimensions but can be of the order of 10^{-5} oersted	Accuracies of 0.05% have been obtained
Hall-effect transducer: Magnetic field impressed on a plate carrying an electric current generates a potential difference in a direction perpendicular to both the current and the magnetic field.	Magnetic field	1–20,000 gauss	Not applicable	Depends on plate thickness and current; of the order of -1×10^{-8} volt-cm/amp-gauss for bismuth	Can be calibrated within 1%

Frequency response	Temperature effects	Type of output	Output range or level	Output-impedance characteristics	Application	Remarks
Generally not above 3 Hz for commercial potentiometers	0.002 to 0.15% °C^{-1} due to a change in resistance. Also, some thermoelectric effects depending on types of contacts used	Voltage or current depending on connecting circuit	Wide	Variable	Used for measurement of displacement	Simple, inexpensive, easy to use, many types available commercially
Frequency of applied voltage must be 10 times desired response. Mechanical limitations also	Small influence of temperature may be reduced by using a thermistor circuit	Voltage proportional to input displacement	0.4 to 4.0 mv/0.001-in./v input depending on frequency. Lower frequency produces lower output	Mainly resistive; low to medium impedance, as low as 20 ohms, depending on size	Used for measurement of displacement	Simple, rugged, inexpensive, high output, requires simple accessory equipment. Care must be taken to eliminate stray magnetic fields
Depends strongly on mechanical construction but may go to 50,000 Hz	Not strong if design allows for effects	Capacitance	Usually between 10^{-3} and 10^3 pf change in capacitance over output range	Usually 10^3 to 10^7 ohms	Displacement, area liquid level, pressure, sound-level measurements, and others. Particularly useful where small forces are available for driving the transducer	High output impedance may require careful construction of output circuitry
Depends on external circuitry and mechanical mounting. 20–20 kHz easily obtained; no response to steady-state forces	Wide variation in crystal properties with temperature	Voltage proportional to input force	Wide, depends on crystal size and material; see sensitivity; can have output of several volts	High, of the order of 10^3 megohms	Measurement of force, pressure, sound level (microphone use)	Simple, inexpensive, rugged
Linear 0–500 Hz; current response drops off 15% at 10,000 Hz	Generally not operable above 75 to 100°C	Current	Of the order of 2 μa	High, of the order of 10 megohms	Very useful for counting purposes	Inexpensive, high output
Rise time varies widely with material and incident radiation from 50 μsec to minutes	Response to longer wavelengths increases with reduction in temperature	Current drawn in the external circuit	Depends on incident intensity; see input sensitivity	High, varies from 1 to 10^3 megohms in commercial devices	Widely used for radiant measurements at all wavelengths	Fairly expensive; calls for precise circuitry to utilize its full potential
Rise time of the order of 1 μsec, response into the megacycle range	Variations of 10% over 40°C range depending on external resistance load	Voltage	100–250 mv in normal room light. Selenium cells up to 500 volts at high illumination	3,000 to 10,000 ohms. Capacitance of order of 0.05 farad/cm^2 in selenium cells	Widely used for exposure meters, selenium cells, responds to X-rays	Inexpensive, nonlinear behavior, some aging effects
–3,000 Hz	Small	Voltage	Depends on excitation circuit; see input sensitivity	High, of the order of 1 megohm	Can be used where accurate measurement of displacement is needed	Relatively insensitive to frequency of excitation circuitry
radiofrequency	Small	Voltage	Depends on coil size; see input sensitivity	Depends on coil size	Measurement of magnetic field	Simple, although accurate, coil dimensions must be maintained for high accuracy
High	Large but can be calibrated	Voltage	Millivolts and microvolts	Low, of the order of 100 ohms for a bismuth detector	Measurement of magnetic field	Each transducer usually must be calibrated because of nonuniformity in semiconductors used for construction

4-13 TRANSDUCERS AND ELECTRIC SENSING DEVICES

A large number of devices operate on the principle of transforming an input representing a physical variable into an electric signal. Such devices are called transducers. In the following sections we shall discuss some of the more widely used transducers and their principles of operation. Table 4-4 presents a compact summary of transducer characteristics, and the reader should consult this table throughout the following discussions to maintain an overall perspective. It may be noted that the subsequent paragraphs will be primarily concerned with principles of operation without repeating much of the detailed information contained in Table 4-4. Thus, the verbal discussion and tabular presentation are complementary and should be used jointly.

It should be noted that the discussions of transducers that follow are concerned primarily with electric effects. Other transducers of a more specialized nature (thermocouples, strain gages, pressure transducers, and nuclear radiation transducers) will be discussed in subsequent chapters.

4-14 THE VARIABLE-RESISTANCE TRANSDUCER

The variable-resistance transducer is a very common device, which may be constructed in the form of a moving contact on a slide-wire, a moving contact on a coil of wire, through either linear or angular movement, or a contact that moves through an angular displacement on a solid conductor like a piece of graphite. The device may also be called a *resistance potentiometer* or *rheostat* and is available commercially in many sizes, designs, and ranges. Costs can range from a few cents for a simple potentiometer used as a volume control in a radio circuit to hundreds of dollars for a precision device used for accurate laboratory work.

The variable-resistance transducer fundamentally is a device for converting either linear or angular displacement into an electric signal; however, through mechanical methods it is possible to convert force and pressure to a displacement so that the device may also be useful in force and pressure measurements.

4-15 THE DIFFERENTIAL TRANSFORMER (LVDT)

A schematic diagram of the differential transformer is shown in Fig. 4-39. Three coils are placed in a linear arrangement as shown with a magnetic core which may move freely inside the coils. The construction of the device is indicated in Fig. 4-40. An alternating input voltage is impressed on the center coil, and the output voltage from the two end coils depends on the magnetic coupling between the core and the coils. This coupling is, in turn, dependent on the position of the core. Thus, the output voltage of the device is an indication of the displacement of the core. As long as the core remains

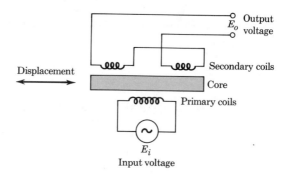

Fig. 4-39 Schematic diagram of a differential transformer.

near the center of the coil arrangement, the output is very nearly linear, as indicated in Fig. 4-41. The linear range of commercial differential transformers is clearly specified, and the devices are seldom operated outside this range. When operating in the linear range, the device is called a linear variable differential transformer (LVDT). Near the null position a slight nonlinearity condition is encountered, as illustrated in Fig. 4-42. It will be noted that Fig. 4-41 considers the phase relationship of the output voltage, while the "V" graph in Fig. 4-42 indicates the absolute magnitude of the output. There is a 180° phase shift from one side of the null position to the other.

The frequency response of LVDTs is primarily limited by the inertia characteristics of the device. In general, the frequency of the applied voltage should be 10 times the desired frequency response.

Commercial LVDTs are available in a broad range of sizes and are widely used for displacement measurements in a variety of applications.

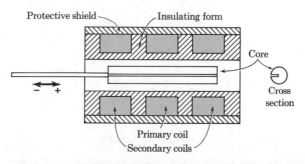

Fig. 4-40 Construction of a commercial linear variable differential transformer (LVDT). (*Courtesy Schaevitz Engineering Company.*)

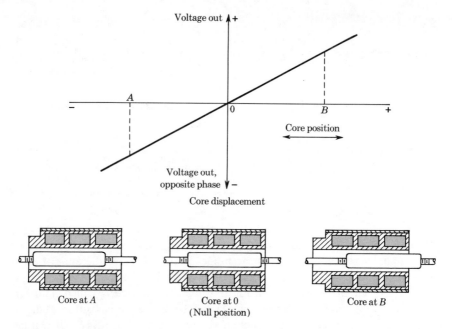

Fig. 4-41   Output characteristics of an LVDT according to Ref. [7].

Force and pressure measurements may also be made after a mechanical conversion.   Table 4-4 indicates the general characteristics of the LVDT. The interested reader should consult Ref. [7] for more detailed information.

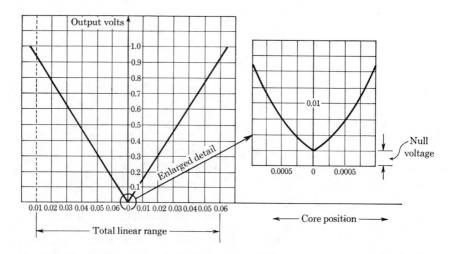

Fig. 4-42   V graph for an LVDT showing slight nonlinear behavior in the null region [7].

4-16 CAPACITIVE TRANSDUCERS

Consider the capacitive transducer shown in Fig. 4-43.   The capacitance (in picofarads) of this arrangement is given by

$$C = 0.225\epsilon \frac{A}{d} \tag{4-40}$$

where d = distance between the plates, in.
   A = *overlapping* area, in.^{2}
   ϵ = dielectric constant

This plate arrangement may be used to measure a change in the distance d through a change in capacitance.   A change in capacitance may also be registered through a change in the overlapping area A resulting from a relative movement of the plates in a lateral direction or a change in the dielectric constant of the material between the plates.   The capacitance may be measured with bridge circuits.   The output impedance of a capacitor is given by

$$Z = \frac{1}{2\pi f C} \tag{4-41}$$

where Z = impedance, ohms
   f = frequency, Hz
   C = capacitance, farads

In general, the output impedance of a capacitive transducer is high; this fact may call for careful design of the output circuitry.
   The capacitive transducer may be used for displacement measurements through a variation of either the spacing distance a or the plate area.   It is commonly used for liquid-level measurements, as indicated in Fig. 4-44. Two electrodes are arranged as shown, and the dielectric constant varies between the electrodes according to the liquid level.   Thus, the capacitance between the electrodes is a direct indication of the liquid level.

Example 4-9   A capacitive transducer is constructed of two 1-in.^{2} plates separated by a distance of 0.01 in. in air.   Calculate the displacement sensitivity of such an arrangement.   The dielectric constant for air is 1.0006.

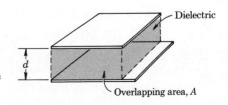

Fig. 4-43 Schematic of a capacitive transducer.

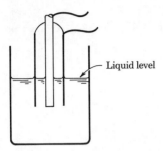

Liquid level

Fig. 4-44 Use of capacitive transducer for liquid-level measurement.

Solution   The sensitivity is found by differentiating Eq. (4-40).

$$S = \frac{\partial C}{\partial d} = -\frac{0.225 \epsilon A}{d^2}$$

Thus,

$$S = -\frac{(0.225)(1.0006)(1)}{(0.01)^2} = -2.25 \times 10^3 \text{ pf/in.}$$

Example 4-10   For the capacitive transducer in Example 4-9 the allowable uncertainty in the spacing measurement is $w_d = \pm 0.0001$ in., while the estimated uncertainty in the plate area is ± 0.005 in.^{2}.   Calculate the tolerable uncertainty in the capacitance measurement in order to achieve the allowable uncertainty in the spacing measurement.

Solution   Solving Eq. (4-40) for d, we have

$$d = 0.225 \frac{\epsilon A}{C} \tag{a}$$

Making use of Eq. (3-2), we obtain

$$\frac{w_d}{d} = \left[\left(\frac{w_c}{C} \right)^2 + \left(\frac{w_A}{A} \right)^2 \right]^{\frac{1}{2}} \tag{b}$$

We have

$$\frac{w_d}{d} = \frac{0.0001}{0.01} = 0.01 \qquad \frac{w_A}{A} = \frac{0.005}{1.0} = 0.005$$

so that

$$\frac{w_c}{C} = 0.00866 = 0.866\%$$

The nominal value of C is

$$C = \frac{(0.225)(1.0006)(1.0)}{0.01} = 22.513 \text{ pf}$$

so that the tolerable uncertainty in C is

$$w_C = (22.513)(0.00866) = \pm 0.195 \text{ pf}$$

4-17 PIEZOELECTRIC TRANSDUCERS

Consider the arrangement shown in Fig. 4-45. A piezoelectric crystal is placed between two plate electrodes. When a force is applied to the plates, a stress will be produced in the crystal and a corresponding deformation. With certain crystals this deformation will produce a potential difference at the surface of the crystal, and the effect is called the *piezoelectric effect*. The induced charge on the crystal is proportional to the impressed force and is given by

$$Q = dF \tag{4-42}$$

where Q is in coulombs, F is in newtons, and the proportionality constant d is called the piezoelectric constant. The output voltage of the crystal is given by

$$E = gtp \tag{4-43}$$

where t is the crystal thickness in meters, p is the impressed pressure in newtons per square meter, and g is called voltage sensitivity and is given by

$$g = \frac{d}{\epsilon} \tag{4-44}$$

Values of the piezoelectric constant and voltage sensitivity for several common piezoelectric materials are given in Table 4-5.

The voltage output depends on the direction in which the crystal slab is cut in respect to the crystal axes. In Table 4-5 an X (or Y) cut means that a perpendicular to the largest face of the cut is in the direction of the x axis (or y axis) of the crystal.

Piezoelectric crystals may also be subjected to various types of shear stresses instead of the simple compression stress shown in Fig. 4-45, but the output voltage is a complicated function of the exact crystal orientation. Piezoelectric crystals are widely used as inexpensive pressure transducers for dynamic measurements and are commonly employed as phonograph pickups. General information on the piezoelectric effect is given by Cady [1].

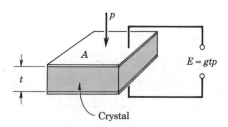

Fig. 4-45 The piezoelectric effect.

Table 4-5   Piezoelectric constants†

Material	Orientation	Charge sensitivity d, $\dfrac{coulombs/m^2}{newtons/m^2}$	Voltage sensitivity g, $\dfrac{volts/m}{newtons/m^2}$
Quartz	X cut; length along Y length longitudinal	2.25×10^{-12}	0.055
	X cut; thickness longitudinal	−2.04	−0.050
	Y cut; thickness shear	4.4	−0.108
Rochelle salt	X cut 45°; length longitudinal	435.0	0.098
	Y cut 45°; length longitudinal	−78.4	−0.29
Ammonium dihydrogen phosphate	Z cut 0°; face shear	48.0	0.354
	Z cut 45°; length longitudinal	24.0	0.177
Commercial barium titanate ceramics	To polarization	130–160	0.0106
	To polarization	−56.0	0.0042–0.0053

† According to Lion [3].

Example 4-11   A quartz piezoelectric crystal having a thickness of 2 mm and a voltage sensitivity of 0.055 volt-m/newton is subjected to a pressure of 200 psi.   Calculate the voltage output.

Solution   We calculate the voltage output with Eq. (4-43):

$$p = (200)(6.895 \times 10^3) = 1.38 \times 10^6 \text{ newtons/m}^2$$
$$t = 2 \times 10^{-3} \text{ m}$$

Thus,

$$E = (0.055)(2 \times 10^{-3})(1.38 \times 10^6) = 151.8 \text{ volts}$$

4-18   PHOTOELECTRIC EFFECTS

A photoelectric transducer converts a light beam into a usable electric signal. Consider the circuit shown in Fig. 4-46.   Light strikes the photoemissive

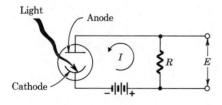

Fig. 4-46   The photoelectric effect.

cathode and releases electrons, which are attracted toward the anode, thereby producing an electric current in the external circuit. The cathode and anode are enclosed in a glass or quartz envelope, which is either evacuated or filled with an inert gas. The photoelectric sensitivity is defined by

$$I = S\Phi \tag{4-45}$$

where I = photoelectric current
Φ = illumination of the cathode
S = sensitivity

The sensitivity is usually expressed in units of amperes per watt or amperes per lumen.

Photoelectric-tube response to different wavelengths of light is influenced by two factors: (1) the transmission characteristics of the glass-tube envelope and (2) the photoemissive characteristics of the cathode material. Photoemissive materials are available which will respond to light over a range of 0.2 to 0.8 μ. Most glasses transmit light in the upper portion of this range, but many do not transmit below about 0.4 μ. Quartz, however, transmits down to 0.2 μ. Various noise effects are present in photoelectric tubes, and the interested reader should consult the discussion by Lion [3] for more information.

Photoelectric tubes are quite useful for measurement of light intensity. Inexpensive devices can be utilized for counting purposes through periodic interruption of a light source.

4-19 PHOTOCONDUCTIVE TRANSDUCERS

The principle of the photoconductive transducer is shown in Fig. 4-47. A voltage is impressed on the semiconductor material as shown. When light strikes the semiconductor material, there is a decrease in the resistance, thereby producing an increase in the current indicated by the meter. A variety of substances are used for photoconductive materials, and a rather

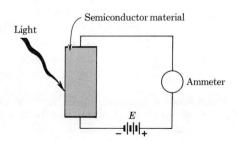

Fig. 4-47 Schematic diagram of a photoconductive transducer.

detailed discussion of the pertinent literature on the subject is given in Refs. [3] and [10].

Photoconductive transducers enjoy a wide range of applications and are useful for measurement of radiation at all wavelengths. It must be noted, however, that extreme experimental difficulties may be encountered when operating with long-wavelength radiation.

The responsivity R_v of a detector is defined as

$$R_v = \frac{\text{rms output voltage}}{\text{rms power incident upon the detector}} \tag{4-46}$$

The *noise equivalent power* (NEP) is defined as the minimum-radiation input that will produce a signal-to-noise ratio of unity. The *detectivity D* is defined as

$$D = \frac{R_v}{\text{rms noise-voltage output of cell}} \tag{4-47}$$

The detectivity is the reciprocal of NEP. A normalized detectivity D^* is defined as

$$D^* = (A\,\Delta f)^{\frac{1}{2}} D \tag{4-48}$$

where A is the area of the detector and Δf is a noise-equivalent bandwidth. The units of D^* are usually $(\text{cm})(\text{Hz})^{\frac{1}{2}}(\text{watt})^{-1}$, and the term is used in describing the performance of detectors so that the particular surface area and bandwidth will not affect the result. Figures 4-48 and 4-49 illustrate the performance of several photoconductive detectors over a range of wavelengths. In these figures the wavelength is expressed in microns, where $1\mu = 10^{-6}$ m. The symbols represent lead sulfide (PbS), lead selenide (PbSe), lead telluride (PbTe), indium antimonide (InSb), and gold- and antimony-doped germanium (Ge: Au, Sb). For these figures D^* is a monochromatic detectivity for an incident radiation, which is chopped at 900 Hz and a 1-Hz bandwidth.

The lead-sulfide cell is very widely used for detection of thermal radiation in the wavelength band of 1 to 3 μ. By cooling the detector, more favorable response at higher wavelengths can be achieved up to about 4 or 5 μ. For measurements at longer wavelengths the indium-antimonide detector is preferred, but it has a lower detectivity than lead sulfide. Some of the applications of these cells are discussed in Chap. 12.

Example 4-12 Calculate the incident radiation at 2 μ that is necessary to produce a signal-to-noise (S/N) ratio of 40 db with a lead-sulfide detector at room temperature, having an area of 1 mm².

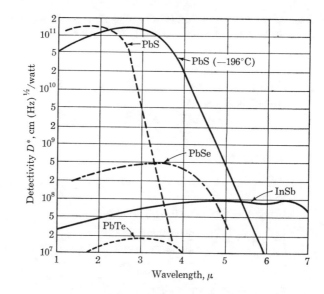

Fig. 4-48 Absolute spectral response of typical detectors at room temperature according to Ref. [11].

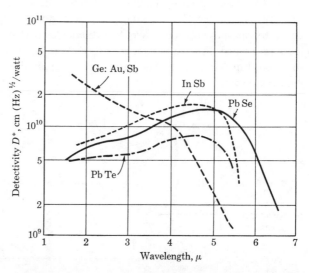

Fig. 4-49 Absolute spectral response of typical detectors cooled to liquid-nitrogen temperatures according to Ref. [11].

Solution   We first insert Eqs. (4-46) and (4-47) in Eq. (4-48) to express the detectivity in terms of voltages and incident power.   Thus,

$$D^* = (A \, \Delta f)^{\frac{1}{2}} \frac{E_{\text{out}}}{E_{\text{noise}}} \frac{1}{P_{\text{incident}}} \tag{a}$$

From Fig. 4-48, $D^* = 1.5 \times 10^{11}$ (cm)(Hz)$^{\frac{1}{2}}$/(watt) for $\Delta f = 1$ Hz.   For a S/N ratio of 40 db, we have

$$40 = 20 \log \frac{E_{\text{out}}}{E_{\text{noise}}}$$

so that

$$\frac{E_{\text{out}}}{E_{\text{noise}}} = 100$$

Using $A = 10^{-2}$ cm^2, Eq. (*a*) yields

$$1.5 \times 10^{11} = \frac{(10^{-2})^{\frac{1}{2}}(1)(100)}{P_{\text{incident}}}$$

Thus,

$$P_{\text{incident}} = 6.7 \times 10^{-11} \text{ watt}$$

4-20 PHOTOVOLTAIC CELLS

The photovoltaic-cell principle is illustrated in Fig. 4-50.   The sandwich construction consists of a metal base plate, a semiconductor material, and a thin transparent metallic layer.   This transparent layer may be in the form of a sprayed, conducting lacquer.   When light strikes the barrier between the transparent metal layer and the semiconductor material, a voltage is generated as shown.   The output of the device is strongly dependent on the load resistance R.   The open-circuit voltage approximates a logarithmic function, but more linear behavior may be approximated by decreasing the load resistance.

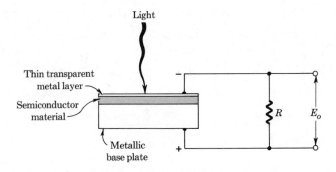

Fig. 4-50   Schematic diagram of a photovoltaic cell.

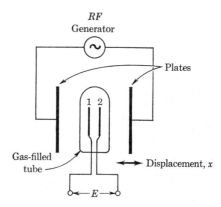

Fig. 4-51 Schematic diagram of an ionization-displacement transducer.

Perhaps the most widely used application of the photovoltaic cell is the light exposure meter in photographic work. The logarithmic behavior of the cell is a decided advantage in such applications because of its sensitivity over a broad range of light intensities.

4-21 IONIZATION TRANSDUCERS

A schematic of the ionization transducer is shown in Fig. 4-51. The tube contains a gas at low pressure while the RF generator impresses a field on this gas. As a result of the RF field, a glow discharge is created in the gas and the two electrodes 1 and 2 detect a potential difference in the gas plasma. The potential difference is dependent on the electrode spacing and the capacitive coupling between the RF plates and the gas. When the tube is located at the central position between the plates, the potentials on the electrodes are the same; but when the tube is displaced from this central position, a dc potential difference will be created. Thus, the ionization transducer is a useful device for measuring displacement. Some of the basic operating characteristics are given in Table 4-4, and a detailed description of the output characteristics is given by Lion [4].

4-22 MAGNETOMETER SEARCH COIL

A schematic of the magnetometer search coil is shown in Fig. 4-52. A flat coil with N turns is placed in the magnetic field as shown. The length of the coil is L, and the cross-sectional area is A. The magnetic field strength H and the magnetic flux density B are in the direction shown, where

$$B = \mu H \tag{4-49}$$

and μ is the magnetic permeability. The voltage output of the coil E is

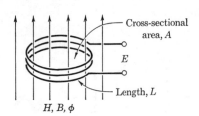

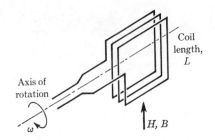

Fig. 4-52 Schematic of a magneto-meter search coil.

Fig. 4-53 Use of a rotating search coil for measurement of steady magnetic fields.

given by

$$E = NA \cos \alpha \frac{dB}{dt} \qquad (4\text{-}50)$$

where α is the angle formed between the direction of the magnetic field and a line drawn perpendicular to the plane of the coil. The total flux through the loop is

$$\phi = A \cos \alpha B \qquad (4\text{-}51)$$

so that

$$E = N \frac{d\phi}{dt} \qquad (4\text{-}52)$$

Note that the voltage output of the device is dependent on the rate of change of the magnetic field and that a stationary coil placed in a steady magnetic field will produce a zero-voltage output. The search coil is thus a transducer that transforms a magnetic field signal into a voltage.

In order to perform a measurement of a steady magnetic field it is necessary to provide some movement of the search coil. A typical method is to use a rotating coil as shown in Fig. 4-53. The rms value of the output voltage for such a device is

$$E_{\text{rms}} = \frac{1}{\sqrt{2}} NAB\omega \qquad (4\text{-}53)$$

where ω is the angular velocity of rotation. Oscillating coils are also used. The accuracy of the search coil device depends on the accuracy with which the dimensions of the coil are known. The coil should be small enough that the magnetic field is constant over its area.

In the above equations the magnetic flux density is expressed in webers per square meter, the area is in square meters, the time is in seconds, the magnetic flux is in webers, the magnetic field strength (magnetic intensity) is in amperes per meter, and the magnetic permeability for free space is $4\pi \times 10^{-7}$ henry/m. An alternate set of units uses B in gauss, H in oersteds, A in centimeters squared, and μ in abhenrys per centimeter. The magnetic permeability for free space in this instance is unity.

Example 4-13   A rotating search coil has 10 turns with a cross-sectional area of 5 cm². It rotates at a constant speed of 100 rpm. The output voltage is 40 mv. Calculate the magnetic field strength.

Solution   According to Eq. (4-53),

$$B = \frac{\sqrt{2}\,E_{rms}}{NA\omega} = \frac{\sqrt{2}(0.04)}{(10)(5 \times 10^{-4})[(100)(2\pi)/60]}$$

$$= 1.08 \text{ weber/m}^2$$

$$H = \frac{B}{\mu} = \frac{1.08}{4\pi \times 10^{-7}}$$

$$= 8.6 \times 10^5 \text{ amp/m}$$

4-23   HALL-EFFECT TRANSDUCERS

The principle of the Hall effect is indicated in Fig. 4-54. A conductor or semiconductor plate of thickness t is connected as shown so that an external current I passes through the material. When a magnetic field is impressed on the plate in a direction perpendicular to the surface of the plate, there will be a potential E_H generated as shown. This potential is called the Hall voltage and is given by

$$E_H = K_H \frac{IB}{t} \tag{4-54}$$

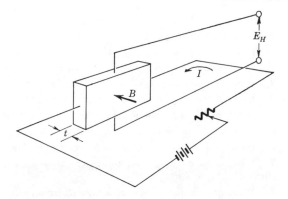

Fig. 4-54   The Hall effect.

Table 4-6   Hall coefficients for different materials†

Material	Field strength, gauss	Temp., °C	K_H, $\dfrac{volt\text{-}cm}{amp\text{-}gauss}$
As	4,000–8,000	20	4.52×10^{-11}
C	4,000–11,000	Room	-1.73×10^{-10}
Bi	1,130	20	-1×10^{-8}
Cu	8,000–22,000	20	-5.2×10^{-13}
Fe	17,000	22	1.1×10^{-11}
n-Ge	100–8,000	25	-8.0×10^{-5}
Si	20,000	23	4.1×10^{-8}
Sn	4,000	Room	-2.0×10^{-14}
Te	3,000–9,000	20	5.3×10^{-7}

† According to Lion [3].

where I is in amperes, B is in gauss, and t is in centimeters.   The proportionality constant is called the Hall coefficient and has the units of volt-cm/amp-gauss.   Typical values of K_H for several materials are given in Table 4-6.

Example 4-14   A Hall-effect transducer is used for the measurement of a magnetic field of 5,000 gauss.   A 2-mm slab of bismuth is used with a current of 3 amp. Calculate the voltage output of the device.

Solution   We use Eq. (4-54) and the data of Table 4-6.

$$E_H = K_H \frac{IB}{t}$$

$$= \frac{(-1 \times 10^{-8})(3)(5,000)}{(2 \times 10^{-1})}$$

$$= -7.5 \times 10^{-4} \text{ volt}$$

REVIEW QUESTIONS

1. How does an electrodynamometer differ from a D'Arsonval meter?   What advantage does it have over the D'Arsonval meter?

2. How do measurements of ac currents differ from measurements of dc currents?

3. How are very high-frequency currents measured?

4. What are the applications for an electrostatic voltmeter?

5. How can an ammeter be converted to a voltmeter?

6. How is the current-sensitive input circuit dependent on internal-meter impedance?

7. What is meant by a ballast circuit?   How does it differ from a voltage-divider circuit?

8. What are the advantages of a high meter impedance when used with a voltage-divider circuit?

9. What is meant by loading error?

10. Why does the internal resistance of the galvanometer not influence the reading in a potentiometer circuit?

11. Differentiate between a bridge operated on the *null* or *deflection* principle.

12. What is meant by a *voltage-sensitive* deflection-bridge circuit?

13. Under what circumstances would a cathode-follower or emitter-follower amplifier circuit be used?

14. How does the decibel notation for voltage level differ from that for power level?

15. Why is a VTVM useful for electrical measurements?

16. What is a Lissajous diagram?

17. What is the upper-frequency limit permissible for (*a*) stylus-type oscillographs and (*b*) light-beam oscillographs?

PROBLEMS

4-1. Expand Eq. (4-10) in series form, and indicate the relation which may be used to obtain a linear approximation for the current.   Also, show the error which results from this approximation.

4-2. Expand the equation for voltage in the ballast circuit [Eq. (4-12)] in series form, and indicate the relation which may be used for a linear approximation.   Show the error which results from this approximation.

4-3. Derive an expression for the sensitivity of the circuit in Fig. 4-11, defined by

$$S = \frac{di}{dR}$$

Find the condition for maximum sensitivity.

4-4. Obtain a linear approximation for the sensitivity of the ballast circuit of Fig. 4-12. Under what conditions would this relation apply?   Estimate the error in the approximation.

4-5. A Wheatstone bridge circuit has resistance arms of 400, 40, 602, and 6,000 ohms taken sequentially around the bridge.   The galvanometer has a resistance of 110 ohms and is connected between the junction of the 40- and 602-ohm resistors to the junction of the 400- and 6,000-ohm resistors.   The battery has an emf of 3 volts and negligible internal resistance.   Calculate the voltage across the galvanometer and the galvanometer current.

4-6. It is known that a certain resistor has a resistance of approximately 800 ohms. This resistor is placed in a Wheatstone bridge the other three arms of which have resistances of exactly 800 ohms.   A 4-volt battery with negligible internal resistance is used in the circuit.   The galvanometer resistance is 100 ohms, and the indicated galvanometer current is 0.08 μa.   Calculate the resistance of the unknown resistor.

4-7. Two galvanometers are available for use with a Wheatstone bridge having equal ratio arms of 100 ohms.   One galvanometer has a resistance of 100 ohms and a sensitivity of 0.05 μa/mm, whereas the other has a resistance of 200 ohms and a sensitivity of 0.01 μa/mm.   A 4-volt battery is used in the circuit, and it has negligible internal resistance.   An unknown resistance of approximately 500 ohms is to be measured with the bridge.   Calculate the deflection of each galvanometer for an error of 0.05 percent in the determination.   State assumptions necessary to make this calculation.

4-8. Two known ratio arms of a Wheatstone bridge are 4,000 and 400 ohms.   The bridge is to be used to measure a resistance of 100 ohms. Two galvanometers are available: one with a resistance of 50 ohms and a sensitivity of 0.05 μa/mm and one with a resistance of 500 ohms and a sensitivity of 0.2 μa/mm. Which galvanometer would you prefer to use? Why? Assume that the galvanometer is connected from the junction of the ratio arms to the opposite corner of the bridge.

4-9. A Wheatstone bridge is constructed with ratio arms of 60 and 600 ohms. A 4-volt battery with negligible internal resistance is used and is connected from the junction of the ratio arms to the opposite corner. A galvanometer having a resistance of 50 ohms and a sensitivity of 0.05 μa/mm is connected between the other corners. When the adjustable arm reads 200 ohms, the galvanometer deflection is 30 mm. What is the value of the unknown resistance?

4-10. The four arms of a Wheatstone bridge have resistances of 500, 1,000, 600, and 290 ohms taken in sequence around the bridge. The battery of 3 volts connects between the 1,000- and 600-ohm resistors and the 500- and 290-ohm resistors. A galvanometer with a resistance of 50 ohms and a sensitivity of 0.05 μa/mm is connected across the other two terminals. The galvanometer is shunted by a resistance of 30 ohms. Calculate the galvanometer deflection. Repeat the calculation for a series resistance of 30 ohms connected in the galvanometer circuit instead of the shunt arrangement.

4-11. Suppose an *end resistor* is added to the circuit of Fig. 4-13 so that there results

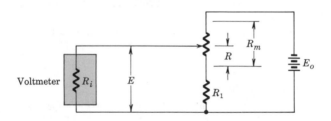

where R_1 is the end resistor. Derive an expression for the voltage output and loading error of such an arrangement. What advantage does it offer over the circuit in Fig. 4-13? Suppose the end resistor were attached to the other end of the variable resistance. What would be the advantage in this circumstance? What would be the advantage if an end resistor were placed on each end of the variable resistor?

4-12. Design a bandpass filter to operate between the limits of 500 and 2,000 Hz, with a load resistance of 16 ohms.

4-13. Design a low-pass filter with a cutoff frequency of 500 Hz with a load resistance of 1,000 ohms.

4-14. Design a high-pass filter with a cutoff frequency of 1,000 Hz with a load resistance of 1,000 ohms.

4-15. A voltage of 500 volts is impressed on a 150-kilohm resistor. The impedance of the voltage source is 10 kilohm. Two meters are used to measure the voltage across the 150-kilohm resistor: a volt-ohmmeter with an internal impedance of 1,000 ohms/volt and a VTVM with an impedance of 11 megohms. Calculate the voltage indicated by each of these devices.

4-16. Plot the gain of the filter circuit of Example 4-6.

4-17. Five 1-in.^{2} plates are arranged as shown. The plate spacings are 0.01 in. The arrangement is to be used for a displacement transducer by observing the change in capacitance with the distance x.   Calculate the sensitivity of the device in picofarads per inch.   Assume that the plates are separated by air.

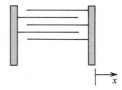

4-18.   Calculate the voltage-displacement sensitivity for the LVDT whose characteristics are shown in Fig. 4-42.

4-19.   The piezoelectric crystal of Example 4-11 is used for a pressure measurement at a nominal 100 psi.   The uncertainty in the voltage measurement is ± 0.5 volt, and the uncertainty in the crystal thickness is ± 0.0003 in.   Estimate the uncertainty in the pressure measurement.

4-20.   A rotating search coil like that shown in Fig. 4-53 has a nominal area of 1 cm² with 50 turns of small-diameter wire.   The rotational speed is nominally 180 rpm.   Calculate the voltage output when the coil is placed in a magnetic field of 1 weber/m².

4-21.   An amplifier has a stated power bandwidth of $+0$, -1 db over a frequency range of 10 Hz to 30 kHz.   If the 0-db level is 35 watts, what is the power that would occur at the lower limit of this specification?

4-22.   An amplifier is stated to produce a 35-watt output into an 8-ohm load when driven with an input voltage of 2.2 mv.   The hum and noise is stated as -65 db referenced to 10 mv at the input.   What is the voltage output of the noise when operating into the 8-ohm load?

4-23.   Suppose the amplifier of Prob. 4-22 requires an input of 180 mv to produce the 35-watt output and has a hum and noise level of -75 db referenced to 0.25 volt at the input.   What is the voltage output of the noise under these conditions?   Calculate the relative decibel level of this noise and the noise calculated in Prob. 4-22.

4-24.   An emf of 100 volts is connected across a load of 10,000 ohms.   In order to measure this voltage, a meter having an internal impedance of 100 kilohms is connected across the load.   Calculate the percent error that will result from the loading by the meter.

4-25.   In Prob. 4-24 calculate the number of decibels below the true voltage indicated by the meter.

4-26.   A certain transducer has an internal resistance of 10 kilohms.   The output of the device is to be measured with a voltage-divider circuit using a voltage source of 50 volts and a meter with an impedance of 100 kilohms.   Calculate the maximum loading error between the ranges of 10 and 90 percent of the measured variables.

4-27.   A simple ballast circuit is used to measure the output of a pressure pickup.   The circuit is designed so that the internal resistance is six times the total transducer resistance. A source of 100 volts is used to energize the circuit.   Calculate the voltage output at 25, 50, 60, and 80 percent of full load on the transducer.   What error would result if the voltage output were assumed to vary linearly with the load?

4-28.   A simple Wheatstone bridge has arms of $R_1 = 121$ ohms, $R_2 = 119$ ohms, and $R_3 = 121$ ohms.   What is the value of R_4 for balance?   If $R_4 = 122$ ohms and the bridge is driven with a source at 100 volts, what is the output voltage?

4-29. A lead-sulfide detector is used to measure a radiation signal at 3.5 μ. What reduction in incident power, measured in decibels, is afforded when the detector is cooled to $-196°C$, assuming a constant S/N ratio?

4-30. An indium-antimonide detector is used to sense radiant energy at 5 μ. What incident flux on a 4-mm² detector at room temperature is necessary to produce a S/N ratio of 45 db?

4-31. A current-sensitivity input circuit is used to measure the output of a transducer having a total resistance of 500 ohms. A 12-volt source is used, and the internal-meter impedance is 7,000 ohms. What is the current indication for 25, 50, 75, and 100 percent of full scale? What error would result at each of these readings if the current output were assumed to vary linearly with the transducer resistance?

4-32. A simple voltage-divider circuit is used to measure the output of a transducer whose total resistance is 1,000 ohms $\pm$ 1 percent. The internal resistance of the voltmeter is 10,000 ohms $\pm$ 5 percent, and the meter is calibrated to read within $\frac{1}{2}$ percent of the true voltage at a full scale of 30 volts. The source voltage is 24 volts $\pm$ 0.05 volts. Calculate the percent full-scale reading of the transducer for voltage readings of 6, 12, 18, and 24 volts. Estimate the uncertainty for each of these readings.

4-33. Suppose the transducer of Prob. 4-32 is connected to an end resistor of 500 ohms, as shown in Prob. 4-11. Calculate the percent full-scale reading of the transducer for voltage readings of 12, 15, 18, and 24 volts. Estimate the uncertainty for each of these readings, assuming the end resistor is accurate within 1 percent.

REFERENCES

1. Cady, W. G.: "Piezoelectricity," McGraw-Hill Book Company, New York, 1946.
2. Frank, E.: "Electrical Measurement Analysis," McGraw-Hill Book Company, New York, 1959.
3. Lion, K. S.: "Instrumentation in Scientific Research," McGraw-Hill Book Company, New York, 1959.
4. Lion, K. S.: Mechanic-electric Transducer, *Rev. Sci. Instr.*, col. 27, no. 2, pp. 222–225, 1956.
5. Sweeney, R. J.: "Measurement Techniques in Mechanical Engineering," John Wiley & Sons, Inc., New York, 1953.
6. ———, "Frequency and Time Standards," Application Note 52, Hewlett-Packard Co., Palo Alto, Calif.
7. ———, "Notes on Linear Variable Differential Transformers," Bulletin AA-1a, Schaevitz Engineering Co., Camden, N.J.
8. ———, "The Radio Amateur's Handbook," 36th ed., Amateur Radio Relay League, West Hartford, Conn., 1959.
9. ———, "Reference Data for Radio Engineers," International Telephone and Telegraph Corp., New York, 1956.
10. International Conference on Photoconductivity, *J. Phys. Chem. Solids*, vol. 22, pp. 1–409, December, 1961.
11. Beyen, W., P. Bratt, W. Engeler, L. Johnson, H. Levinstein, and A. MacRaw: Infrared Detectors Today and Tomorrow, *Proc. IRIS*, vol. 4, no. 1, March, 1959.
12. Bartholomew, D.: "Electrical Measurements and Instrumentation," Allyn and Bacon, Inc., Boston, Mass., 1963.
13. Doebelin, E. O.: "Measurement Systems: Analysis and Design," McGraw-Hill Book Company, New York, 1966.

5
Displacement and Area Measurements

5-1 INTRODUCTION

Many of the transducers discussed in Chap. 4 represent excellent devices for measurement of displacement. In this chapter we wish to examine the general subject of dimensional and displacement measurements and indicate some of the techniques and instruments that may be utilized for such purposes, making use, where possible, of the information in the preceding sections.

Dimensional measurements are categorized as determinations of the size of an object, while a *displacement measurement* implies the measurement of the movement of a point from one position to another. An area measurement on a standard geometric figure is a combination of appropriate dimensional measurements through a correct analytical relationship. The determination of areas of irregular geometric shapes usually involves a mechanical, graphical, or numerical integration.

Displacement measurements may be made under both steady and transient conditions. Transient measurements fall under the general class of subjects discussed in Chap. 11. The present chapter is concerned only with static measurements.

5-2 DIMENSIONAL MEASUREMENTS

The standard units of length have been discussed in Chap. 2. All dimensional measurements are eventually related to these standards. Simple dimensional measurements with an accuracy of ± 0.01 in. may be made with graduated metal machinists' scales or wood scales which have accurate engraved markings. For large dimensional measurements metal tapes are used to advantage. The primary errors in such measurement devices, other than readability errors, are usually the result of thermal expansion or contraction of the scale. On long metal tapes used for surveying purposes this can represent a substantial error, especially when used under extreme temperature conditions. It may be noted, however, that thermal expansion effects represent fixed errors and may easily be corrected when the measurement temperature is known.

Vernier calipers represent a convenient modification of the metal scale to improve the readability of the device. The caliper construction is shown in Fig. 5-1, and an expanded view of the vernier scale is shown in Fig. 5-2. The caliper is placed on the object to be measured and the fine adjustment rotated until the jaws fit tightly against the workpiece. The increments along the primary scale are 0.025 in. The vernier scale shown is used to read to 0.001 in. so that it has 25 equal increments (0.001 is $\frac{1}{25}$ of 0.025) and a total length of $\frac{24}{25}$ times the length of the primary scale graduations. Consequently, the vernier scale does not line up exactly with the primary scale, and the ratio of the last coincident number on the vernier to the total vernier length will equal the fraction of a whole primary scale division indicated by the index position. In the example shown in Fig. 5-2 the reading would be $2.350 + (\frac{14}{25})(0.025) = 2.364$ in.

The micrometer calipers shown in Fig. 5-3 represent a more precise measurement device than the vernier calipers. Instead of using the vernier scale arrangement a calibrated screw thread and circumferential scale divisions are used to indicate the fractional part of the primary scale divisions. In order to obtain the maximum effectiveness of the micrometer, care must be exerted to ensure that a consistent pressure is maintained on the workpiece. The spring-loaded ratchet device on the handle enables the operator

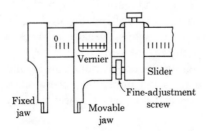

Fig. 5-1 Vernier caliper.

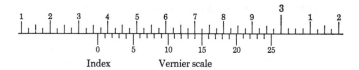

Fig. 5-2 Expanded view of vernier scale.

to maintain such a condition. When properly used, the micrometer can be employed for the measurement of dimensions within 0.0001 in.

Dial indicators are devices that perform a mechanical amplification of the displacement of a pointer or follower in order to measure displacements within about 0.001 in. The construction of such indicators provides a gear rack, which is connected to a displacement-sensing shaft. This rack engages a pinion which in turn is used to provide a gear-train amplification of the movement. The output reading is made on a circular dial.

Example 5-1 A 100-ft steel tape is used for a surveying instrument in the summer such that the tape temperature is 105°F. A measurement indicates 84.56 ± 0.002 ft. The linear thermal coefficient of expansion is $6.47 \times 10^{-6}/°F$ at 60°F. Calculate the true distance measurement.

Solution The indicated tape length would be the true value if the measurement were taken at 60°F. At the elevated temperature the tape has expanded and consequently reads too small a distance. The actual length of the 100-ft tape at 105°F is

$$L(1 + \alpha \, \Delta T) = [1 + (6.47 \times 10^{-6})(45)](100)$$

$$= 100.029115 \text{ ft}$$

Such a true length would be indicated as 100 ft. The true reading for the above situation is thus

$$(84.56)(1.00029115) = 84.585 \text{ ft}$$

5-3 GAGE BLOCKS

Gage blocks represent industrial dimension standards. They are small steel blocks about $\frac{3}{8} \times 1\frac{3}{8}$ in. with highly polished parallel surfaces. The thickness

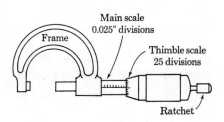

Fig. 5-3 Micrometer calipers.

of the blocks is specified in accordance with the following tolerances:

Grade of block	Tolerance, $\mu in.$†
AA	2
A	4
B	8

Gage blocks are available in a range of thicknesses that make it possible to stack them in a manner such that with a set of 81 blocks any dimension between 0.100 and 8.000 in. can be obtained in increments of 0.0001 in. The blocks are stacked through a process of wringing. With surfaces thoroughly clean the metal surfaces are brought together in a sliding fashion while a steady pressure is exerted. The surfaces are sufficiently flat so that when the wringing process is correctly executed, they will adhere together as a result of molecular attraction. The adhesive force may be as great as 30 times atmospheric pressure.

Because of their high accuracy, gage blocks are frequently used for calibration of other dimensional measurement devices. For very precise measurements they may be used for direct dimensional comparison tests with a machined item. A discussion of the methods of producing gage-block standards is given in Ref. [3]. The literature of manufacturers of gage blocks furnishes an excellent source of information on the measurement techniques which are employed in practice.

5-4 OPTICAL METHODS

An optical method for measuring dimensions very accurately is based on the principle of light interference. The instrument based on this principle is called an interferometer and is used for the calibration of gage blocks and other dimensional standards. Other optical instruments in wide use are various types of microscopes and telescopes, including the conventional surveyor's transit which is employed for measurement of large distances.

Consider the two sets of light beams shown in Fig. 5-4. In Fig. 5-4a the two beams are in phase so that the brightness at point P is augmented when they intersect. In Fig. 5-4b the beams are out of phase by half a wavelength so that a cancellation is observed, and the light waves are said to *interfere* with each other. This is the essence of the interference principle. The effect of the cancellation is brought about by allowing two light waves from a single source to travel along paths of different lengths. When the difference in the distance is an integral multiple of wavelength, there will be a reinforcement of the waves while there will be a cancellation when the difference in the distances is an odd multiple of half-wavelengths.

† Tolerances are for blocks less than 1 in. thick; for greater thickness the same tolerances are per inch.

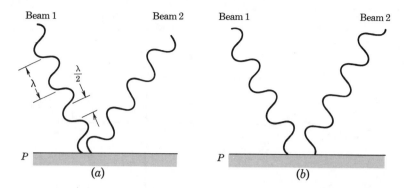

Fig. 5-4 Interference principle.   (*a*) Beams in phase; (*b*) beams out of phase.

Now let us apply the interference principle to dimensional measurements.   Consider the two parallel plates shown in Fig. 5-5.   One plate is a transparent, strain-free glass accurately polished flat within a few microinches.   The other plate has a reflecting metal surface.   The glass plate is called an optical flat.   Parallel light beams A and B are projected on the plates from a suitable collimating source.   The separation distance between the plates d is assumed to be quite small.   The reflected beam A intersects the incoming beam B at the point P.   Since the reflected beam has traveled further than beam B by a distance of $2d$, it will create an interference at point P if this incremental distance is an odd multiple of $\lambda/2$.   If the distance $2d$ is an even multiple of $\lambda/2$, the reflected beam will augment beam B.   Thus, for $2d = \lambda/2$, $3\lambda/2$, etc., the screen S will detect no reflected light.   Now consider the same two plates, but let them be tilted slightly so that the distance between the plates is a variable.   Now, if one views the reflected light beams, alternate light and dark regions will appear on the screen indicating the variation in the plate spacing.   The dark lines or regions are called fringes,

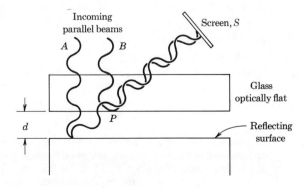

Fig. 5-5 Application of interference principle.

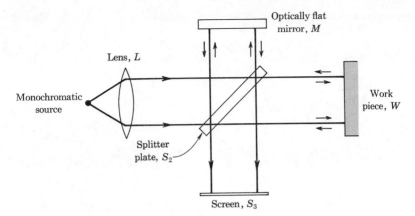

Fig. 5-6 Schematic of interferometer.

and the change in the separation distance between the positions of two fringes
corresponds to

$$\Delta(2d) = \frac{\lambda}{2} \tag{5-1}$$

The interference principle offers a convenient means for measuring small
surface defects and for calibrating gage blocks.   The use of a tilted optical
flat as in Fig. 5-5 is an awkard method of utilizing the principle, however.
For practical purposes the interferometer, as indicated schematically in Fig.
5-6, is employed.   Monochromatic light from the source is collimated by the
lens L onto the splitter plate S_2, which is a half-silvered mirror that reflects
half of the light toward the optically flat mirror M and allows transmission of
the other half toward the workpiece W.   Both beams are reflected back and
recombined at the splitter plate S_2 and then transmitted to the screen.
Fringes may appear on the screen resulting from differences in the optical
path lengths of the two beams.   If the instrument is properly constructed,
these differences will arise from dimensional variations of the workpiece.
The interferometer is primarily used for calibration of gage blocks and other
applications where extremely precise absolute dimensional measurements
are required.   For detailed information on experimental techniques used in
interferometry the reader should consult Refs. [3] and [5].   The use of the
interferometer for fluid-flow measurements will be discussed in Chap. 7.

Example 5-2   A mercury light source employs a green filter such that the wavelength
is 5460 A.   This light is collimated and directed onto two tilted surfaces like
those shown in Fig. 5-5.   At one end the surfaces are in precise contact.   Between
the point of contact and a distance of 3.000 in., five interference fringes are observed.   Calculate the separation distance between the two surfaces and the tilt
angle at this position.

Solution The five fringe lines correspond to $\lambda/2, 3\lambda/2, \ldots, 9\lambda/2$; that is, for the fifth fringe line

$$2d = \frac{9\lambda}{2}$$

We have $\lambda = 5460 \times 10^{-8}$ cm $= 2.15 \times 10^{-5}$ in. so that

$$d = \tfrac{9}{4}(2.15 \times 10^{-5}) = 48.4 \ \mu\text{in.}$$

The tilt angle is

$$\phi = \tan^{-1}\frac{48.4 \times 10^{-6}}{3.000} = \frac{48.4 \times 10^{-6}}{3.000} = 16.1 \times 10^{-6} \text{ rad}$$

Mechanical displacement may also be measured with the aid of the electric transducers discussed in Chap. 4. The LVDT, for example, can be used to sense displacements as small as 1 μin. Resistance transducers are primarily of value for measurement of fairly large displacements because of their poor resolution. Capacitance and piezoelectric transducers, on the other hand, provide high resolution and are suitable for dynamic measurements.

5-5 PNEUMATIC DISPLACEMENT GAGE

Consider the system shown in Fig. 5-7. Air is supplied at a constant pressure p_1. The flow through the orifice and through the outlet of diameter d_2 is governed by the separation distance x between the outlet and the workpiece. The change in flow with x will be indicated by a change in the pressure downstream from the orifice p_2. Thus, a measurement of this pressure may be taken as an indication of the separation distance x. For purposes of analysis we assume incompressible flow. (See Sec. 7-3 for a discussion of the validity of this assumption.) The volumetric flow through an orifice may be represented by

$$Q = CA\sqrt{\Delta p} \qquad\qquad\qquad (5\text{-}2)$$

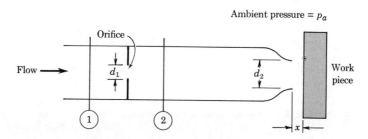

Fig. 5-7 Pneumatic displacement device.

where C = discharge coefficient

   A = flow area of the orifice

   Δp = pressure differential across the orifice

There are *two* orifices in the situation depicted in Fig. 5-7, the obvious one and the orifice formed by the flow restriction between the outlet and the workpiece. We shall designate the area of the first orifice A_1 and that of the second A_2. Then, Eq. (5-2) becomes

$$Q = C_1 A_1 \sqrt{p_1 - p_2} = C_2 A_2 \sqrt{p_2 - p_a} \qquad (5\text{-}3)$$

where p_a is the ambient pressure and is assumed constant. Equation (5-3) may be rearranged to give

$$r = \frac{p_2 - p_a}{p_1 - p_a} = \frac{1}{1 + (A_2/A_1)^2} \qquad (5\text{-}4)$$

where it is assumed that the discharge coefficients C_1 and C_2 are equal. We may now observe that

$$A_1 = \frac{\pi d_1{}^2}{4} \qquad (5\text{-}5)$$

$$A_2 = \pi d_2 x \qquad (5\text{-}6)$$

Thus we see the relation between the pressure ratio r and the workpiece displacement x. Graneek and Evans [1] have shown experimentally that the relation between r and the area ratio A_2/A_1 is very nearly linear for $0.4 < r < 0.9$ and that

$$r = 1.10 - 0.50 \frac{A_2}{A_1} \qquad (5\text{-}7)$$

for this range. Introducing Eqs. (5-5) and (5-6), we have

$$r = \frac{p_2 - p_a}{p_1 - p_a} = 1.10 - 2.00\left(\frac{d_2}{d_1{}^2}\right)x \qquad \text{for } 0.4 < r < 0.9 \qquad (5\text{-}8)$$

The pneumatic displacement gage is mainly used for small displacement measurements.

Example 5-3   A pneumatic displacement gage like the one shown in Fig. 5-7 has $d_1 = 0.030$ in. and $d_2 = 0.062$ in. The supply pressure is 10.0 psig, and the differential pressure $p_2 - p_a$ is measured with a water manometer which may be read with an uncertainty of 0.05 in. H_2O. Calculate the displacement range for which Eq. (5-8) applies and the uncertainty in this measurement, assuming that the supply pressure remains constant.

Solution   We have

$$\frac{d_2}{d_1{}^2} = \frac{0.062}{(0.030)^2} = 68.8$$

When $r = 0.4$, we have, from Eq. (5-8),

$$x = \frac{1.10 - 0.4}{(2.00)(68.8)} = 0.0509 \text{ in.}$$

When $r = 0.9$, $x = 0.0145$ in.
     Utilizing Eq. (3-2) as applied to Eq. (5-8), we have

$$w_r = \left[\left(\frac{\partial r}{\partial x} \right)^2 w_x{}^2 \right]^{\frac{1}{2}} = \pm \left| \frac{\partial r}{\partial x} \right| w_x$$

Furthermore,

$$w_r = \frac{w_{\Delta p}}{p_1 - p_a} \qquad \frac{\partial r}{\partial x} = -(2.00)(68.8) = -137.6$$

The uncertainty in the measurement of $p_2 - p_a$ is

$$w_{\Delta p} = (0.05)(0.0361) = 1.805 \times 10^{-3} \text{ psig}$$

Thus, the uncertainty in x is given by

$$w_x = \frac{1.805 \times 10^{-3}}{(137.6)(10.00)} = 1.313 \times 10^{-6} \text{ in.} = 1.313 \ \mu\text{in.}$$

From this example we see that the pneumatic gage can be quite sensitive, even with modest pressure-measurement facilities at hand.

In addition to its application as a steady-state–displacement-measurement device, the pneumatic gage may be employed as a dynamic sensor in conjunction with properly designed fluidic circuits.   By periodically interrupting the discharge, the device may serve as a periodic signal generator for fluidic circuits.   Different interruption techniques may be employed to generate square-, triangle-, or sine-wave signals [7].   The device may also be employed for fluidic reading of coded information on plates or cards that pass under the outlet jet [8 and 9].   The rapidity with which such readings may be made depends on the dynamic response of the gage and its associated connecting lines and pressure transducers.   Studies [11] have shown that the device can produce good frequency response for signals up to 500 Hz.

5-6   AREA MEASUREMENTS

There are many applications that require a measurement of a plane area. Graphical determinations of the area of survey plots from maps, the integration of a function to determine the area under a curve, and analyses of experimental data plots all may rely on a measurement of a plane area. There are also many applications for the measurement of surface areas, but such measurements are considerably more difficult to perform.

5-7 THE PLANIMETER

The planimeter is a mechanical integrating device that is used for measurement of plane areas. Consider the schematic representation shown in Fig. 5-8. The point O is fixed while the tracing point T is moved around the periphery of the figure whose area is to be determined. The wheel W is mounted on the arm BT so that it is free to rotate when the arm undergoes an angular displacement. The wheel has engraved graduations and a vernier scale so that its exact number of revolutions may be determined as the tracing point moves around the curve. The planimeter and area are placed on a flat, relatively smooth surface so that the wheel W will only slide when the arm BT undergoes an axial translational movement. Thus the wheel registers zero angular displacement when an axial translational movement of arm BT is experienced. Let the length of the tracing arm BT be L and the distance from point B to the wheel be a. The diameter of the wheel is D. The distance OB is taken as R. Now suppose the arm BT is rotated an angle $d\theta$ and the arm OB through an angle $d\phi$ as a result of movement of the tracing point. The area swept out by the arms BT and OB is

$$dA = \tfrac{1}{2}L^2 \, d\theta + LR \sin \beta \, d\phi + \tfrac{1}{2}(R^2 + L^2 - 2aL) \, d\phi \qquad (5\text{-}9)$$

where β is the angle between the two arms. Similarly, the distance traveled

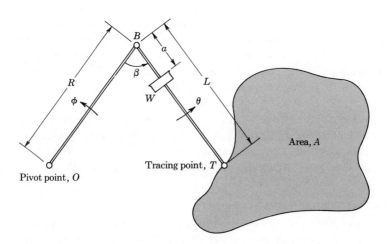

Fig. 5-8 Schematic of a polar planimeter.

by the rim of the wheel owing to rotation is

$$ds = a \, d\theta + R \sin \beta \, d\phi \tag{5-10}$$

Equations (5-9) and (5-10) may now be integrated for a complete traverse of the tracing point around the area.   There results

$$A = \tfrac{1}{2}L^2 \int d\theta + LR \int \sin \beta \, d\phi + \tfrac{1}{2}(R^2 + L^2 - 2aL) \int d\phi \tag{5-11}$$

and

$$s = a \int d\theta + R \int \sin \beta \, d\phi \tag{5-12}$$

Since the tracing arm comes back to its original position,

$$\int d\theta = 0$$

$$s = R \int \sin \phi \, d\phi$$

and

$$A = Ls + \tfrac{1}{2}(R^2 + L^2 - 2aL)(\phi_2 - \phi_1) \tag{5-13}$$

If the pole point O is outside the area, as shown in Fig. 5-8, $\phi_2 = \phi_1$.   If it is inside the area, $\phi_2 - \phi_1 = 2\pi$.   Note that when the pole is outside the area, the distances R and a do not enter into the calculation of the area.   The last term in Eq. (5-13) represents the area of the *zero circle*, which is the area the tracing point would sweep out when the pivot point is inside the area and the wheel reading is zero.

The instrument described above is called a *polar planimeter*.   Typical commercial devices have a wheel circumference of 2.50 in., and models are available with both fixed- and adjustable-length tracing arms.   The area of the zero circle is frequently given by the manufacturer but may also be determined as in Example 5-4.   The construction of the polar planimeter is indicated in Fig. 5-9.

The polar planimeter is not generally suitable for the measurement of long narrow areas because the pivot point must be fixed.   The *roller planimeter* shown in Fig. 5-10 is suitable for these measurements.   The two rollers R are connected through a bevel-gear arrangement to the table A.   The wheel W is in contact with the table and rotates in accordance with the number of revolutions of the table.   The tracing point T is connected to the mounting for the wheel so that its movement causes the wheel to move inward and outward on the rotating table.   The rotation of the wheel

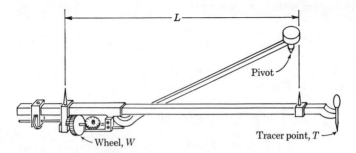

Fig. 5-9 Construction of a polar planimeter.

depends upon its distance from the center of the table so that it is, in turn, dependent on the displacement of the tracing point in a direction parallel to the axis of the two rollers.   Thus, the total indication of the wheel is a function of the distance moved by the rollers and the perpendicular displacement of the tracing point and is consequently a function of the area traversed.   The device may be used for a large class of area measurements since it can move in two directions.

Example 5-4 A planimeter with a fixed tracing arm 4.00 in. long has a wheel circumference of 2.50 in.   The instruction sheet for the instrument is lost so that the area of the zero circle is not known.   To determine the area of the zero circle, a certain area is measured with the pole both inside and outside the area.   With the pole outside the area the wheel records 2.55 revolutions.   With the pole inside the area the wheel records -13.01 revolutions.   Calculate the area of the zero circle and the reading of the planimeter when it is used to trace out a circle 9 in. in radius.

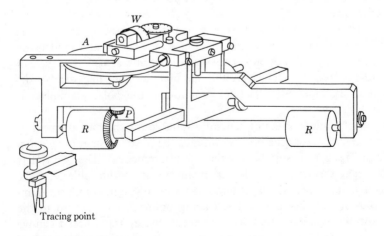

Fig. 5-10 Roller planimeter.

Solution   From Eq. (5-13) we have

$$A = Ls + A_0$$

where A_0 is the area of the zero circle.   The true area is the value of Ls when the pole is outside the area.   Thus,

$$A = (4.00)(2.55)(2.50) = 25.50 \text{ in.}^2$$

and

$$A_0 = 25.50 - (4.00)(-13.01)(2.50) = 155.6 \text{ in.}^2$$

When the planimeter is used to measure a 9-in.-radius circle, we have

$$s = \frac{A - A_0}{L} = \frac{\pi(9)^2 - 155.6}{4.00} = 24.717$$

The number of wheel revolutions is thus

$$n = \frac{s}{2.50} = \frac{24.717}{2.50} = 9.887$$

5-8 GRAPHICAL AND NUMERICAL METHODS FOR AREA MEASUREMENT

A very simple method of plane-area measurement is to place the figure on coordinate paper and count the number of squares enclosed by the figure. An appropriate scale factor is then applied to determine the area.   Numerical integration is commonly applied to determine the area under an irregular curve.   Perhaps the two most common methods are the trapezoidal rule and Simpson's rule.   Consider the area shown in Fig. 5-11.   The area under the curve is

$$A = \int y \, dx \qquad\qquad (5\text{-}14)$$

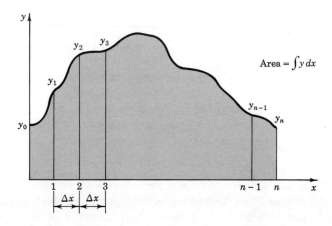

Fig. 5-11   Plane-area determination.

If the figure is divided into equal increments Δx along the x axis, the trapezoidal rule gives for the area

$$A = \left(\frac{y_0 + y_n}{2} + \sum_{i=1}^{n-1} y_i\right) \Delta x \tag{5-15}$$

When the area is divided into an *even* number of increments, Simpson's rule gives

$$A = \frac{\Delta x}{3} \left\{y_0 + y_n + \sum_{i=1}^{n-1} y_i[3 + (-1)^{i+1}]\right\} \tag{5-16}$$

The trapezoidal rule is obtained by joining the ordinates of the curve with straight lines, while the result given by Eq. (5-16) is obtained by joining three points at a time with a parabola. Both of the above equations are special cases of a general class of equations called the *Newton-Cotes integration formulas*. A derivation of these formulas is based on an approximation of the actual curve with a polynomial which agrees with it at $n + 1$ equally spaced points. The general form of the integration formulas is derived in Ref. [2] and may be written in the following way:

$$\int_{x_1}^{x_2} f(x)\, dx = \Delta x \sum_{k=0}^{n} C_k f(x_k) \tag{5-17}$$

where $\Delta x = \dfrac{x_2 - x_1}{n}$ \hfill (5-18)

$$x_k = x_1 + k\, \Delta x \tag{5-19}$$

and the coefficients C_k are given by

$$C_k = \int_0^n \frac{s(s-1)\cdots(s-k+1)(s-k-1)\cdots(s-n)}{k(k-1)\cdots(k-k+1)(k-k-1)\cdots(k-n)}\, ds \tag{5-20}$$

The variable s is defined according to

$$s = \frac{x - x_1}{n} \tag{5-21}$$

Note that the Newton-Cotes formulas use n increments in the independent variable and that this requires a matching of a polynomial at $n + 1$ points in the region that is to be integrated. A discussion of the errors involved in the above formulas is given by Hildebrand [2].

5-9 SURFACE AREAS

Consider the general three-dimensional surface shown in Fig. 5-12. The surface is described by the function

$$z = f(x, y)$$

and the surface area is given in Ref. [4] as

$$A = \int \int \left[\left(\frac{\partial z}{\partial x} \right)^2 + \left(\frac{\partial z}{\partial y} \right)^2 + 1 \right]^{\frac{1}{2}} dx \, dy \tag{5-22}$$

If the function z is known and well-behaved, the integral in Eq. (5-22) may be evaluated directly. Let us consider the case where the function is not given but specific values of z are known for incremental changes in x and y. The increments in x and y are denoted by Δx and Δy, while the value of z is denoted by $z_n{}^p$ where the subscript n refers to the x increments and the superscript p refers to the y increment. We thus have the approximations

$$\frac{\partial z}{\partial x} \approx \frac{z_{n+1}^p - z_n{}^p}{\Delta x}$$

$$\frac{\partial z}{\partial y} \approx \frac{z_n^{p+1} - z_n{}^p}{\Delta y}$$

The integral in Eq. (5-22) is now replaced by the double sum

$$A = \sum_p \sum_n \left[\left(\frac{z_{n+1}^p - z_n{}^p}{\Delta x} \right)^2 + \left(\frac{z_n^{p+1} - z_n{}^p}{\Delta y} \right)^2 + 1 \right]^{\frac{1}{2}} \Delta x \, \Delta y \tag{5-23}$$

The surface area may be determined by performing this numerical summation.

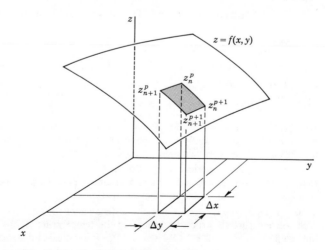

Fig. 5-12 Surface-area determination.

PROBLEMS

5-1. A 12-in. steel scale is graduated in increments of 0.01 in. and is accurate when used at a temperature of 60°F. Calculate the error in an 11-in. measurement when the ambient temperature is 100°F. Should the person using the scale be concerned about this error? Why?

5-2. Calculate the temperature error in a 76-ft measurement with a steel surveyor's tape at $-10°F$ when the tape is accurate at 60°F.

5-3. Show that the spacing distance d in Fig. 5-5 can be represented by

$$d = \frac{2n - 1}{4} \lambda$$

where n is the number of fringe lines.

5-4. A pneumatic displacement gage is designed according to the arrangement in Fig. 5-7. An air supply pressure of 20 psig is available, and displacements are to be measured over a range of 0.050 in. The orifice diameter is 0.025 in. Calculate the maximum displacement which may be measured in the linear range of operation and the outlet tube diameter d_2.

5-5. A planimeter is used to measure the area of an irregular plane figure. The wheel circumference is 2.50 in., and the tracing arm is adjustable. With a tracing-arm length of 2.00 in. and the pole outside the area the wheel indicates 9.52 revolutions. With a tracing-arm length of 4.00 in. and the pole outside the area the wheel indicates 4.70 revolutions. Do the two readings correspond to the same area? If not, how do you explain the differences?

5-6. Plot the equation

$$y = 3 + 4x - 6x^2 + 2.8x^3 - 0.13x^4$$

for the range $0 < x < 5$. Determine the area under the curve by counting squares and also by numerical integration using the trapezoidal relation and Simpson's rule. Calculate the error in each of these three cases by comparing the results with that obtained analytically.

5-7. Use the trapezoidal rule and Simpson's method to perform the integration

$$A = \int_0^\pi \sin x \, dx$$

Use 4, 8, and 12 increments of x and calculate the error for each case.

5-8. Consider the sphere given by

$$x^2 + y^2 + z^2 = 25$$

Using the summation of Eq. (5-23), calculate the surface area bounded by $x = \pm 1$ and $y = \pm 1$. Use $\Delta x = \Delta y = 0.5$. Determine the error in the calculation by comparing it with the true value as calculated from Eq. (5-22).

5-9. By suitable numerical integration determine the surface area and volume of a right circular cone having a height of 5 in. and a base diameter of 6 in. Compare the result with that obtained by an exact calculation.

5-10. A steel tape is used to measure a distance of 83 ft at 50°F. Assuming that the tape will measure the true distance when the temperature is 70°F, what would be the indicated distance at 50°F? What would be the indicated distance at 100°F?

5-11. A mercury light source and green filter (5460 A) are used with an interferometer to determine the distance between an optically flat glass plate and a metal plate, as shown in Fig. 5-5. For this arrangement nine interference fringes are observed. Calculate the separation distance between the two surfaces.

5-12. A pneumatic displacement gage is to be used to measure displacements between 0.005 and 0.01 in. The output of the device is a hypodermic needle having a diameter of 0.020 in. The measurement is to be performed with an uncertainty of 1 μin. Specify values of the orifice diameter, upstream pressure, and allowable uncertainty in the pressure measurement to accomplish this objective.

5-13. The planimeter of Example 5-4 is used to measure the area of an irregular-shaped figure. With the pole outside the area, the wheel records 5.23 revolutions. What is the area of the figure? What would be the reading if the pole were placed inside the figure?

5-14. The pneumatic displacement gage of Example 5-3 is operated under the same conditions as given, but it is discovered that the supply pressure has a random fluctuation of ± 0.07 psia during the measurements. Calculate the uncertainty in the dimensional measurement under these new conditions, assuming that the uncertainty in $p_2 - p_a$ remains at 0.05 in. H_2O.

REFERENCES

1. Graneek, M., and J. C. Evans: A Pneumatic Calibrator of High Sensitivity, *Engineer*, p. 62, July 13, 1951.
2. Hildebrand, F. B.: "Introduction to Numerical Analysis," McGraw-Hill Book Company, New York, 1956.
3. Peters, C. G., and W. B. Emerson: Interference Methods for Producing and Calibrating End Standards, *J. Res. Natl. Bur. Std.*, vol. 44, p. 427, 1950.
4. Sokolnikoff, I. S., and R. M. Redheffer: "Mathematics of Physics and Modern Engineering," McGraw-Hill Book Company, New York, 1958.
5. ———, Metrology of Gauge Blocks, *Natl. Bur. Std. (U.S.)*, *Circ.* 581, April, 1957.
6. Fish, V. T., and G. M. Lance: An Accelerometer for Fluidic Control Systems, ASME Paper No. 67-WA/FE-29.
7. Goldstein, S.: A New Type of Pneumatic Triangle-Wave Oscillator, ASME Paper No. 66-WA/AUT-4.
8. Rosenbaum, H. M., and J. S. Gant: A Pneumatic Tape Reader, 2d Cranfield Fluidics Conference, E2-13, January, 1967.
9. Gant, G. C.: A Fluidic Digital Displacement Indicator, 2d Cranfield Fluidics Conference, E1-1, January, 1967.
10. Eckerlin, H. M., and D. A. Small: A Method of Air Gauge Circuit Analysis, *Advan. Fluidics*, ASME, 1967.
11. Moses, H. L., D. A. Small, and G. A. Cotta: Response of a Fluidic Air Gauge, *J. Basic Eng.*, pp. 475–478, September, 1969.

6
Pressure Measurement

6-1 INTRODUCTION

Pressure is represented as a force per unit area. As such, it has the same units as stress and may, in a general sense, be considered as a type of stress. For our purposes we shall designate the force per unit area exerted by a fluid on a containing wall as the pressure. The forces that arise as a result of strains in solids are designated as stresses and discussed in Chap. 10. Thus, our discussion of pressure measurement is one which is restricted to fluid systems. *Absolute pressure* refers to the absolute value of the force per unit area exerted on the containing wall by a fluid. *Gage pressure* represents the difference between the absolute pressure and the local atmospheric pressure. *Vacuum* represents the amount by which the atmospheric pressure exceeds the absolute pressure. From these definitions we see that the absolute pressure may not be negative and the vacuum may not be greater than the local atmospheric pressure. The three terms are illustrated graphically in Fig. 6-1. It is worthwhile to mention that local fluid pressure may be dependent upon many variables; elevation, flow velocity, fluid density, and temperature are parameters that are of frequent importance.

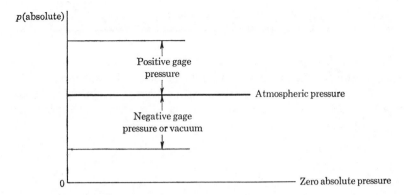

Fig. 6-1 Relationship between pressure terms.

In the engineering system of units pressure is usually expressed in pounds per square inch absolute (psia). Gage pressure carrying the same units is designated with the symbol psig. Pressure is frequently expressed in terms of the height of a column of mercury which it will support at a temperature of 68°F. At standard atmospheric pressure of 14.696 psia this height is 760 mm with the density of mercury taken as 13.5951 g/cm³. Some other units that are commonly used are:

1 microbar = 1 dyne per square centimeter = 2.089 lb_f/ft²
1 millimeter = 1333.22 microbars
1 micron = 1 μ = 10^{-6} meter Hg = 10^{-3} millimeter Hg
1 torr = 1 millimeter Hg

Fluid pressure results from a momentum exchange between the molecules of the fluid and a containing wall. The total momentum exchange is dependent upon the total number of molecules striking the wall per unit time and the average velocity of the molecules. For an ideal gas, it may be shown that the pressure is given by

$$p = \tfrac{1}{3}nmv_{\mathrm{rms}}^2 \tag{6-1}$$

where
n = molecular density
m = molecular mass
v_{rms} = root-mean-square molecular velocity

It may also be shown that

$$v_{\mathrm{rms}} = \sqrt{\frac{3kT}{m}} \tag{6-2}$$

where T = absolute temperature of the gas

$k = 1.3803 \times 10^{-23}$ joule/molecule-°K (Boltzmann's constant)

Equation (6-1) is a kinetic-theory interpretation of the ideal-gas law. An expression for the pressure in a liquid would not be so simple.

The mean free path is defined as the average distance a molecule travels between collisions. For an ideal gas whose molecules act approximately like billiard balls

$$\lambda = \frac{\sqrt{2}}{8\pi r^2 n} \qquad (6\text{-}3)$$

where r is the effective radius of the molecule and λ is the mean free path. It is clear that the mean free path increases with a decrease in the gas density. At standard atmospheric pressure and temperature the mean free path is quite small, of the order of 10^{-5} cm. At a pressure of 1 μ, however, the mean free path would be of the order of 1 cm. At very low pressures the mean free path may be significantly greater than a characteristic dimension of the containing vessel. For air the relation for mean free path reduces to

$$\lambda = 8.64 \times 10^{-7} \frac{T}{p} \text{ ft} \qquad (6\text{-}3a)$$

where T is in degrees Rankine and p is in lb_f/ft^2.

A variety of devices are available for pressure measurement, as we shall see in the following sections. Static, i.e., steady-state, pressure is not difficult to measure with good accuracy. Dynamic measurements, however, are much more perplexing because they are influenced strongly by the characteristics of the fluid which is studied as well as the construction of the measurement device. In many instances a pressure instrument that gives very accurate results for a static measurement may be entirely unsatisfactory for dynamic measurements. We shall discuss some of the factors that are important for good dynamic response in conjunction with the exposition associated with the different types of pressure-measurement devices.

6-2 DYNAMIC RESPONSE CONSIDERATIONS

The transient response of pressure-measuring instruments is dependent on two factors: (1) the response of the transducer element that senses the pressure and (2) the response of the pressure-transmitting fluid and the connecting tubing, etc. This latter factor is frequently the one which determines the overall frequency response of a pressure-measurement system, and, eventually, direct calibration must be relied upon for determining this response. An estimate of the behavior may be obtained with the following analysis.

Consider the system shown in Fig. 6-2. The fluctuating pressure has a frequency of ω and an amplitude of p_0 and is impressed on the tube of length L and radius r. At the end of this tube is a chamber of volume V where the connection to the pressure-sensitive transducer is made. The mass of fluid vibrates under the influence of fluid friction in the tube which tends to dampen the motion. If the conventional formula for laminar friction resistance in tube flow is used to represent this friction, the resulting expression for the pressure amplitude ratio is

$$\left|\frac{p}{p_0}\right| = \frac{1}{\{[1 - (\omega/\omega_n)^2]^2 + 4h^2(\omega/\omega_n)^2\}^{\frac{1}{2}}} \tag{6-4}$$

In this equation, p is the amplitude of the pressure signal impressed on the transducer. The natural frequency ω_n is given by

$$\omega_n = \sqrt{\frac{3\pi r^2 c^2}{4LV}} \tag{6-5}$$

and the damping ratio h is

$$h = \frac{2\mu}{\rho c r^3}\sqrt{\frac{3LV}{\pi}} \tag{6-6}$$

In the above formulas, c represents the velocity of sound in the fluid, μ is the dynamic viscosity of the fluid, and ρ is the fluid density. The phase angle for the pressure signal is

$$\phi = \tan^{-1}\frac{-2h(\omega/\omega_n)}{1 - (\omega/\omega_n)^2} \tag{6-7}$$

The velocity of sound for air may be calculated from

$$c = 49.1\ T^{\frac{1}{2}}\ \text{ft/sec}$$

where T is the absolute temperature in degrees Rankine. When the tube diameter is very small as in a capillary, it is possible to produce a very large

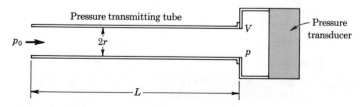

Fig. 6-2 Schematic of pressure-transmitting system.

damping ratio so that Eq. (6-4) will reduce to the following for frequencies below the natural frequencies:

$$\left| \frac{p}{p_0} \right| = \frac{1}{[1 + 4h^2(\omega/\omega_n)^2]^{\frac{1}{2}}} \tag{6-8}$$

If the transmitting fluid is a gas, the entire system can act as a Helmholtz resonator with a resonant frequency of

$$\omega = \left[\frac{\pi r^2 c^2}{V(L + \frac{1}{2}\sqrt{\pi^2 r^2})} \right]^{\frac{1}{2}} \tag{6-9}$$

More complete information on the dynamic response of pressure-measurement systems is given in Refs. [1], [7], and [11].

From both Eqs. (6-4) and (6-8) it is evident that a capillary tube may be used for effective damping of pressure signals. The tube is then said to act as an acoustical filter. The similarity of this system to that described in Sec. 2-6 and the discussion of the seismic instrument in Sec. 11-3 are to be noted. It should be noted also that the actual dynamic response for tube systems is strongly frequency-dependent, and the preceding formulas must be accepted with possible modification in high-frequency ranges. Tijdeman [18] and others [14 to 17] discuss experimental and analytical solutions which are available for such problems. In a dynamic pressure-measurement application one must also consider the frequency response of the pressure transducer and its movement in the overall measurement system. In general, one should try to design the system so that the natural frequency of the transducer is substantially greater than the signal frequency to be measured. Dynamic pressure measurements are particularly applicable to sound-level determinations, as discussed in Chap. 11.

Example 6-1 A small tube, 0.02 in. in diameter, is connected to a pressure transducer through a volume of 0.2 in.^{3}. The tube has a length of 3 in. Air at 14.7 psia and 70°F is the pressure transmitting fluid. Calculate the natural frequency for this system and the damping ratio.

Solution We shall use Eq. (6-9) for this calculation. For air,

$$\rho = \frac{p}{RT} = \frac{(14.7)(144)}{(53.35)(530)} = 0.0752 \text{ lb}_m/\text{ft}^3$$

$$c = (49.1)(530)^{\frac{1}{2}} = 1127 \text{ ft/sec}$$

$$\mu = 0.0440 \text{ lb}_m/\text{hr-ft}$$

Thus,

$$\omega_n = \left\{ \frac{\pi(0.01)^2(1127)^2(1728)}{(12)^2(0.2)[\frac{3}{12} + (0.5)(\pi)(0.01/12)]} \right\}^{\frac{1}{2}}$$

$$= 95.5 \text{ Hz}$$

The damping ratio is calculated with Eq. (6-6):

$$h = \frac{(2)(0.0440)(1728)}{(3600)(0.0752)(1127)(0.01)^3} \left[(3)\left(\frac{3}{12}\right)\left(\frac{0.2}{1728}\right)\pi \right]^{\frac{1}{2}}$$

$$= 2.57$$

Example 6-2 Calculate the attenuation of a 40-Hz pressure signal in the system of Example 6-1.

Solution For this calculation we employ Eq. (6-4). We have

$$\frac{\omega}{\omega_n} = \frac{40}{95.5} = 0.418$$

so that

$$\left|\frac{p}{p_0}\right| = \frac{1}{\{[1 - (0.418)^2]^2 + (4)(2.57)^2(0.418)^2\}^{\frac{1}{2}}}$$

$$= 0.435$$

The calibration of pressure transducers for dynamic measurement applications is rather involved and discussed in detail in Ref. [19].

6-3 MECHANICAL PRESSURE-MEASUREMENT DEVICES

Mechanical devices offer the simplest means for pressure measurement. In this section we shall examine the principles of some of the more important arrangements.

The fluid manometer is a widely used device for measurement of fluid pressures under steady-state conditions. Consider first the *U-tube manometer* shown in Fig. 6-3. The difference in pressure between the unknown pressure p and the atmosphere is determined as a function of the differential height h. The density of the fluid transmitting the pressure p is ρ_f, and the density of the

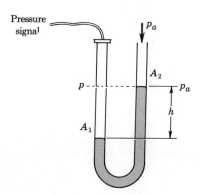

Fig. 6-3 U-tube manometer.

manometer fluid is designated as ρ_m. A pressure balance of the two columns dictates that

$$p_a + \frac{g}{g_c} h \rho_m = p + \frac{g}{g_c} h \rho_f \tag{6-10}$$

or

$$p - p_a = \frac{g}{g_c} h(\rho_m - \rho_f) \tag{6-11}$$

Equation (6-11) gives the basic principle of the U-tube manometer. It is to be noted that the distance h is measured parallel to the gravitational force and that the differential pressure $p - p_a$ is measured at the location designated by the dashed line. If the location of the pressure source is at a different elevation from this point, there could be an appreciable error in the pressure determination, depending upon the density of the transmitting fluid.

A *well-type manometer* operates in the same manner as the U-tube manometer except that the construction is as shown in Fig. 6-4. In this case the pressure balance of Eq. (6-10) still yields

$$p - p_a = \frac{g}{g_c} h(\rho_m - \rho_f)$$

This equation is seldom used, however, because the height h is not the fluid displacement which is normally measured. Typically, the well-type manometer is filled to a certain level at zero-pressure differential conditions. A measurement is then made of the displacement of the small column from this zero level. Designating this displacement by h', we have

$$h' A_2 = (h - h') A_1 \tag{6-12}$$

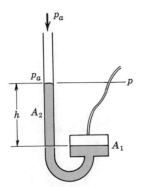

Fig. 6-4 Well-type manometer.

since the volume displacements are the same on both sides of the manometer. Inserting Eq. (6-12) in (6-10) gives

$$p - p_a = \frac{g}{g_c} h' \left(\frac{A_2}{A_1} + 1 \right)(\rho_m - \rho_f) \tag{6-13}$$

Commercial well-type manometers have the scale for the manometer column graduated so that the user need not apply the area correction factor to the indicated displacement h'. Thus, for an area ratio of $A_2/A_1 = 0.03$, a true reading of 10.0 in. for h' would be indicated as 10.3 as a result of the special scale graduation. The indicated value is then substituted for h in Eq. (6-11).

Manometers may be oriented in an inclined position to lengthen the scale and improve readability, or special optical sightglasses and vernier scales may be employed to provide more accurate location and indication of the manometer-fluid height than could be obtained with the naked eye. When mercury is the manometer fluid, variable-reluctance pickups may be used to accurately sense the fluid height. Special metal floats may also afford such a convenience with less dense fluids which are nonconductive.

When a well-type manometer is arranged as in Fig. 6-5, it is commonly called a barometer. The top of the column is evacuated while the well is exposed to atmospheric pressure. The height h is thus a measure of the absolute atmospheric pressure. When $p_a = 14.696$ psia, the height of a column of mercury at 68°F would be 760 mm.

6-4 DEAD-WEIGHT TESTER

The dead-weight tester is a device used for balancing a fluid pressure with a known weight. Typically, it is a device which is used for static calibration of pressure gages and is seldom employed for an actual pressure measurement. Our discussion will be concerned only with the use of the dead-weight tester as a calibration device.

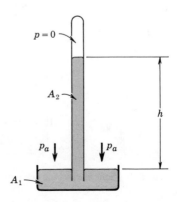

Fig. 6-5 Manometer used as a barometer.

Consider the schematic in Fig. 6-6. The apparatus is set up for calibration of the pressure gage G. The chamber and cylinder of the tester are filled with a clean oil by first moving the plunger to its most forward position and then slowly withdrawing it while the oil is poured in through the opening for the piston. The gage to be tested is installed and the piston inserted in the cylinder. The pressure exerted on the fluid by the piston is now transmitted to the gage when the valve is opened. This pressure may be varied by adding weights to the piston or by using different piston-cylinder combinations of varying areas. The viscous friction between the piston and the cylinder in the axial direction may be substantially reduced by rotating the piston-weight assembly while the measurement is taken. As the pressure is increased, it may be necessary to advance the plunger to account for compression of the oil and any entrapped gases in the apparatus. High-pressure–dead-weight testers have a special lever system which is used to apply large forces to the piston.

The accuracies of dead-weight testers are limited by two factors: (1) the friction between the cylinder and the piston and (2) the uncertainty in the area of the piston. The friction is reduced by rotation of the piston and use of long enough surfaces to ensure negligible flow of oil through the annular space between the piston and the cylinder. The area upon which the weight force acts is not the area of the piston nor the area of the cylinder; it is some effective area between these two which depends on the clearance spacing and the viscosity of the oil. The smaller the clearance, the more closely the effective area will approximate the cross-sectional area of the piston. It can

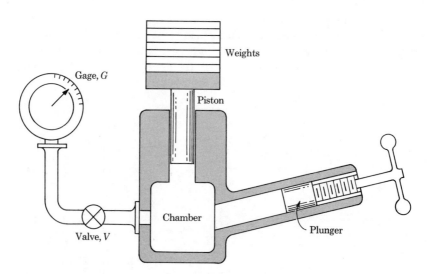

Fig. 6-6 Schematic of a dead-weight tester.

be shown† that the percentage error due to the clearance varies according to

$$\text{Percent error} \sim \frac{(\rho\,\Delta p)^{\frac{1}{2}}b^3}{\mu DL} \tag{6-14}$$

where ρ = density of the oil

Δp = pressure differential on the cylinder

b = clearance spacing

μ = viscosity

D = piston diameter

L = piston length

At high pressures there can be an elastic deformation of the cylinder which increases the clearance spacing and thereby increases the error of the tester.

6-5 BOURDON-TUBE PRESSURE GAGE

Bourdon-tube pressure gages enjoy a wide range of application where consistent, inexpensive measurements of static pressure are desired. They are commercially available in many sizes (1 to 16 in. diameter) and accuracies. The Heise gage‡ is an extremely accurate bourdon-tube gage having an accuracy of 0.1 percent of full-scale reading and is frequently employed as a secondary pressure standard in laboratory work.

The construction of a bourdon-tube gage is shown in Fig. 6-7. The bourdon tube itself is usually an elliptical cross-sectional tube having a "C"-shape configuration. When the pressure is applied to the inside of the tube, an elastic deformation results, which, ideally, is proportional to the pressure. The degree of linearity depends on the quality of the gage. The end of the gage is connected to a spring-loaded linkage, which amplifies the displacement and transforms it to an angular rotation of the pointer. The linkage is constructed so that the mechanism may be adjusted for optimum linearity and minimum hysteresis, as well as to compensate for wear which may develop over a period of time. Electrical-resistance strain gages (Sec. 10-7) may also be installed on the bourdon tube to sense the elastic deformation.

6-6 DIAPHRAGM AND BELLOWS GAGES

Diaphragm and bellows gages represent similar types of elastic deformation devices useful for many pressure-measurement applications. Consider first the flat diaphragm subjected to the differential pressure $p_1 - p_2$, as shown in

† See, for example, Ref. [10], p. 105.
‡ Manufactured by Heise Gage Company, Newton, Conn.

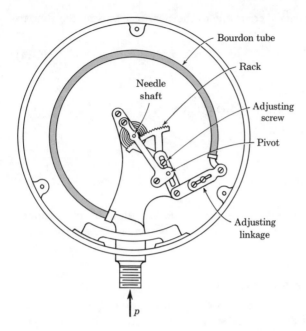

Fig. 6-7 Schematic of a bourdon-tube pressure gage.

Fig. 6-8. The diaphragm will be deflected in accordance with this pressure differential and the deflection sensed by an appropriate displacement transducer. Electrical-resistance strain gages may also be installed on the diaphragm, as shown in Fig. 6-9. The output of these gages is a function of the local strain, which, in turn, may be related to the diaphragm deflection and pressure differential. The deflection generally follows a linear variation with Δp when the deflection is less than one-third the diaphragm thickness. Figure 6-10 compares the deflection characteristics of three diaphragm

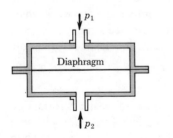

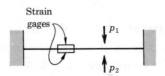

Fig. 6-8 Schematic of a dia-phragm gage.

Fig. 6-9 Diaphragm gage using electrical-resistance strain gages.

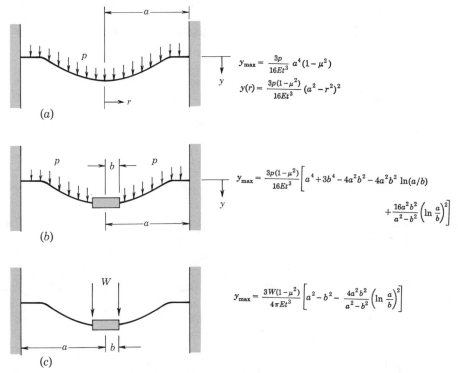

$$y_{max} = \frac{3p}{16Et^3} a^4 (1 - \mu^2)$$

$$y(r) = \frac{3p(1 - \mu^2)}{16Et^3} (a^2 - r^2)^2$$

(a)

$$y_{max} = \frac{3p(1 - \mu^2)}{16Et^3} \left[a^4 + 3b^4 - 4a^2b^2 - 4a^2b^2 \ln(a/b) \right. $$
$$\left. + \frac{16a^2b^2}{a^2 - b^2} \left(\ln \frac{a}{b} \right)^2 \right]$$

(b)

$$y_{max} = \frac{3W(1 - \mu^2)}{4\pi Et^3} \left[a^2 - b^2 - \frac{4a^2b^2}{a^2 - b^2} \left(\ln \frac{a}{b} \right)^2 \right]$$

(c)

Fig. 6-10 Deflection characteristics of three diaphragm arrangements according to Ref. [9]. (a) Edges fixed, uniform load over entire surface; (b) outer edge fixed and supported, inner edge fixed, uniform load over entire actual surface; (c) outer edge fixed and supported, inner edge fixed, uniform load along inner edge.

arrangements as given by Roark [9]. Note that the first two diaphragms have uniform pressure loading over the entire surface of the disk, while the third type has a load which is applied at the center boss. In all three cases it is assumed that the outer edge of the disk is rigidly fixed and supported. To facilitate linear response over a larger range of deflections than that imposed by the one-third-thickness restriction, the diaphragm may be constructed from a corrugated disk, as shown in Fig. 6-11. This type of diaphragm is most suitable for those applications where a mechanical device is used for sensing the deflection of the diaphragm. Larger deflections are usually necessary with a mechanical amplification device than for electric

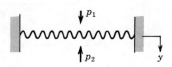

Fig. 6-11 Corrugated-disk diaphragm.

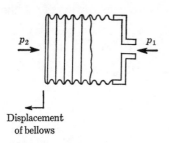

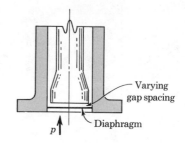

Fig. 6-12 Schematic of a bellows pressure gage.

Fig. 6-13 Capacitance pressure gage.

transducers. A good summary of the properties of corrugated diaphragms is given in Ref. [12].

The bellows gage is depicted schematically in Fig. 6-12. A differential pressure force causes a displacement of the bellows, which may be converted to an electric signal or undergo a mechanical amplification to permit display of the output on an indicator dial. The bellows gage is generally unsuitable for transient measurements because of the larger relative motion and mass involved. The diaphragm gage, on the other hand, may be quite stiff, involves rather small displacements, and is suitable for high-frequency pressure measurements.

The deflection of a diaphragm under pressure may be sensed by a capacitance variation as shown in Fig. 6-13. Such pressure pickups are well-suited for dynamic measurements since the natural frequency of diaphragms can be rather high. The capacitance pickup, however, involves

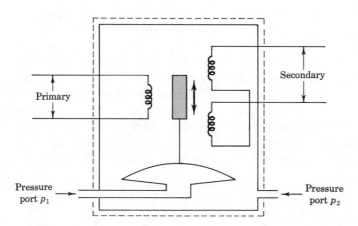

Fig. 6-14 Schematic of LVDT-diaphragm differential pressure gage. (*Courtesy Sanborn Company.*)

low sensitivity, and special care must be exerted in the construction of readout circuitry. A schematic diagram of a LVDT-diaphragm differential pressure gage is shown in Fig. 6-14. Commercial models of this type of gage permit measurement of pressures as low as 0.000035 psi.

The natural frequency of a circular diaphragm fixed at its perimeter is given by Hetenyi [5] as

$$f = \frac{10.21}{a^2} \sqrt{\frac{g_c E t^2}{12(1 - \mu^2)\rho}} \text{ Hz} \tag{6-15}$$

where E = modulus of elasticity, psi

$\quad t$ = thickness, in.

$\quad a$ = radius of the diaphragm, in.

$\quad \rho$ = density of the material, $\text{lb}_m/\text{in.}^3$

$\quad g_c$ = dimensional conversion constant, expressed in units of inches $(g_c = 385.9 \text{ lb}_m\text{-in.}/\text{lb}_f\text{-sec}^2)$

$\quad \mu$ = Poisson's ratio

Equation (6-15) may be simplified to the following relation for steel diaphragms:

$$f = 1.934 \times 10^6 \frac{t}{\pi a^2} \tag{6-16}$$

where t and a are in inches.

Example 6-3 A diaphragm pressure gage is to be constructed of spring steel ($E = 29 \times 10^6$ psi, $\mu = 0.3$) 2 in. in diameter and is to be designed to measure a maximum pressure of 200 psig. Calculate the thickness of the gage required so that the maximum deflection is one-third this thickness. Calculate the natural frequency of this diaphragm.

Solution Using the relation from Fig. 6-10, we have

$$y_{max} = \tfrac{1}{3}t = \frac{3 \, \Delta p}{16 E t^3} a^4 (1 - \mu^2)$$

$$t^4 = \frac{(9)(200)(1.0)^4}{(16)(29 \times 10^6)} [1 - (0.3)^2] = 3.53 \times 10^{-6} \text{ in.}^4$$

$$t = 0.0433 \text{ in.}$$

We may calculate the natural frequency from Eq. (6-15)

$$f = \frac{10.21}{(1)^2} \left[\frac{(385.9)(29 \times 10^6)(0.0433)^2(1,728)}{(12)(1 - 0.09)(490)} \right]^{\frac{1}{2}}$$

$$= 26,600 \text{ Hz}$$

6-7 THE BRIDGMAN GAGE†

It is known that the resistance of fine wires changes with the pressure according to a linear relationship.

$$R = R_1(1 + b \, \Delta p) \tag{6-17}$$

R_1 is the resistance at 1 atm, b is the pressure coefficient of resistance, and Δp is the gage pressure. The effect may be used for measurement of pressures as high as 100,000 atm [4]. A pressure transducer based on this principle is called a Bridgman gage. A typical gage employs a fine wire of Manganin (84 percent Cu, 12 percent Mn, 4 percent Ni) wound in a coil and enclosed in a suitable pressure container. The pressure coefficient of resistance for this material is about $1.7 \times 10^{-7} \, \text{psi}^{-1}$. The total resistance of the wire is about 100 ohms, and conventional bridge circuits are employed for measuring the change in resistance. Such gages are subject to aging over a period of time so that frequent calibration is required; however, when properly calibrated the gage can be used for high-pressure measurement with an accuracy of 0.1 percent. The transient response of the gage is exceedingly good. The resistance wire itself can respond to variations in the megacycle range. Of course, the overall frequency response of the pressure-measurement system would be limited to much lower values because of the acoustic response of the transmitting fluid. Many of the problems associated with high-pressure measurement are discussed more fully in Refs. [2], [6], and [13].

6-8 LOW-PRESSURE MEASUREMENT

The science of low-pressure measurement is a rather specialized field which requires considerable care on the part of the experimentalist. The purpose of our discussion is to call attention to the more prominent types of vacuum instruments and describe the physical principles upon which they operate. For those readers requiring more specialized information, we refer them to the excellent monograph by Dushman and Lafferty [3]. The reader may also consult this reference for information on the various techniques for producing and maintaining a vacuum.

For moderate vacuum measurements the bourdon gage, manometers, and various diaphragm gages may be employed. Our discussion in this section, however, is concerned with the measurement of low pressures which are not usually accessible to the conventional gages. In this sense, we are primarily interested in absolute pressures below 1 torr.

6-9 THE MCLEOD GAGE‡

The McLeod gage is a modified mercury manometer which is constructed as shown in Fig. 6-15. The movable reservoir is lowered until the mercury

† P. W. Bridgman, *Proc. Natl. Acad. Sci., U.S.*, vol. 3, p. 10, 1917.
‡ H. McLeod, *Phil. Mag.*, vol. 48, p. 110, 1874.

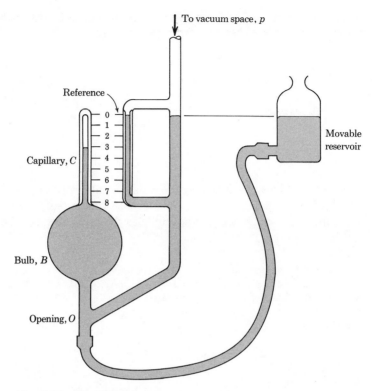

Fig. 6-15 The McLeod gage.

column drops below the opening O. The bulb B and capillary C are then at the same pressure as that of the vacuum source p. The reservoir is subsequently raised until the mercury fills the bulb and rises in the capillary to a point where the level in the reference capillary is located at the zero point. The volume of the capillary per unit length is denoted by a so that the volume of the gas in the capillary is

$$V_c = ay \tag{6-18}$$

where y is the length of the capillary occupied by the gas.

We designate the volume of the capillary, bulb, and tube down to the opening as V_B. The pressure of the gas in the capillary is thus

$$p_c = p \frac{V_B}{V_c} \tag{6-19}$$

Now, the pressure indicated by the capillary is

$$p_c - p = y \tag{6-20}$$

where we are expressing the pressure in terms of the height of the mercury column. Combining Eqs. (6-18) to (6-20) gives

$$p = \frac{ay^2}{V_B - ay} \tag{6-21}$$

For most cases $ay \ll V_B$ and

$$p = \frac{ay^2}{V_B} \tag{6-22}$$

Commercial McLeod gages have the capillary calibrated directly in microns. The McLeod gage is insensitive to condensed vapors that may be present in the sample and are generally applicable to the pressure range from 10^{-2} to $10^2 \mu$.

Example 6-4   A McLeod gage has $V_B = 100$ cm³ and a capillary diameter of 1 mm. Calculate the pressure indicated by a reading of 3.00 cm. What error would result if Eq. (6-22) were used instead of Eq. (6-21)?

Solution   We have

$$V_c = \frac{\pi(1)^2}{4}(30.0) = 23.6 \text{ mm}^3$$

$$V_B = 10^5 \text{ mm}^3$$

From Eq. (6-21),

$$p = \frac{(23.6)(30.0)^2}{10^5 - 23.6} = 0.212 \text{ torr} = 212 \mu$$

The fractional error in using Eq. (6-22) would be

$$\text{Error} = \frac{ay}{V_B} = 2.36 \times 10^{-4}$$

or a negligibly small value.

6-10 PIRANI THERMAL-CONDUCTIVITY GAGE†

At low pressures the effective thermal conductivity of gases decreases with pressure. The Pirani gage is a device that measures the pressure through the change in thermal conductance of the gas. The gage is constructed as shown in Fig. 6-16. An electrically heated filament is placed inside the vacuum space. The heat loss from the filament is dependent upon the thermal conductivity of the gas and the filament temperature. The lower the pres-

† M. Pirani, *Verhandl. deut. physik. Ges.*, vol. 8, p. 686, 1906.

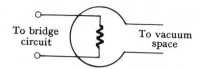

Fig. 6-16 Schematic of Pirani gage.

sure, the lower the thermal conductivity and consequently the higher the filament temperature for a given electric-energy input. The temperature of the filament could be measured by a thermocouple, but in the Pirani-type gage the measurement is made by observing the variation in resistance of the filament material (tungsten, platinum, etc.). The resistance measurement may be performed with an appropriate bridge circuit. The heat loss from the filament is also a function of the ambient temperature, and, in practice, *two* gages are connected in series, as shown in Fig. 6-17, to compensate for possible variations in the ambient conditions. The measurement gage is evacuated, and both it and the sealed gage are exposed to the same environment conditions. The bridge circuit is then adjusted (through resistance R_2) to produce a null condition. When the test gage is now exposed to the particular pressure conditions the deflection of the bridge from the null position will be compensated for changes in environment temperature.

Pirani gages require an empirical calibration and are not generally suitable for use at pressures much below 1 μ. The upper limit is about 1 torr. For higher pressures the thermal conductance changes very little with pressure. It must be noted that the heat loss from the filament is also a function of the conduction losses to the filament supports and radiation losses to the surroundings. The lower limit of applicability of the gage is the point where these effects overshadow the conduction into the gas. The transient response of the Pirani gage is poor. The time necessary for the establishment of thermal equilibrium may be of the order of several minutes at low pressures.

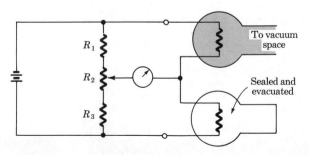

Fig. 6-17 Pirani-gage arrangement to compensate for change in ambient temperatures.

6-11 THE KNUDSEN GAGE†

Consider the arrangement shown in Fig. 6-18. Two vanes V along with the mirror M are mounted on the thin-filament suspension. Near these vanes are two heated plates P, each of which is maintained at a temperature T. The separation distance between the vanes and plates is *less than* the mean free path of the surrounding gas. Heaters are installed so that the temperature of the plates is higher than that of the surrounding gas. The vanes are at the temperature of the gas T_g. The molecules striking the vanes from the hot plates have a higher velocity than those leaving the vanes because of the difference in temperature. Thus, there is a net momentum imparted to the vanes which may be measured by observing the angular displacement of the mirror, similar to the technique used in a lightbeam galvanometer. The total momentum exchange with the vanes is a function of molecular density, which, in turn, is related to the pressure and temperature of the gas. An expression for the gas pressure may thus be derived in terms of the temperatures and the measured force. For small temperature differences $T - T_g$ it may be shown that this relation is [3]

† M. Knudsen, *Ann. Physik*, vol. 32, p. 809, 1910.

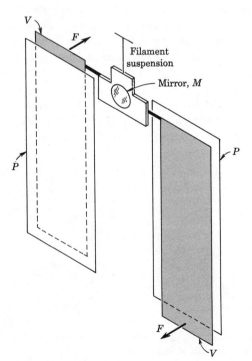

Fig. 6-18 Schematic of Knudsen gage.

$$p = 4F \frac{T_g}{T - T_g} \tag{6-23}$$

where the pressure is in dynes per square centimeter when the force is in dynes. The temperatures are in degrees Kelvin.

The Knudsen gage furnishes an absolute measurement of the pressure, which is independent of the molecular weight of the gas. It is suitable for use between 10^{-5} and $10 \, \mu$ and may be used as a calibration device for other gages in this region.

6-12 THE IONIZATION GAGE

Consider the arrangement shown in Fig. 6-19, which is similar to the ordinary triode vacuum tube. The heated cathode emits electrons, which are accelerated by the positively charged grid. As the electrons move toward the grid, they produce ionization of the gas molecules through collisions. The plate is maintained at a negative potential so that the positive ions are collected there, producing the plate current i_p. The electrons and negative ions are collected by the grid, producing the grid current i_g. It is found that the pressure of the gas is proportional to the ratio of plate current to grid current.

$$p = \frac{1}{S} \frac{i_p}{i_g} \tag{6-24}$$

where the proportionality constant S is called the sensitivity of the gage. A typical value for nitrogen is $S = 20 \, \text{torr}^{-1}$, but the exact value must be determined by calibration of the particular gage. The value of S is a function of the tube geometry and the type of gas.

Conventional ionization gages are suitable for measurements between 1.0 and $10^{-5} \, \mu$, and the current output is usually linear in this range. At higher pressures there is the danger of burning out the cathode. Special types of ionization gages are suitable for measurements of pressures as low as $10^{-12} \, \text{torr}$. Very precise experimental techniques are required, however, in order to perform measurements at these high vacuums. The interested reader should consult Ref. [3] for additional information.

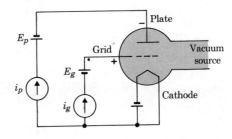

Fig. 6-19 Schematic of ionization gage.

6-13 THE ALPHATRON†

The Alphatron is a radioactive ionization gage, which is shown schematically in Fig. 6-20. A small radium source serves as an α-particle emitter. These particles ionize the gas inside the gage enclosure, and the degree of ionization is determined by measuring the voltage output E_0. The degree of ionization is a direct linear function of pressure for a rather wide range of pressures from 10^{-3} to 10^3 torr. The output characteristics, however, are different for each type of gas used. The lower pressure limit of the gage is determined by the length of the mean free path of the alpha particles as compared with the enclosure dimensions. At very low pressures the mean free path becomes so large that very few collisions are probable in the gage and hence the ionization level is very small. The Alphatron has the advantages that it may be used at atmospheric pressure as well as high vacuum and that there is no heated filament to contend with as in the conventional ionization gage. Consequently, there is no problem of accidentally burning out a filament because of an inadvertent exposure of the gage to high (above 10^{-1} torr) pressures.

6-14 SUMMARY

Figure 6-21 gives a convenient summary of the pressure ranges for which the gages discussed are normally employed in practice.

REVIEW QUESTIONS

1. Distinguish between gage pressure, absolute pressure, and vacuum.
2. To transmit a high-frequency-pressure signal one should select
    (*a*) A short small-diameter tube
    (*b*) A short large-diameter tube

† National Research Corp., Cambridge, Mass.

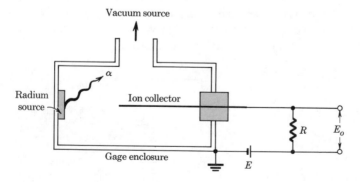

Fig. 6-20   Schematic of Alphatron gage.

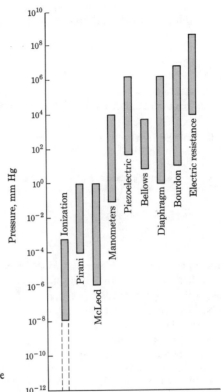

Fig. 6-21 Summary of applicable
range of pressure gages.

   (c) A long small-diameter tube
   (d) A long large-diameter tube

3. What are the advantages of the manometer pressure-measurement device?

4. What is the advantage of a well-type manometer?

5. What are some advantages of the bourdon-tube, diaphragm, and bellows gages?

6. Describe the principle of operation of a McLeod gage.

7. Describe the Pirani gage.

8. When is the Knudsen gage used?

9. Describe the ionization gage. How does it differ from the Pirani gage? What disadvantages does it have?

PROBLEMS

6-1. A mercury barometer is constructed like that shown in Fig. 6-5. The column is a glass tube 0.250 in. ID and 0.375 in. OD, and the well is a glass dish 1.50 in. ID. Calculate the percent error which would result if an area correction factor were not used.

6-2. Derive an expression for the radius of a simple diaphragm as shown in Fig. 6-10 using the following restrictions:

(a) The maximum deflection is one-third the thickness.

(b) The maximum deflection must be 100 times as great as the uncertainty in the deflection measurement w_y.

Assume that the maximum pressure differential Δp is given as well as w_y.

6-3. Determine the factor to convert pressure in inches of water to pounds per square foot.

6-4. The effective radius of an air molecule is about 1.85×10^{-8} cm. Calculate the mean free path at 70°F and the following pressures: 1 atm, 1 torr, 1 μ, 1 in. H_2O, and $10^{-3}\,\mu$.

6-5. Reduce Eq. (6-3) to an expression for mean free path in terms of pressure in microns, temperature in degrees Kelvin, molecular weight of the gas, and effective molecular diameter.

6-6. A dynamic pressure measurement is to be made with an apparatus similar to that shown in Fig. 6-2. The appropriate dimensions are:

$$L = 2.50 \text{ in.}$$
$$r = 0.005 \text{ in.}$$
$$V = 0.10 \text{ in.}^3$$

The fluid is air at 70°F and 14.7 psia. Plot the pressure amplitude ratio versus ω/ω_n according to both Eq. (6-4) and Eq. (6-8).

6-7. Plot the error in Eq. (6-8) versus r^2L/V.

6-8. Calculate the resonant frequency of the system in Prob. 6-6, assuming that it acts as a Helmholtz resonator.

6-9. A well-type manometer uses a special bromide fluid having a specific gravity of 2.95. The well has a diameter of 3.00 in., and the tube has a diameter of 0.200 in. The manometer is to be used to measure a differential pressure in a water-flow system. The scale placed alongside the tube has no correction factor for the area ratio of the manometer. Calculate the value of a factor that may be multiplied by the manometer reading in inches to find the pressure differential in pounds per square inch.

6-10. A vacuum gage is to use an LVDT-diaphragm combination like that shown in Fig. 6-14. The LVDT has a sensitivity of 0.0001 in., and the diaphragm is to be constructed of steel ($E = 29 \times 10^6$, $\mu = 0.3$) with a diameter of 6 in. Calculate the diaphragm thickness in accordance with the restriction that the maximum deflection not exceed one-third this thickness. What is the lowest pressure which may be sensed by this instrument?

6-11. Calculate the natural frequency of the diaphragm in Prob. 6-10.

6-12. A Bridgman gage uses a coil of Manganin wire having a nominal resistance of 100 ohms at atmospheric pressure. The gage is to be used to measure a pressure of 1,000 psig with an uncertainty of 0.1 percent. What is the allowable uncertainty in the resistance measurement?

6-13. Suppose the Bridgman gage of Prob. 6-12 is connected to the bridge circuit of Fig. 4-17 so that the gage is R_1 and all resistances are equal to 100 ohms at a pressure of 1 atm. The battery voltage is 4.0 volts, and the detector is a high-impedance voltage-measuring device. The bridge is assumed to be in balance at $p = 1$ atm. Calculate the voltage output of the bridge at $p = 1,000$ psig.

6-14. Rework Example 6-2 assuming the diameter of the capillary to be 0.2 mm.

6-15. A Knudsen gage is to be designed to operate at a maximum pressure of 1.0 μ. For this application the spacing of the vane and plate is to be less than 0.3 mean free path at this pressure. Calculate the force on the vanes at pressures of 1.0 and 0.01 μ when the gas temperature is 20°C and the temperature difference is 50.0°K.

6-16. A capacitance-diaphragm pressure gage as shown in Fig. 6-13 is to be used to measure pressure differentials as high as 1,000 psi at frequencies as high as 15,000 Hz. The diameter of the diaphragm is not to exceed 0.500 in. Calculate the thickness and diameter of a diaphragm to accomplish this (the natural frequency should be at least 30,000 Hz). Choose a suitable gap spacing, and estimate the capacitance-pressure sensitivity of the device. Assume the dielectric constant is that of air.

6-17. A bourdon-tube pressure gage having an internal volume of 1.0 in.^{3} is used for measuring pressure in a fluctuating air system having frequencies as high as 100 Hz. Design an acoustical filter which will attenuate all frequencies above 20 Hz by 99 percent. Plot the frequency response of this filter.

6-18. An acoustic filter is to be designed to attenuate sharp pressure transients in air above 50 Hz. The volume of air contained in the pressure-transducer cavity is 0.6 in.^{3}, and a capillary tube connects the cavity to the pressure source. If the 50-Hz frequency is to be attenuated by 50 percent, determine the capillary length and diameter for system natural frequencies of (*a*) 50 Hz, (*b*) 100 Hz, and (*c*) 500 Hz.

6-19. A U-tube manometer uses tubes of 0.250 and 0.500 in. diameters for the two legs. When subjected to a certain pressure, the difference in height of the two fluid columns is 10.0 in. Hg. What would have been the reading if both tubes were the same diameter? The measurement is performed on air.

6-20. A pressure signal is fed through a line having an inside diameter of 0.062 in. and a length of 5 ft. The line is connected to a pressure transducer having a volume of approximately 0.3 in.^{3}. Air at 100 psia and 200°F is the transmitting fluid. Calculate the natural frequency and damping ratio for this system.

6-21. The manometer of Prob. 6-19 uses a fluid having a specific gravity of 1.85. The sensing fluid is water. What is the pressure difference when the difference in heights of the columns is 5.0 in.? Assume that both legs of the manometer are filled with water.

6-22. A diaphragm-pressure gage is constructed of spring steel to measure a pressure differential of 1,000 psi. The diameter of the diaphragm is 0.5 in. Calculate the diaphragm thickness so that the maximum deflection is one-third the thickness. What is the natural frequency of this diaphragm?

6-23. A Bridgman gage is to be used to measure a pressure of 10,000 psi using a Manganin element having a resistance of 100 ohms at atmospheric pressure. Calculate the resistance of the gage under high-pressure conditions. If the gage is one leg of a bridge whose other legs all have values of exactly 100 ohms, calculate the voltage output of the bridge for a constant-voltage source of 24 volts.

6-24. A McLeod gage is available which has a volume V_B of 150 cm^3 and a capillary diameter of 0.3 mm. Calculate the gage reading for a pressure of 30 μ.

6-25. What is the approximate range of mean free paths for air over the range of pressures for which the Knudsen gage is applicable?

6-26. A special high-pressure U-tube manometer is constructed to measure pressure differential in air at 2,000 psia and 70°F. When an oil having a specific gravity of 0.83 is used as the fluid, calculate the differential pressure in psia that would be indicated by a 5.3-in. reading.

6-27. A diaphragm like that shown in Fig. 6-10c has $a = 1.0$ in., $b = 0.125$ in., and $t = 0.048$ in. and is constructed of spring steel. It is subjected to a total loading of 600 lb$_f$. Calculate the deflection.

REFERENCES

1. Arons, A. B., and R. H. Cole: Design and Use of Piezo-electric Gages for Measurement of Large Transient Pressures, *Rev. Sci. Instr.*, vol. 21, pp. 31–38, 1950.
2. Bridgman, P. W.: "The Physics of High Pressure," The Macmillan Company, New York, 1931.
3. Dushman, S., and J. M. Lafferty: "Scientific Foundations of Vacuum Technique," 2d ed., John Wiley & Sons, Inc., New York, 1962.
4. Hall, H. T.: Some High Pressure—High Temperature Apparatus Design Considerations, *Rev. Sci. Instr.*, vol. 29, p. 267, 1958.
5. Hetenyi, M. (ed.): "Handbook of Experimental Stress Analysis," John Wiley & Sons, Inc., New York, 1950.
6. Howe, W. H.: The Present Status of High Pressure Measurements, *ISA Journal*, vol. 2, pp. 77 and 109, 1955.
7. Iberall, A. S.: Attenuation of Oscillatory Pressures in Instrument Lines, *Trans. ASME*, vol. 72, p. 689, 1950.
8. Neubert, H. K. P.: "Instrument Transducers," Oxford University Press, Fair Lawn, N.J., 1963.
9. Roark, R. J.: "Formulas for Stress and Strain," 3d ed., McGraw-Hill Book Company, New York, 1954.
10. Sweeney, R. J.: "Measurement Techniques in Mechanical Engineering," John Wiley & Sons, Inc., New York, 1953.
11. Taback, I.: The Response of Pressure Measuring Systems to Oscillating Pressure, *NACA Tech. Note* 1819, February, 1949.
12. Wildhack, W. A., R. F. Dresslea, and E. C. Lloyd: Investigation of the Properties of Corrugated Diaphragms, *Trans. ASME*, vol. 79, pp. 65–82, 1957.
13. Giardini, A. A. (ed.): "High Pressure Measurements," Butterworth & Co. (Publishers), Ltd., London, 1963.
14. Karam, J. T., and M. E. Franke: The Frequency Response of Pneumatic Lines, *J. Basic Eng., Trans. ASME*, ser. D, vol. 90, pp. 371–378, 1967.
15. Nichols, N. B.: The Linear Properties of Pneumatic Transmission Lines, *Trans. Instr. Soc. Am.*, vol. 1, pp. 5–14, 1962.
16. Watts, G. P.: The Response of Pressure Transmission Lines, Preprint No. 13.3-1-65, 20th Annual ISA Conf. and Exhibit, Los Angeles, Calif., October 4–7, 1965.
17. Bergh, H., and H. Tijdeman: Theoretical and Experimental Results for the Dynamic Response of Pressure Measuring Systems, *NLR Tech. Rept.* F. 238, 1965.
18. Tijdeman, H.: Remarks on the Frequency Response of Pneumatic Lines, *J. Basic Eng.*, pp. 325–328, June, 1969.
19. Schweppe, J. L., et al.: Methods for the Dynamic Calibration of Pressure Transducers, *Natl. Bur. Std. (U.S.), Monograph* 67, 1963.

7
Flow Measurement

7-1 INTRODUCTION

The measurement of fluid flow is important in applications ranging from measurements of blood-flow rates in a human artery to the measurement of the flow of liquid oxygen in a rocket. Many research projects and industrial processes depend upon a measurement of fluid flow to furnish important data for analysis. In some cases extreme precision is called for in the flow measurement, while in other instances only crude measurements are necessary. The selection of the proper instrument for a particular application is governed by many variables, including cost. For many industrial operations the accuracy of a fluid-flow measurement is directly related to profit. A simple example is the gasoline pump at the neighborhood service station; another example is the water meter at home. It is easy to see how a small error in flow measurement on a large natural gas or oil pipeline could make a difference of thousands of dollars over a period of time. Thus, the laboratory scientist is not the only person who is concerned with accurate flow measurement; the engineer in industry is also vitally interested because of the impact which flow measurements may have on the profit and loss statement of the company.

Flow-rate-measurement devices frequently require accurate pressure and temperature measurements in order to calculate the output of the instrument. Chapters 6 and 8 consider these associated measurement topics in detail, and the reader should consult the appropriate sections from time to time to relate specific pressure and temperature measurement devices to the material in the present chapter. We may remark at this time, however, that the overall accuracy of many of the most widely used flow-measurement devices is governed primarily by the accuracy of some pressure or temperature measurement.

Our objective in this chapter is to present a broad discussion of flow measurements and indicate the principles of operation of a number of devices that are commonly used. We shall also give the calculation methods that are connected with some of these devices and discuss some methods of flow visualization. In concluding the chapter a tabular comparison of the various methods will be presented, pointing out their range of applicability and expected accuracies.

7-2 POSITIVE-DISPLACEMENT METHODS

The flow rate of a nonvolatile liquid like water may be measured through a direct-weighing technique. The time necessary to collect a quantity of the liquid in a tank is measured, and an accurate measurement is then made of the weight of liquid collected. The average flow rate is thus calculated very easily. Improved accuracy may be obtained by using longer or more precise time intervals or more precise weight measurements. The direct-weighing technique is frequently employed for calibration of water and other liquid flowmeters and thus may be taken as a standard calibration technique. Obviously, it is not suited for transient flow measurements.

Positive-displacement flowmeters are generally used for those applications where consistently high accuracy is desired under steady flow conditions. A typical positive-displacement device is the home water meter shown

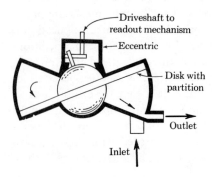

Fig. 7-1 Schematic of a nutating-disk meter.

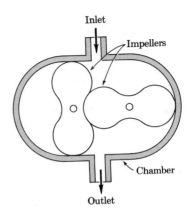

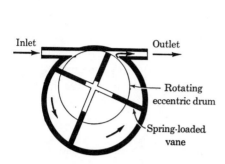

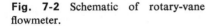

Fig. 7-2 Schematic of rotary-vane flowmeter.

Fig. 7-3 Schematic of lobed-impeller flowmeter.

schematically in Fig. 7-1. This meter operates on the nutating-disk principle. Water enters the left side of the meter and strikes the disk, which is eccentrically mounted. In order for the fluid to move through the meter, the disk must "wobble" or nutate about the vertical axis since both the top and the bottom of the disk remain in contact with the mounting chamber. A partition separates the inlet and outlet chambers of the disk. As the disk nutates, it gives direct indication of the volume of liquid which has passed through the meter. The indication of the volumetric flow is given through a gearing and register arrangement which is connected to the nutating disk. The nutating-disk meter may give reliable flow measurements within 1 percent.

Another type of positive-displacement device is the rotary-vane meter shown in Fig. 7-2. The vanes are spring-loaded so that they continuously maintain contact with the casing of the meter. A fixed quantity of fluid is trapped in each section as the eccentric drum rotates, and this fluid eventually finds its way out the exit. An appropriate register is connected to the shaft of the eccentric drum to record the volume of the displaced fluid. The uncertainties of rotary-vane meters are of the order of 0.5 percent, and the meters are relatively insensitive to viscosity since the vanes always maintain good contact with the inside of the casing.

The lobed-impeller meter shown in Fig. 7-3 may be used for either gas- or liquid-flow measurements. The impellers and case are carefully machined so that accurate fit is maintained. In this way, the incoming fluid is always trapped between the two rotors and is conveyed to the outlet as a result of their rotation. The number of revolutions of the rotors is an indication of the volumetric flow rate.

Example 7-1 A lobed-impeller flowmeter is used for measurement of the flow of nitrogen at 20 psia and 100°F. The meter has been calibrated so that it indicates the volumetric flow with an accuracy of $\pm$ one-half of one percent from 1,000 to 3,000 cfm. The uncertainties in the gas pressure and temperature measurements are ± 0.025 psi and ± 1.0°F, respectively. Calculate the uncertainty in a mass flow measurement at the given pressure and temperature conditions.

Solution The mass flow is given by

$$\dot{m} = \rho Q$$

where the density of nitrogen is given by

$$\rho = \frac{p}{R_{N_2}T}$$

Using Eq. (3-2), we obtain the following equation for the uncertainty in the mass flow as

$$\frac{w_{\dot{m}}}{\dot{m}} = \left[\left(\frac{w_Q}{Q} \right)^2 + \left(\frac{w_p}{p} \right)^2 + \left(\frac{w_T}{T} \right)^2 \right]^{\frac{1}{2}}$$

Using the given data,

$$\frac{w_{\dot{m}}}{\dot{m}} = \left[(0.005)^2 + \left(\frac{0.025}{20} \right)^2 + \left(\frac{1}{560} \right)^2 \right]^{\frac{1}{2}} = 5.05 \times 10^{-3}$$

or 0.505 percent. Thus, the uncertainties in the pressure and temperature measurements do not appreciably influence the overall uncertainty in the mass flow measurements.

7-3 FLOW-OBSTRUCTION METHODS

Several types of flowmeters fall under the category of obstruction devices. Such devices are sometimes called head meters because a head-loss or pressure-drop measurement is taken as an indication of the flow rate. Let us first consider some of the general relations for obstruction meters. We shall then examine the applicability of these relations to specific devices.

Consider the one-dimensional flow system shown in Fig. 7-4. The continuity relation for this situation is

$$\dot{m} = \rho_1 A_1 u_1 = \rho_2 A_2 u_2 \tag{7-1}$$

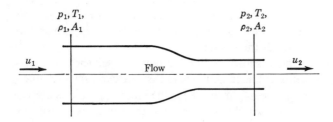

Fig. 7-4 General one-dimensional flow system.

where u is the velocity. If the flow is adiabatic and frictionless and the fluid is incompressible, the familiar Bernoulli equation may be written

$$\frac{p_1}{\rho_1} + \frac{u_1{}^2}{2g_c} = \frac{p_2}{\rho_2} + \frac{u_2{}^2}{2g_c} \tag{7-2}$$

where now $\rho_1 = \rho_2$. Solving Eqs. (7-1) and (7-2) simultaneously gives for the pressure drop

$$p_1 - p_2 = \frac{u_2{}^2 \rho}{2g_c}\left[1 - \left(\frac{A_2}{A_1}\right)^2 \right] \tag{7-3}$$

and the volumetric flow rate may be written

$$Q = A_2 u_2 = \frac{A_2}{\sqrt{1 - (A_2/A_1)^2}}\sqrt{\frac{2g_c}{\rho}(p_1 - p_2)} \tag{7-4}$$

where $Q = $ ft³/sec
$\quad\ A = $ ft²
$\quad\ \rho = $ lb$_m$/ft³
$\quad\ p = $ lb$_f$/ft²
$\quad\ g_c = 32.17$ lb$_m$-ft/lb$_f$-sec²

Thus, we see that a channel like the one shown in Fig. 7-4 could be used for a flow measurement by simply measuring the pressure drop $(p_1 - p_2)$ and calculating the flow from Eq. (7-4). No such channel, however, is frictionless, and some losses are always present in the flow. The volumetric flow rate calculated from Eq. (7-4) is the ideal value, and it is usually related to the actual flow rate and an empirical *discharge coefficient* C by the following relation:

$$\frac{Q_{\text{actual}}}{Q_{\text{ideal}}} = C \tag{7-5}$$

The discharge coefficient is not a constant and may depend strongly on the flow Reynolds number and the channel geometry.

When the flow of an ideal gas is considered, the following equation of state applies:

$$p = \rho RT \tag{7-6}$$

where T is the absolute temperature and R is the gas constant for the particular gas. For reversible adiabatic flow the steady-flow energy equation is

$$c_p T_1 + \frac{u_1{}^2}{2g_c} = c_p T_2 + \frac{u_2{}^2}{2g_c} \tag{7-7}$$

where c_p is the specific heat at constant pressure and is assumed constant for an ideal gas. When Eqs. (7-1), (7-6), and (7-7) are combined, there results

$$\dot{m}^2 = 2g_c A_2{}^2 \frac{\gamma}{\gamma - 1} \frac{p_1{}^2}{RT_1} \left[\left(\frac{p_2}{p_1}\right)^{2/\gamma} - \left(\frac{p_2}{p_1}\right)^{(\gamma+1)/\gamma} \right] \tag{7-8}$$

where the velocity of approach, i.e., the velocity at section 1, is assumed to be very small. This relationship may be simplified to

$$\dot{m} = \sqrt{\frac{2g_c}{RT_1}} \, A_2 \left[p_2 \, \Delta p - \left(\frac{1.5}{\gamma} - 1\right) \Delta p^2 + \cdots \right]^{\frac{1}{2}} \tag{7-9}$$

with $\Delta p = p_1 - p_2$ and $\gamma = c_p/c_v$ is the ratio of specific heats for the gas. Equation (7-9) is valid for $\Delta p < p_1/4$. When $\Delta p < p_1/10$, a further simplification may be made to give

$$\dot{m} = A_2 \sqrt{\frac{2g_c p_2}{RT_1} (p_1 - p_2)} \tag{7-10}$$

Note that Eq. (7-10) reduces to Eq. (7-4) when the relation for density from Eq. (7-6) is substituted. Thus, for small values of Δp compared with p_1, the flow of a compressible fluid may be approximated by the flow of an incompressible fluid.

Three typical obstruction meters are shown in Fig. 7-5. The venturi offers the advantages of high accuracy and small pressure drop, while the orifice is considerably lower in cost. Both the flow nozzle and the orifice have relatively high permanent pressure drop. Flow-rate calculations for all three devices are made on the basis of Eq. (7-4) with appropriate empirical

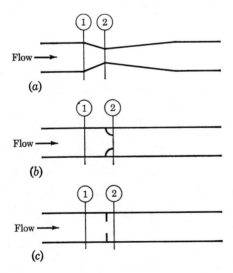

Fig. 7-5 Schematic of three typical obstruction meters. (a) Venturi; (b) flow nozzle; (c) orifice.

constants defined as follows:

$$M = \text{velocity of approach factor} = \frac{1}{\sqrt{1 - (A_2/A_1)^2}} \qquad (7\text{-}11)$$

$$K = \text{flow coefficient} = CM \qquad (7\text{-}12)$$

$$\beta = \text{diameter ratio} = \frac{d}{D} = \sqrt{\frac{A_2}{A_1}} \qquad (7\text{-}13)$$

When flow measurements of a compressible fluid are made, an additional parameter, the *expansion factor* Y, is used. For venturis and nozzles this factor is given by

$$Y_a = \left[\left(\frac{p_2}{p_1}\right)^{2/\gamma} \frac{\gamma}{\gamma - 1} \frac{1 - (p_2/p_1)^{(\gamma-1)/\gamma}}{1 - (p_2/p_1)} \frac{1 - (A_2/A_1)^2}{1 - (A_2/A_1)^2(p_2/p_1)^{2/\gamma}} \right]^{\frac{1}{2}} \qquad (7\text{-}14)$$

while for orifices an empirical expression for Y is given as

$$Y_1 = 1 - \left[0.41 + 0.35\left(\frac{A_2}{A_1}\right)^2 \right] \frac{p_1 - p_2}{\gamma p_1} \qquad (7\text{-}15)$$

when either flange taps or vena contracta taps are used. For orifices with pipe taps the following relation applies:

$$Y_2 = 1 - [0.333 + 1.145(\beta^2 + 0.7\beta^5 + 12\beta^{13})] \frac{p_1 - p_2}{\gamma p_1} \qquad (7\text{-}16)$$

The empirical expansion factors given by Eqs. (7-15) and (7-16) are accurate within ± 0.5 percent for $0.8 < p_2/p_1 < 1.0$.

We thus have the following semiempirical equations, which are conventionally applied to venturis, nozzles, or orifices:

Venturis, incompressible flow

$$Q_{\text{act}} = CMA_2\sqrt{\frac{2g_c}{\rho}} \sqrt{p_1 - p_2} \qquad (7\text{-}17)$$

Nozzles and orifices, incompressible flow

$$Q_{\text{act}} = KA_2\sqrt{\frac{2g_c}{\rho}} \sqrt{p_1 - p_2} \qquad (7\text{-}18)$$

The use of the flow coefficient instead of the product CM is merely a matter of convention. When compressible fluids are used, the above equations are modified by the factor Y and the fluid density is evaluated at inlet conditions. We then have

Venturis, compressible flow

$$\dot{m}_{\text{act}} = YCMA_2\sqrt{2g_c\rho_1(p_1 - p_2)} \qquad (7\text{-}19)$$

Nozzles and orifices, compressible flow

$$\dot{m}_{act} = YKA_2\sqrt{2g_c\rho_1(p_1 - p_2)} \tag{7-20}$$

Detailed tabulations of the various coefficients have been made in Ref. [1], some of which are presented in Figs. 7-9 through 7-15. Examples 7-2 and 7-3 illustrate the use of these charts for practical calculations.

7-4 PRACTICAL CONSIDERATIONS FOR OBSTRUCTION METERS

The construction of obstruction meters has been standardized by the American Society of Mechanical Engineers [1], [2]. The recommended proportions of venturi tubes are shown in Fig. 7-6. Note that the pressure taps are connected to manifolds which surround the upstream and throat portions of the tube. These manifolds receive a sampling of the pressure all around the sections so that a good average value is obtained. The discharge coefficients for such venturi tubes are shown in Fig. 7-9, with the tolerance limits indicated by the dashed lines. In general, the discharge coefficient is smaller for pipes less than 2 in. in diameter, and the approximate behavior is indicated in Fig. 7-10. More precise values of the discharge coefficient for a venturi may be obtained by direct calibration, in which case accuracies of ± 0.5 percent may

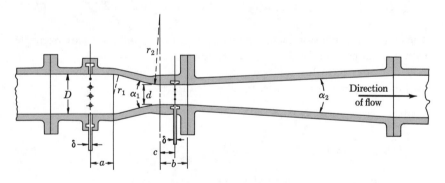

$D =$ Pipe diameter inlet and outlet
$d =$ Throat diameter as required
$a = 0.25D$ to $0.75D$ for $4'' \leq D \leq 6''$, $0.25D$ to $0.50D$ for $6'' < D \leq 32''$
$b = d$
$c = d/2$
$\delta = 3/16$ in. to $1/2$ in. according to D. Annular pressure chamber
      with at least four piezometer vents
$r_2 = 3.5d$ to $3.75d$
$r_1 = 0$ to $1.375D$
$\alpha_1 = 21° \pm 2°$
$\alpha_2 = 5°$ to $15°$

Fig. 7-6 Recommended proportions of venturi tubes according to Ref. [1].

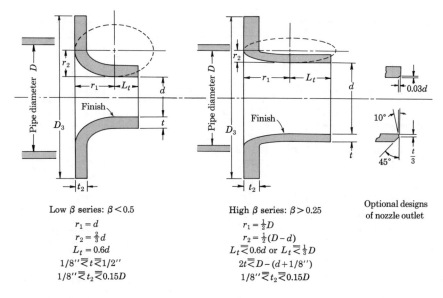

Low β series: $\beta < 0.5$

$r_1 = d$
$r_2 = \frac{2}{3}d$
$L_t = 0.6d$
$1/8'' \overline{\geqslant} t \overline{\geqslant} 1/2''$
$1/8'' \overline{\geqslant} t_2 \overline{\geqslant} 0.15D$

High β series: $\beta > 0.25$

$r_1 = \frac{1}{2}D$
$r_2 = \frac{1}{2}(D-d)$
$L_t \overline{\geqslant} 0.6d$ or $L_t \overline{\leqslant} \frac{1}{3}D$
$2t \overline{\geqslant} D - (d + 1/8'')$
$1/8'' \overline{\geqslant} t_2 \overline{\geqslant} 0.15D$

Optional designs
of nozzle outlet

Fig. 7-7 Recommended proportions of the ASME long-radius flow nozzle according to Ref. [1].

be obtained fairly easily. The recommended dimensions for ASME flow nozzles are shown in Fig. 7-7, and the discharge coefficients are shown in Fig. 7-11.

The recommended installations for concentric, thin-plate orifices are

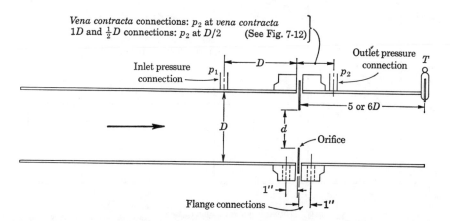

Fig. 7-8 Recommended location of pressure taps for use with concentric, thin-plate, square-edged orifices according to Ref. [1].

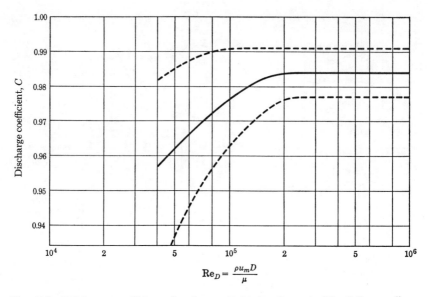

Fig. 7-9 Discharge coefficients for the venturi tube shown in Fig. 7-6 according to Ref. [1]. Values are applicable for $0.25 < \beta < 0.75$ and $D > 2$ in.

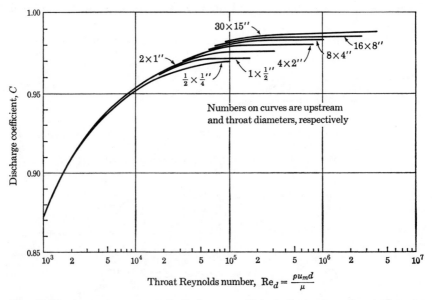

Fig. 7-10 Approximate venturi discharge coefficients for various throat diameters according to Ref. [15].

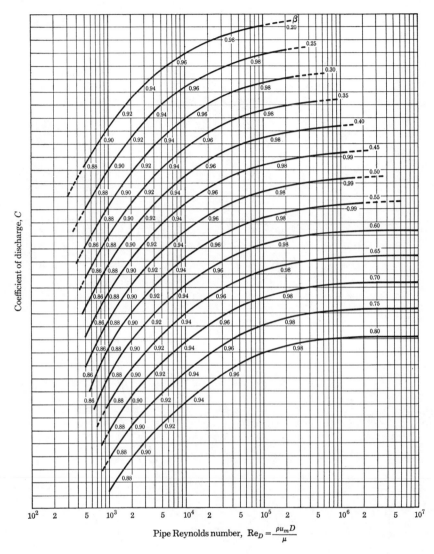

Fig. 7-11 Discharge coefficients for ASME long-radius nozzles shown in Fig. 7-7 according to Ref. [1].

shown in Fig. 7-8. Note that three standard sets of pressure-tap locations are used.

1. Both pressure taps are installed in the flanges as shown.
2. The inlet pressure tap is located one pipe diameter upstream, and the

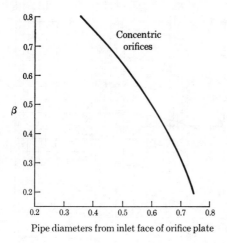

Pipe diameters from inlet face of orifice plate

Fig. 7-12 Location of outlet pressure connections for orifices with vena contracta taps according to Ref. [1].

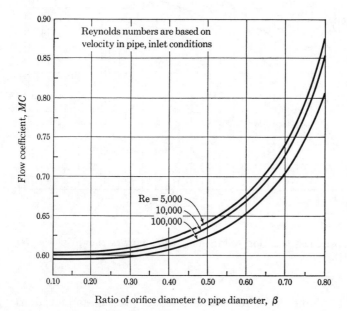

Ratio of orifice diameter to pipe diameter, β

Fig. 7-13 Flow coefficients for concentric orifices in pipes. Pressure taps one diameter upstream and one-half diameter downstream. Applicable for $1.25 < D < 3.00$ in. (*From Ref. [15].*)

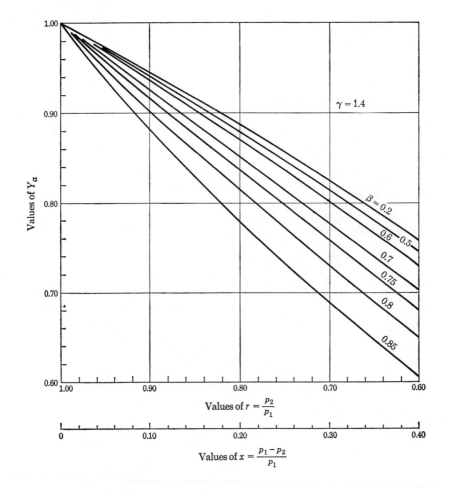

Fig. 7-14 Adiabatic expansion factors for use with venturis and flow nozzles as calculated from Eq. (7-14). (*From Ref.* [2].)

outlet pressure tap is located one-half diameter downstream of the orifice.

3. The inlet pressure tap is located one pipe diameter upstream, and the outlet pressure tap is located at the vena contracta of the orifice as given by Fig. 7-12.

Figure 7-13 gives the values of the orifice flow coefficient for pipe sizes $1\frac{1}{4}$ to 3 in. with pressure taps located according to case 2 above. Flow coefficients for other cases are given in Ref. [1].

The various flow coefficients are plotted as a function of Reynolds number defined by

$$\text{Re} = \frac{\rho u_m d}{\mu} \tag{7-21}$$

where ρ = fluid density
       μ = dynamic viscosity
       u_m = mean flow velocity
       d = diameter *at the particular section for which the Reynolds number is specified*

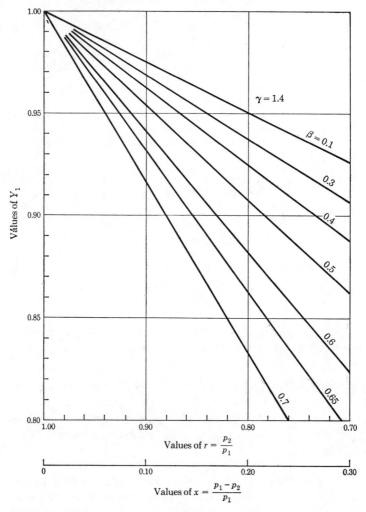

Fig. 7-15 Expansion factors for square-edged orifices with pipe taps as calculated from Eq. (7-16). (*From Ref.* [2].)

Note that some charts, viz., Fig. 7-13, base the Reynolds number on upstream conditions, while others, viz., Fig. 7-10, base it on throat conditions. The product ρu_m may be calculated from the mass flow according to

$$\dot{m} = \rho u_m A_c \qquad (7\text{-}22)$$

where A_c is the cross-sectional area for the flow where u_m is measured. For a circular cross section, $A_c = \pi d^2/4$.

Example 7-2 A venturi tube is to be used to measure a maximum flow rate of water of 50 gpm at 70°F. The throat Reynolds number is to be at least 10^5 at these flow conditions. A differential pressure gage is selected, which has an accuracy of 0.25 percent of full scale, and the upper scale limit is to be selected to correspond to the maximum flow rate. Determine the size of the venturi and the maximum range of the differential pressure gage, and estimate the uncertainty in the mass flow measurement at nominal flow rates of 50 and 25 gpm. Use either Fig. 7-9 or Fig. 7-10 to determine the discharge coefficient.

Solution The properties of water are

$$\rho = 62.4 \text{ lb}_m/\text{ft}^3 \qquad \mu = 2.36 \text{ lb}_m/\text{hr-ft}$$

From the given maximum flow rate and throat Reynolds number we may calculate the maximum allowable throat diameter:

$$\text{Re}_d = \frac{\rho u_m d}{\mu} = \frac{\dot{m} d}{(\pi d^2/4)\mu} = \frac{4\dot{m}}{\pi d \mu} = 10^5$$

The maximum flow rate is

$$\dot{m} = (50)(8.33)(60) = 2.5 \times 10^4 \text{ lb}_m/\text{hr}$$

so that

$$d_{\max} = \frac{(4)(2.5 \times 10^4)}{\pi(10^5)(2.36)} = 0.135 \text{ ft} = 1.62 \text{ in.}$$

We shall select a venturi with a 1.0-in. throat diameter since we have a discharge-coefficient curve for this size in Fig. 7-10. The upstream pipe diameter is taken as 2.0 in. From Fig. 7-10 we estimate the discharge coefficient for this size venturi as 0.976 for $8 \times 10^4 < \text{Re}_d < 3 \times 10^5$. The uncertainty in this coefficient will be taken as ± 0.002 since Fig. 7-10 is a general set of curves. With this selection of venturi size, the maximum throat Reynolds number becomes

$$(\text{Re}_d)_{\max} = (10^5)\left(\frac{1.62}{1.0}\right) = 1.62 \times 10^5$$

The minimum Reynolds number is thus half this value or 8.1×10^4. The maximum pressure differential may be calculated with Eq. (7-17).

$$Q_{\text{act}} = CMA_2\sqrt{\frac{2g_c}{\rho}}\sqrt{\Delta p} \qquad (7\text{-}17)$$

or

$$\frac{(50)(231)}{(60)(1728)} = \frac{(0.976)\pi(1.0)^2}{(4)(144)\sqrt{1 - (\frac{1}{2})^2}}\sqrt{\frac{(2)(32.2)}{62.4}}\sqrt{\Delta p}$$

This yields

$\Delta p = 948$ psf $= 6.58$ psi

Let us assume that a differential pressure gage is at our disposal having a maximum range of 1,000 psf. In accordance with the problem statement the uncertainty in the pressure reading would be

$w_{\Delta p} = \pm 2.5$ psf

When the flow is reduced to 25 gpm, the pressure differential will be *one-fourth* of that at 50 gpm. To estimate the uncertainty in the flow measurement we shall assume that the dimensions of the venturi are known exactly, as well as the density of the water. For the calculation we utilize Eq. (3-2). The quantities of interest are:

$$\frac{\partial Q}{\partial C} = MA_2 \sqrt{\frac{2g_c}{\rho}} \sqrt{\Delta p}$$

$$\frac{\partial Q}{\partial \Delta p} = \frac{CMA_2}{2\sqrt{\Delta p}} \sqrt{\frac{2g_c}{\rho}}$$

$w_c = \pm 0.002$

Thus,

$$\frac{w_Q}{Q} = \left[\left(\frac{w_c}{C}\right)^2 + \frac{1}{4}\left(\frac{w_{\Delta p}}{\Delta p}\right)^2\right]^{\frac{1}{2}}$$

For $Q = 50$ gpm,

$$\frac{w_Q}{Q} = \left[\left(\frac{0.002}{0.976}\right)^2 + \frac{1}{4}\left(\frac{2.5}{948}\right)^2\right]^{\frac{1}{2}}$$

$$= 0.002435 \qquad \text{or} \quad 0.2435\%$$

For $Q = 25$ gpm,

$$\frac{w_Q}{Q} = \left[\left(\frac{0.002}{0.976}\right)^2 + \frac{1}{4}\left(\frac{2.5}{984/4}\right)^2\right]^{\frac{1}{2}}$$

$$= 0.00566 \qquad \text{or} \quad 0.566\%$$

Example 7-3 An orifice with pressure taps one diameter upstream and one-half diameter downstream is installed in a 2.00-in.-diam pipe and used to measure the same flow of water as in Eaxmple 7-2. For this orifice, $\beta = 0.50$. The differential pressure gage has an accuracy of 0.25 percent of full scale, and the upper scale limit is selected to correspond to the maximum flow rate. Determine the range of the pressure gage and the uncertainty in the flow-rate measurements at nominal flow rates of 50 and 25 gpm. Assume that the uncertainty in the flow coefficient is ± 0.002.

Solution We first calculate the pipe Reynolds numbers. Using the properties from Example 7-2,

$$\text{Re}_d = \frac{(2.5 \times 10^4)(4)}{\pi(2.0/12)(2.36)} = 8.09 \times 10^4 \qquad \text{at 50 gpm}$$

$$\text{Re}_d = 4.05 \times 10^4 \qquad\qquad\qquad\qquad \text{at 25 gpm}$$

From Fig. 7-13 the flow coefficient is estimated as

$K = 0.625$ at 50 gpm
$K = 0.630$ at 25 gpm

The volumetric flow is

$$Q = \frac{(50)(231)}{(60)(1728)} = 0.1115 \text{ ft}^3/\text{sec} \qquad \text{at 50 gpm}$$

$$Q = 0.0558 \text{ ft}^3/\text{sec} \qquad\qquad\qquad \text{at 25 gpm}$$

The nominal values of the differential pressure are then calculated from Eq. (7-18) as

$$0.1115 = (0.625) \frac{\pi(1)^2}{(4)(144)} \sqrt{\frac{(2)(32.2)}{(62.4)}} \sqrt{\Delta p} \qquad \text{at 50 gpm}$$

$\Delta p = 1{,}037 \text{ psf} = 7.21 \text{ psi}$ at 50 gpm
$\Delta p = \phantom{1{,}0}255 \text{ psf} = 1.77 \text{ psi}$ at 25 gpm

A suitable differential pressure gage might be one with a maximum range of 1,200 psf. The same equation for uncertainty applies in this problem as in Example 7-2 except that the flow coefficient K is used instead of the discharge coefficient. Thus,

$$\frac{w_Q}{Q} = \left[\left(\frac{w_K}{K}\right)^2 + \frac{1}{4}\left(\frac{w_{\Delta p}}{\Delta p}\right)^2\right]^{\frac{1}{2}}$$

with $w_K = 0.002$ and $w_{\Delta p} = (0.0025)(1{,}200) = 3.0 \text{ psf}$

For $Q = 50$ gpm,

$$\frac{w_Q}{Q} = \left[\left(\frac{0.002}{0.625}\right)^2 + \frac{1}{4}\left(\frac{3.0}{1{,}037}\right)^2\right]^{\frac{1}{2}}$$

$$= 0.00351 \qquad \text{or } 0.351\%$$

For $Q = 25$ gpm,

$$\frac{w_Q}{Q} = \left[\left(\frac{0.002}{0.630}\right)^2 + \frac{1}{4}\left(\frac{3.0}{255}\right)^2\right]^{\frac{1}{2}} = 0.00669 \qquad \text{or } 0.669\%$$

7-5 THE SONIC NOZZLE

All the obstruction meters discussed above may be used with gases. When the flow rate is sufficiently high, the pressure differential becomes quite large and eventually sonic flow conditions may be achieved at the minimum flow area. Under these conditions the flow is said to be "choked," and the flow rate takes on its maximum value for the given inlet conditions. For an ideal gas with constant specific heats it may be shown that the pressure ratio for this choked condition, assuming isentropic flow, is

$$\left(\frac{p_2}{p_1}\right)_{\text{critical}} = \left(\frac{2}{\gamma + 1}\right)^{\gamma/(\gamma - 1)} \tag{7-23}$$

This ratio is called the critical pressure ratio. Inserting this ratio in Eq. (7-8) gives for the mass flow rate

$$\dot{m} = A_2 p_1 \sqrt{\frac{2g_c}{RT_1} \left[\frac{\gamma}{\gamma + 1} \left(\frac{2}{\gamma + 1} \right)^{2/(\gamma - 1)} \right]^{\frac{1}{2}}} \tag{7-24}$$

Equation (7-24) is frequently applied to a nozzle when it is known that the pressure ratio p_2/p_1 is less than the critical value given by Eq. (7-23). Under these conditions the ideal flow is dependent only on the inlet stagnation conditions p_1 and T_1. These conditions are usually easy to measure so that the sonic nozzle offers a convenient method for measuring gas flow rates. It may be noted, however, that a large pressure drop must be tolerated with the method. *Upstream stagnation conditions* must be used for p_1 and T_1 in the calculation.

The ideal sonic-nozzle flow rate given by Eq. (7-24) must be modified by an appropriate discharge coefficient which is a function of the geometry of the nozzle and other factors. There may be several complicating conditions, but discharge coefficients of about 0.97 are usually observed. A comprehensive survey of critical flow nozzles is presented by Arnberg [3], and the interested reader should consult this discussion for more information.

The flow obstruction devices discussed above require the use of wall pressure taps. (Other devices also require the use of such taps.) Measurements with wall pressure taps can be subject to several influencing factors, which are discussed in detail by Rayle [19]. In general, the diameter of the pressure tap should be small in comparison with the diameter of the pipe.

Example 7-4 A sonic nozzle is to be used to measure a flow of air at 300 psia and 100°F in a 3-in.-diam pipe. The nominal flow rate is 1 lb$_m$/sec. Calculate the throat diameter (nozzle size) such that critical flow conditions are just obtained.

Solution We use Eq. (7-24) for this calculation with $\gamma = 1.4$ for air. The only unknown in this equation is A_2. Thus, we have

$$\dot{m} = C A_2 p_1 \sqrt{\frac{2g_c}{RT_1} \left[\frac{\gamma}{\gamma + 1} \left(\frac{2}{\gamma + 1} \right)^{2/(\gamma - 1)} \right]^{\frac{1}{2}}}$$

$$1 = A_2 (300)(144) \sqrt{\frac{(2)(32.2)}{(53.35)(560)} \left[\left(\frac{4.1}{2.4} \right) \left(\frac{2}{2.4} \right)^{2/0.4} \right]^{\frac{1}{2}}}$$

and $A_2 = 0.001078$ ft$^2 = 0.1551$ in.^{2}

The diameter at the throat is

$$d = \sqrt{\frac{4}{\pi} (0.1551)} = 0.444 \text{ in.}$$

For the above calculation we have taken the given pressure and temperature as stagnation properties. The temperature would most likely be measured with a stagnation probe so that 100°F is probably the stagnation temperature. The pressure would probably be measured by a static tap in the side of the pipe upstream from the nozzle so that a static-pressure measurement is most likely the one which will be available. If the upstream pipe diameter is large enough, the static pressure will be very nearly equal to the stagnation pressure and the error in the above calculation will be small. Let us examine the above situation, assuming that the 300 psia is a static-pressure measurement. The mass flow upstream is

$$\dot{m} = \frac{p_{1_s}}{RT_{1_s}} A_1 u_1 \tag{a}$$

where the subscript s denotes static properties. The velocity upstream may be written in terms of the stagnation temperature as

$$u_1 = \sqrt{2g_c c_p (T_{1_0} - T_{1_s})} \tag{b}$$

Combining Eqs. (a) and (b), we have

$$\dot{m} = \frac{p_{1_s}}{RT_{1_s}} A_1 \sqrt{2g_c c_p (T_{1_0} - T_{1_s})} \tag{c}$$

Taking $p_{1_s} = 300$ psia and $T_{1_0} = 100°F = 560°R$, we may solve Eq. (c) for T_{1_s}. The result is

$$T_{1_s} \approx 560°R \tag{d}$$

or the upstream velocity is so small that the stagnation properties are very nearly equal to the static properties. This result may be checked by calculating the upstream velocity from Eq. (a) using the result from Eq. (d). We obtain

$$u_1 = 14.1 \text{ ft/sec}$$

The pressure difference $(p_{1_0} - p_{1_s})$ corresponding to this velocity would be only 0.031 psia, while the temperature difference $(T_{1_0} - T_{1_s})$ would be 0.017°F. Both of these values are negligible. It may be noted, however, that if the upstream pipe diameter were considerably smaller, say, 1.0 in. diameter, it might be necessary to correct for a difference between the measured static pressure and the stagnation pressure that must be used in Eq. (7.24).

7-6 FLOW MEASUREMENT BY DRAG EFFECTS

The rotameter is a very commonly used flow-measurement device and is shown schematically in Fig. 7-16. The flow enters the bottom of the tapered vertical tube and causes the bob or "float" to move upward. The bob will rise to a point in the tube such that the drag forces are just balanced by the weight and buoyancy forces. The position of the bob in the tube is then taken as an indication of the flow rate. The device is sometimes called an area meter because the elevation of the bob is dependent on the annular area between it and the tapered glass tube; however, the meter operates on the physical principle of drag so that we choose to classify it in this category. A force balance on the bob gives

$$F_d + \rho_f V_b \frac{g}{g_c} = \rho_b V_b \frac{g}{g_c} \tag{7-25}$$

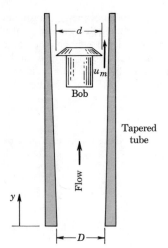

Fig. 7-16   Schematic of a rotameter.

where ρ_f and ρ_b are the densities of the fluid and bob, V_b is the total volume of the bob, g is the acceleration of gravity, and F_d is the drag force, which is given by

$$F_d = C_d A_b \frac{\rho_f u_m^2}{2g_c} \tag{7-26}$$

C_d is a drag coefficient, A_b is the frontal area of the bob, and u_m is the mean flow velocity in the annular space between the bob and the tube.

Combining Eqs. (7-25) and (7-26) gives

$$u_m = \left[\frac{1}{C_d} \frac{2gV_b}{A_b} \left(\frac{\rho_b}{\rho_f} - 1\right)\right]^{\frac{1}{2}} \tag{7-27}$$

or

$$Q = A u_m = A\left[\frac{1}{C_d} \frac{2gV_b}{A_b} \left(\frac{\rho_b}{\rho_f} - 1\right)\right]^{\frac{1}{2}} \tag{7-28}$$

where A is the annular area and is given by

$$A = \frac{\pi}{4} [(D + ay)^2 - d^2] \tag{7-29}$$

D is the diameter of the tube at inlet, d is the maximum bob diameter, y is the vertical distance from the entrance, and a is a constant indicating the tube taper.

The drag coefficient is dependent on the Reynolds number and hence on the fluid viscosity; however, special bobs may be used that have an essentially constant drag coefficient and thus offer the advantage that the meter reading will be essentially independent of viscosity. It may be noted that for many practical meters the quadratic area relation given by Eq. (7-29) becomes nearly linear for the actual dimensions of tube and bob that are used. Assuming such a linear relation, the equation for mass flow would become

$$\dot{m} = C_1 y \sqrt{(\rho_b - \rho_f)\rho_f} \qquad (7\text{-}30)$$

where C_1 is now an appropriate meter constant.

It is frequently advantageous to have a rotameter that gives an indication that is independent of fluid density; i.e., we wish to have

$$\frac{\partial \dot{m}}{\partial \rho_f} = 0$$

Performing the indicated differentiation, we obtain

$$\rho_b = 2\rho_f \qquad (7\text{-}31)$$

and the mass flow is given by

$$\dot{m} = \frac{C_1 y \rho_b}{2} \qquad (7\text{-}32)$$

Thus, by special bob construction the meter may be used to compensate for density changes in the fluid. The error in Eq. (7-32) is less than 0.2 percent for a fluid-density deviation of 5 percent from that given in Eq. (7-31). A photograph of a commercial rotameter is given in Fig. 7-17.

A popular type of flow-measurement device is the turbine meter shown in Fig. 7-18. As the fluid moves through the meter, it causes a rotation of the small turbine wheel. In the turbine-wheel body a permanent magnet is enclosed so that it rotates with the wheel. A reluctance pickup attached to the top of the meter detects a pulse for each revolution of the turbine wheel. Since the volumetric flow is proportional to the number of wheel revolutions, the total pulse output may be taken as an indication of total flow. The pulse rate is proportional to flow rate, and the transient response of the meter is very good. A flow coefficient K for the turbine meter is defined so that

$$Q = \frac{f}{K} \qquad (7\text{-}33)$$

where f is the pulse frequency. The flow coefficient is dependent on flow rate

Fig. 7-17 Photograph of a commercial rotameter. (*Courtesy Fischer and Porter Company.*)

and the kinematic viscosity of the fluid v. A calibration curve for a typical meter is given in Fig. 7-19. It may be seen that this particular meter will indicate the flow accurately within ± 0.5 percent over a rather wide range of flow rates.

7-7 HOT-WIRE ANEMOMETER

The hot-wire anemometer is a device that is most often used in research applications to study varying flow conditions. A fine wire is heated electrically and placed in the flow stream. The heat-transfer rate from the wire has been shown [5] to be

$$q = (a + b\sqrt{\rho u})(T_w - T_\infty) \text{ Btu/(hr)(ft}^2) \tag{7-34}$$

where T_w = wire temperature

T_∞ = fluid temperature

u = fluid velocity

a, b = constants that are obtained by a calibration of the device

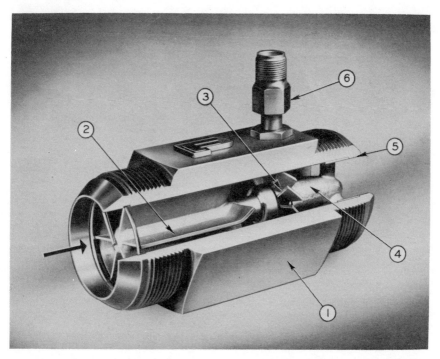

Fig. 7-18 Turbine flowmeter. (1) Meter body; (2) inlet-straightening vanes; (3) turbine blades; (4) downstream afterbody to maintain smooth flow; (5) pipe or tube connector; (6) reluctance pickup. (*Courtesy Fischer and Porter Company.*)

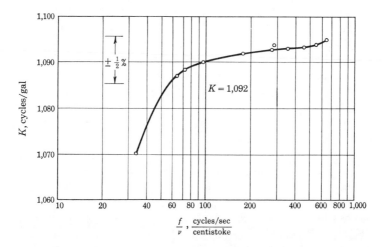

Fig. 7-19 Calibration curve for 1-in. turbine flowmeter of the type shown in Fig. 7-18. Calibration was performed with water.

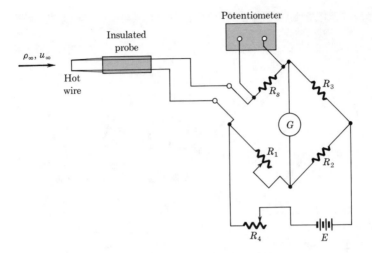

Fig. 7-20 Schematic of hot-wire flow-measurement circuit.

The heat-transfer rate must also be given by

$$q = i^2 R_w = i^2 R_0[1 + \alpha(T_w - T_0)] \tag{7-35}$$

where i = electric current

R_0 = resistance of the wire at the reference temperature T_0

α = temperature coefficient of resistance

For measurement purposes the hot wire is connected to a bridge circuit as shown in Fig. 7-20.   The current is determined by measuring the voltage drop across the standard resistance R_s, and the wire resistance is determined from the bridge circuit.   For steady-state measurements the null condition may be used, while an oscillograph output may be used for transient measurements.   With i and R_w determined, the flow velocity may be calculated with Eqs. (7-34) and (7-35).   Hot-wire probes have been used extensively for measurement of transient flows, especially measurements of turbulent fluctuations.   Time constants of the order of 1 msec may be obtained with 0.0001-in.-diam platinum or tungsten wires operating in air.   A modification of the hot-wire method consists of a small cylinder that is coated with a thin metallic film a few microns thick.   This film then serves as the variable resistance and is extremely sensitive to fluctuations in the fluid velocity.   Such hot-film probes have been used for measurements involving frequencies as high as 50,000 Hz.

The calibration of hot-wire probes is quite complicated, and the interested reader is referred to the discussion by Kovasznay [6] for more information.   It may be noted, however, that complete commercial setups are available for hot-wire and hot-film measurements.

TURBULENCE MEASUREMENTS

In turbulent fluid flow, whether in a channel or in a boundary layer, the fluid-velocity components exhibit a random oscillatory character, which depends on the average fluid velocity, fluid density, viscosity, and other variables.   The instantaneous-velocity components are expressed as

$$u = \bar{u} + u' \qquad (x \text{ component})$$
$$v = \bar{v} + v' \qquad (y \text{ component})$$
$$w = \bar{w} + w' \qquad (z \text{ component})$$

where the bar quantities are the integrated average-velocity components and the primed quantities represent the fluctuations from the average velocity. Figure 7-21 illustrates a typical plot of turbulent fluctuations.   One measure of the *intensity of turbulence* is the root mean square of the fluctuations. For the x component

$$u'_{\text{rms}} = (\overline{u'^2})^{\frac{1}{2}} = \left[\frac{1}{T}\int_0^T u'^2 \, dt\right]^{\frac{1}{2}} \tag{7-36}$$

where T is some time period that is large compared to the turbulence scale. The average kinetic energy (KE) of turbulence per unit mass may be taken as a measure of the total turbulence intensity of the flow stream.   Thus,

$$\frac{\text{Turbulence KE}}{\text{mass}} = \tfrac{1}{2}(\overline{u'^2} + \overline{v'^2} + \overline{w'^2}) \tag{7-37}$$

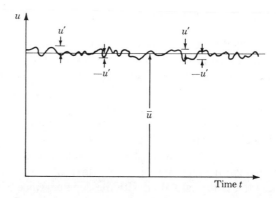

Fig. 7-21   Turbulent fluctuations in direction of flow.

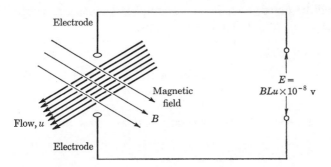

Fig. 7-22 Flow of conducting fluid through a magnetic field.

The hot-wire or hot-film anemometer is very useful for turbulence measurements because it can respond to very rapid changes in flow velocity. Two or more wires at one point in the flow can make simultaneous measurements of the fluctuating components. Various mean values and energy spectra can then be computed from the voltage outputs of the probes. In general, these operations can be performed electronically.

7-8 MAGNETIC FLOWMETERS

Consider the flow of a conducting fluid through a magnetic field as shown in Fig. 7-22. Since the fluid represents a conductor moving in the field, there will be an induced voltage according to

$$E = BLu \times 10^{-8} \text{ volt} \qquad (7\text{-}38)$$

where B = magnetic flux density, gauss
u = velocity of the conductor, cm/sec
L = length of the conductor, cm

The length of the conductor is proportional to the tube diameter, and the velocity is proportional to the mean flow velocity. The two electrodes detect the induced voltage, which may be taken as a direct indication of flow velocity. The construction of a commercial magnetic flowmeter is shown in Fig. 7-23.

Two types of magnetic flowmeters are used commercially. One type has a nonconducting pipe liner and is used for fluids with low conductivities like water. The electrodes are mounted so that they are flush with the non-conducting liner and make contact with the fluid. Alternating magnetic fields are normally used with these meters since the output is low and requires amplification. The second type of magnetic flowmeter is one which is used with high-conductivity fluids, principally liquid metals. A stainless-steel

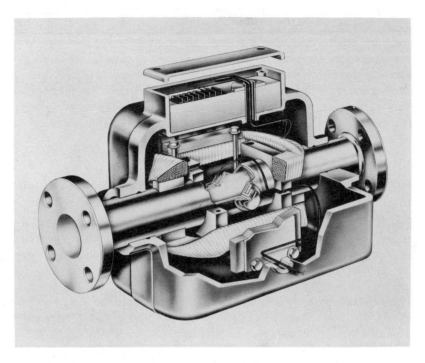

Fig. 7-23 Construction of commercial magnetic flowmeter for measurement of low-conductivity fluids. (*Courtesy Fischer and Porter Company.*)

pipe is employed in this case with the electrodes attached directly to the outside of the pipe and diametrically opposed to each other. The output of this type of meter is sufficiently high that it may be used for direct readout purposes. The meter shown in Fig. 7-23 is one which is used for low-conductivity fluids.

7-9 FLOW-VISUALIZATION METHODS

Fluid flow is a complicated subject with many areas that have not yet yielded to precise analytical techniques. Accordingly, flow-measurement problems are not always simple and precise because of the lack of analytical relations to use for calculation and reduction of experimental data. The interpretation of data involving turbulence or measurements of complicated boundary layer, viscous, and shock-wave effects is not easy. Frequently, the flow may be altered as a result of probes that are inserted to measure the pressure, velocity, and temperature profiles so that the experimentalist is uncertain about the effect he has measured. Flow visualization by optical methods offers the advantage that when properly executed it does not disturb the fluid stream and thus gives the experimentalist an extra tool to use in conjunction with other measurement devices. In some instances flow-visualization techniques may be employed for rather precise measurements of important flow parameters, while in other cases they may serve only to furnish qualitative information regarding the overall flow behavior. In the following paragraphs we shall discuss the principles of some of the basic flow-visualization schemes and indicate their applications. The interested reader should consult Refs. [8] and [17] for a rather complete survey of the subject.

Consider the gaseous flow field shown in Fig. 7-24. The flow is in a direction perpendicular to the figure, i.e., the z direction. An incoming light ray is deflected through an angle ϵ as a result of density gradients in the flow. It may be shown† that the deflection angle for small-density gradients is given as

$$\epsilon = \frac{L}{n_1}\left(\frac{dn}{dy}\right)_{y=y_1} = \frac{L\beta}{\rho_s}\left(\frac{d\rho}{dy}\right)_{y=y_1} \tag{7-39}$$

where L is the width of the flow field, ρ is the local fluid density, ρ_s is a reference density which is usually taken at standard conditions, and n is the index of refraction which for gases may be written as

$$n = 1 + \beta\frac{\rho}{\rho_s} \tag{7-40}$$

β is a dimensionless constant having a value of about 0.000292 for air.

According to Eq. (7-39) the angular deflection of the light ray ϵ is

† See, for example, Ref. [10], p. 156.

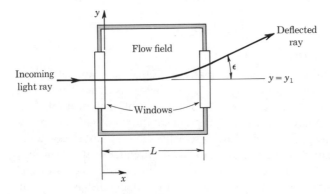

Fig. 7-24 Basic optical effects used for flow visualization.

proportional to the density gradient in the flow. This is the basic optical effect that is used for flow-visualization work. It may be noted that the deflection of the light ray is a measure of the average density gradient integrated over the x coordinate. Thus, the effect is primarily useful for indicating density variations in two dimensions (in this case the y and z dimensions) and will average the variations in the third dimension.

In the following sections we shall discuss several optical methods of flow visualization for use in gas systems. For liquid flow visualization a typical experimental technique is to add a dye to the liquid in order to study the flow phenomena. Another technique which has received considerable attention for use with liquids is the so-called hydrogen-bubble method. The technique consists of using a fine wire, placed in water, as one end of a dc circuit to electrolyze the water. Very small hydrogen bubbles are generated in the liquid. The motion of the bubbles may be studied by illuminating the flow. The application of this method has been given a very complete description by Schraub, Kline et al. [16]. Figure 7-25 indicates the voltage-discharge circuits employed in the technique. A variable-input voltage from 10 to 250 volts is provided and rectified in the diode circuit. A relay is used to periodically charge and discharge the capacitor, and a variable resistor is provided in the charging circuit to tune the system for optimum performance for each wire diameter and flow condition. In some cases it is necessary to add a small amount of sodium sulfate to the water in order to achieve a satisfactory electrolyte concentration.

7-10 THE SHADOWGRAPH

The shadow technique is a method for direct viewing of flow phenomena. Imagine the flow field as shown in Fig. 7-26 with a density gradient in the y direction. The parallel light rays enter the test section as shown. In the

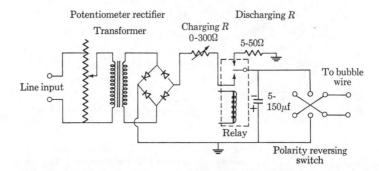

Fig. 7-25 Voltage pulse circuits for hydrogen-bubble method according to [16].

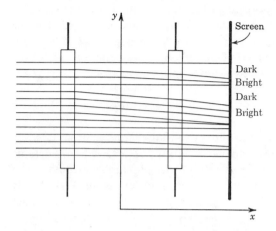

Fig. 7-26 Shadowgraph flow-visualization device.

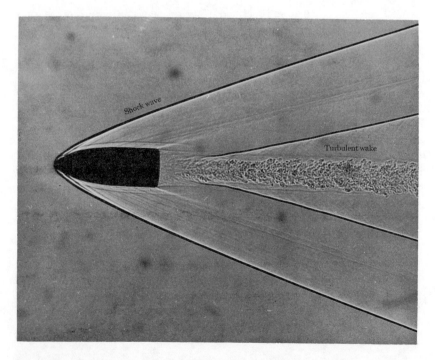

Fig. 7-27 Direct shadowgram of flight of a 0.220-caliber Remington Swift shell. Velocity = 4,100 ft/sec; free-stream pressure is atmospheric. Exposure time was 0.13 μsec. Note clear definition of shock wave and turbulent wake. (*Courtesy ARO, Inc., Arnold Air Force Base, Tenn.*)

regions where there is no density gradient, the light rays will pass straight
through the test section with no deflection.   For the regions where a gradient
exists the rays will be deflected.   The net effect is that the rays will bunch
together after leaving the test section to form bright spots and dark spots.
The illumination will depend on the *relative deflection* of the light rays $d\epsilon/dy$
and hence on $d^2\rho/dy^2$.   The illumination on a screen placed outside the test
section is thus dependent on the second derivative of the density at the
particular point.

The shadowgraph is a very simple optical tool, and its effects may be
viewed in several everyday phenomena using only the naked eye and local
room lighting.   The free-convection boundary layer on a horizontal electric
hot plate is clearly visible when viewed from the edge.   This phenomenon is
visible because of the density gradients that result from the heating of the air
near the hot surface.   It is almost fruitless to try to evaluate local densities
using shadow photography; however, the shadowgraph is exceedingly useful
to view turbulent flow regions, and the method can be used to establish the
location of shock waves with high precision.   Figure 7-27 illustrates shock-
wave and turbulent-flow phenomena as viewed with a shadowgraph.

7-11 THE SCHLIEREN

While the shadowgraph gives an indication of the second derivative of density
in the flow field, the schlieren is a device which indicates the density gradient.
Consider the schematic diagram shown in Fig. 7-28.   Light from a slit
source ab is collimated by the lens L_1 and focused at plane 1 in the test

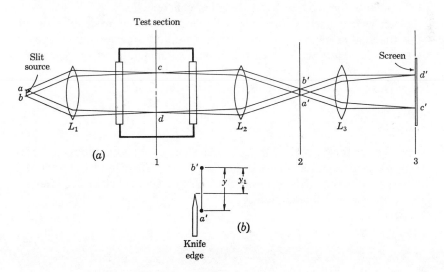

Fig. 7-28   (*a*) Schematic of schlieren flow visualization; (*b*) detail of knife edge.

section.   After passing through lens L_2, there is produced an inverted image of the source at the focal plane 2.   Lens L_3 then focuses the image of the test section on the screen at plane 3.   Now let us consider the imaging process in more detail.   The pencils of light originating at point a occupy a different portion of the various lenses than those originating from point b or any other point in the slit source.   The regions in which these pencils overlap are shown in Fig. 7-28.   Note that *all* pencils of light pass through the image plane cd in the test section and the source image plane $b'a'$.   An image of the test section at $d'c'$ is then uniformly illuminated since the image at $b'a'$ is uniformly illuminated.   This means that all points in the plane $b'a'$ are affected in the same manner by whatever fluid effects may take place in the test section.

If the test section is completely uniform in density, the pencils of light appear as shown in Fig. 7-28; a pencil originating at point c is deflected by the same amount as a pencil originating at point d.   This is consistent with the observation that all pencils originating in plane cd completely fill the image plane $b'a'$.   Now consider the effect of the introduction of an obstruction at plane $b'a'$ under these circumstances.   We immediately conclude that such an obstruction would uniformly decrease the illumination on the screen by a factor proportional to the amount of the area $b'a'$ intercepted.

Suppose now that a density gradient exists at the test section focal plane cd.   This means that all pencils of light originating in this plane would no longer fill the image plane $b'a'$ completely.   If, now, an obstruction is placed at plane $b'a'$, it will intercept more light from some points in the test section plane than from others, resulting in light and dark regions on the screen at plane 3.   The obstruction is called a knife edge, and the resultant variation in illumination on the screen is called the schlieren effect.

Let us examine the variation in illumination in more detail.   In Fig. 7-28b the total height of the source image is y and the portion not intercepted by the knife edge is y_1.   Thus, the general illumination on the screen I is proportional to y_1.   An angular displacement of a pencil of light in plane 1 is ϵ.   This produces a vertical deflection at plane 2 of

$$\Delta y = f_2 \epsilon \tag{7-41}$$

where f_2 is the focal length of lens L_2.   As a result of this deflection, there is a fractional change in illumination on the screen.   The contrast at any point on the screen may be defined as the ratio of the fractional change in illumination to the general illumination or

$$C = \frac{\Delta I}{I} = \frac{\Delta y}{y_1} = \frac{f_2 \epsilon}{y_1} \tag{7-42}$$

The angular deflection is given by Eq. (7-39) so that the expression for contrast may be written

$$C = \frac{f_2 L \beta}{y_1 \rho_s} \left(\frac{d\rho}{dy} \right)_{cd} \tag{7-43}$$

Thus, the contrast on the screen is directly proportional to the density gradient in the flow.   It may be observed that the contrast may be increased by reducing the distance y_1, that is, by intercepting more light at the source image plane.   This also reduces the general illumination so that the contrast may not be increased indefinitely and a compromise must be accepted. Schlieren photographs are used extensively for location of shock waves and complicated boundary-layer phenomena in supersonic flow systems.   Some typical schlieren photographs are shown in Figs. 7-29 and 7-30.

7-12 THE INTERFEROMETER

The Mach-Zehnder interferometer is the most precise instrument for flow visualization.   Consider the schematic representation in Fig. 7-31.   The light source is collimated through lens L_1 onto the splitter plate S_1.   This plate permits half of the light to be transmitted to mirror M_2 while reflecting the other half toward mirror M_1.   Beam 1 passes through the test section while beam 2 travels an alternate path of approximately equal length.   The two beams are brought together again by means of the splitter plate S_2 and eventually focused on the screen.   Now, if the two beams travel paths of different *optical lengths*, either because of the geometry of the system or because of the refractive properties of any element of the optical paths, the two beams will be out of phase and will interfere when they are joined

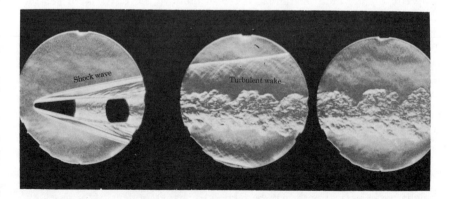

Fig. 7-29 Schlieren photograph of 1.75-in.-diam blunt cone and wake after separation of rear portion.   Velocity = 5,878 ft/sec; free-stream pressure = 94 mm Hg; cone angle = 18°.   (*Courtesy ARO, Inc., Arnold Air Force Base, Tenn.*)

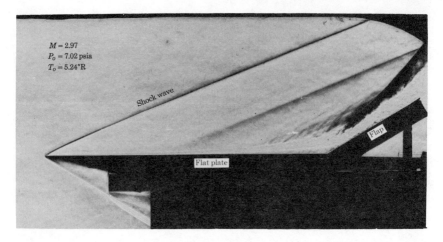

$M = 2.97$
$P_o = 7.02\ \text{psia}$
$T_o = 5.24°\text{R}$

Shock wave

Flap

Flat plate

Fig. 7-30 Schlieren photograph of flow-separation phenomena.   Plate length = 8 in.;
flap angle = 30°.   Note clear definition of shock wave and turbulent flow region at the
intersection of the flap and horizontal plate. (*Courtesy ARO, Inc., Arnold Air Force
Base, Tenn.*)

together at S_2.   There will be alternate bright and dark regions called fringes.
The number of fringes will be a function of the difference in optical path
lengths for the two beams; for a difference in path lengths of one wavelength
there will be one fringe, two fringes for a difference of two wavelengths, and
so on.

The interferometer is used to obtain a direct measurement of density
variations in the test section.   If the density in the test section (i.e., beam 1)
is different from that in beam 2, there will be a change in the refractive
properties of the fluid medium.   If the medium in the test section has the
same optical properties as the medium in beam 2, there will be no fringe
shifts except those resulting from the geometric arrangement of the apparatus.
These fringe shifts may be neutralized by appropriate movements of the
mirrors and splitter plates.   Then, the appearance of fringes on the screen
may be directly related to changes in density in the flow field within the test
section utilizing the following analysis.

The change in optical path in the test section resulting from a change in
refractive index is

$$\Delta L = L(n - n_0) \qquad (7\text{-}44)$$

where L is the thickness of the flow field in the test section.   Using Eq. (7-40),
the change in optical path may be related to change in density for gases by

$$\Delta L = \beta L \frac{\rho - \rho_0}{\rho_s} \qquad (7\text{-}45)$$

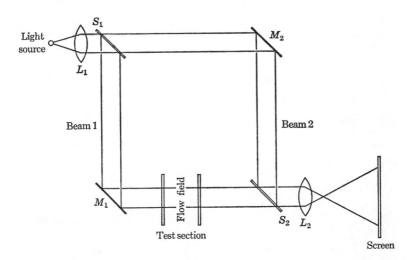

Fig. 7-31   Schematic of Mach-Zehnder interferometer.

The number of fringe shifts N is then given by

$$N = \frac{\Delta L}{\lambda} = \frac{\beta L}{\lambda} \frac{\rho - \rho_0}{\rho_s} \qquad\qquad (7\text{-}46)$$

where λ is the wavelength of the light.   In Eq. (7-46) it is to be noted that $\rho - \rho_0$ represents the change in density from the zero-fringe condition.

The interferometer gives a direct quantitative indication of density changes in the test section, but these changes are represented as integrated values over the entire thickness of the flow field.   It is applicable to a wide range of flow conditions ranging from the low-speed (~ 1 ft/sec) flow in free-convection boundary layers to shock-wave phenomena in supersonic flow. Figures 7-32 and 7-33 show some typical interferometer photographs.   The captions explain the flow phenomena.

7-13 THE LASER ANEMOMETER

We have seen how optical flow-visualization methods offer the advantage that they do not disturb the flow during the measurement process.   The laser anemometer is a device that offers the nondisturbance advantages of optical methods while affording a very precise quantitative measurement of velocities. The device is capable of rapid response and is suitable for measurement of high-frequency turbulence fluctuations.

A schematic of a laser-anemometer system is shown in Fig. 7-34.   The laser beam is focused on a small-volume element in the flow through lens L_1. In order for the device to function, the flow must contain some type of small

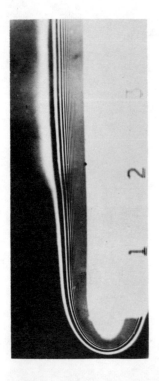

Fig. 7-32 Interferometer photograph of a free-convection boundary layer on a vertical flat plate. Fluid is air. Each fringe line represents a line of constant density or a line of constant temperature. (*Photograph courtesy E. Soehngen.*)

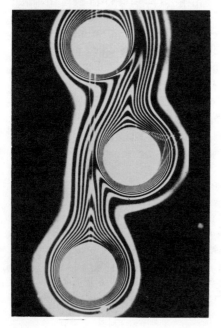

Fig. 7-33 Interferometer photograph of the interaction of free-convection boundary layers on three horizontal heated cylinders. Fluid is air, and each fringe line represents a line of constant temperature. (*Photograph courtesy E. Soehngen.*)

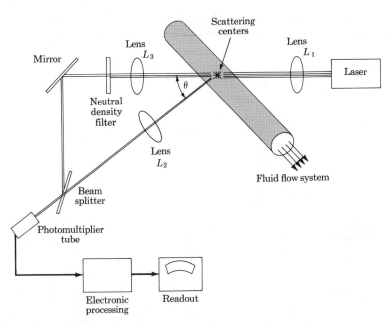

Fig. 7-34 Schematic of laser-anemometer flow-measurement system.

particles to scatter the light, but the particle concentration required is very small. Ordinary tap water, for example, contains enough impurities to scatter the incident beam. Two additional lenses L_2 and L_3 are positioned to receive the laser beam that is transmitted through the fluid (lens L_3) and some portion of the beam that is scattered through the angle θ (lens L_2). The scattered light experiences a Doppler shift in frequency that is directly proportional to the flow velocity. The unscattered portion of the beam is reduced in intensity by the neutral density filter and recombined with the scattered beam through the beam splitter. The laser-anemometer device must be so constructed that the direct and scattered beams travel the same optical path so that an interference will be observed at the photomultiplier tube that is proportional to frequency shift. This shift then gives an indication of the flow velocity. In order to retrieve the velocity data from the photomultiplier signal, rather sophisticated electronic techniques must be employed for signal processing. A spectrum analyzer may be used to determine velocity in steady laminar flow as well as mean velocity and turbulence intensity in turbulent flow.

Laser anemometers that measure more than one velocity component simultaneously have been developed, and the optics and electronic-signal processing for such systems become quite complex and expensive. Even so, the technique offers unusual promise for detailed investigations of turbulence and other flow phenomena, which may not be performed in any other way. Welch and Hines [27] have noted that the sample volume of a focused laser beam can be as small as 10^{-9} in.^{3} with a spatial resolution of the order of a few tens of microns. The interested reader should consult Refs. [20] to [27] for additional information on the construction of laser anemometers and measurements that have been obtained.

Example 7-5 An interferometer is used for visualization of a free-convection boundary layer in air similar to the one shown in Fig. 7-32. For this application the following data were collected:

Plate temperature $T_w = 120°F$
Free-stream air temperature $T_\infty = 70°F$
$\beta = 0.000293$
Depth of test section $L = 18$ in.
Wavelength of light source $\lambda = 5460$ A
Reference density at 70°F
Pressure $= 14.7$ psia

Calculate the number of fringes that will be viewed in the boundary layer.

Solution We use Eq. 7-46 for this calculation. The reference density is the same as the zero-fringe density, so that

$$N = \frac{\beta L}{\lambda} \left(\frac{\rho_w}{\rho_\infty} - 1 \right)$$

For air the density is calculated from the ideal-gas equation of state:

$$\rho = \frac{p}{RT}$$

Thus,

$$N = \frac{\beta L}{\lambda} \left(\frac{T_\infty}{T_w} - 1 \right)$$

$$= \frac{(0.000293)(18.0)(2.54)}{(5460 \times 10^{-8})} \left(\frac{530}{580} - 1 \right)$$

$$= 21.15 \text{ fringes}$$

7-14 SMOKE METHODS

A very simple flow-visualization method utilizes the injection of smoke traces in a gas stream to follow the flow streamlines. The method is primarily of qualitative utility in that direct measurements are difficult to obtain except for certain special phenomena. Figure 7-35 shows an example of a flow system where smoke visualization was used to verify an analytical calculation. In this case, smoke is used to view the complicated secondary flow patterns in a

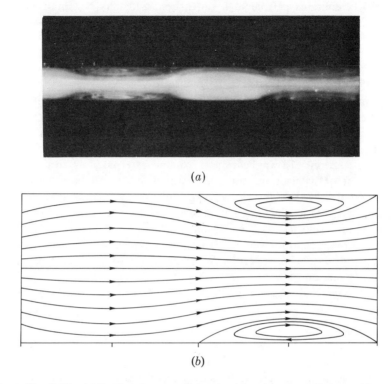

(a)

(b)

Fig. 7-35 (a) Smoke photograph showing secondary flow effects resulting from a standing sound wave in a tube; (b) flow streamlines for the system in (a) as obtained from the theoretical analysis of Ref. [12]. (*Photograph courtesy Dr. T. W. Jackson.*)

channel through which a forced flow is coupled with a standing sound wave. The smoke patterns in Fig. 7-35a agree well with the analytical predictions in Fig. 7-35b.

In order for the smoke filaments to represent streamlines of the flow it is necessary that the individual smoke particles be of sufficiently small mass so that they are carried along freely at the flow velocity. Filtered smoke from burning rotten wood or cigars is generally suitable for smoke studies, as well as smoke from the chemical titanium tetrachloride when it reacts with moisture in air to form hydrochloric acid and titanium oxide. This latter substance, however, is corrosive to many materials used for the construction of containers. Reference [14] discusses the use of titanium tetrachloride for low-speed flow measurements. One of the best fuels for producing nontoxic, noncorrosive, dense smoke is a product called Type-1964 Fog Juice. This fuel has a boiling temperature of approximately

530°F, contains petroleum hydrocarbon, and may be obtained from most theatrical supply houses. Some smoke-generation techniques are discussed in Ref. [28].

7-15 PRESSURE PROBES

A majority of fluid dynamic applications involve a measurement of total flow rate by one or more methods discussed in the previous sections. These measurements ignore the local variation of velocity and pressure in the flow channel and permit an indication of only the total flow through a particular cross section. In applications involving external flow situations, such as aircraft or wind-tunnel tests, an entirely different type of measurement is required. In these instances probes must be inserted in the flow to measure the local static and stagnation pressures. From these measurements the local flow velocity may be calculated. Several probes are available for such measurements, and summaries of the characteristic behavior are given in Refs. [4], [7], [13], and [18]. We shall discuss some of the basic probe types in this section.

The total pressure for isentropic stagnation of an ideal gas is given by

$$\frac{p_0}{p_\infty} = \left(1 + \frac{\gamma - 1}{2} M_\infty^2\right)^{\gamma/(\gamma - 1)} \tag{7-47}$$

where p_0 is the stagnation pressure, p_∞ is the free-stream static pressure, and M_∞ is the free-stream Mach number given by

$$M_\infty = \frac{u_\infty}{a} \tag{7-48}$$

a is the acoustic velocity and may be calculated with

$$a = \sqrt{\gamma g_c R T} \tag{7-49}$$

for an ideal gas. It is convenient to express Eq. (7-47) in terms of the dynamic pressure q defined by

$$q = \tfrac{1}{2}\rho u_\infty^2 = \tfrac{1}{2}\gamma p M_\infty^2 \tag{7-50}$$

Equation (7-47) thus becomes

$$p_0 - p_\infty = \frac{2q}{\gamma M_\infty^2}\left[\left(1 + \frac{\gamma - 1}{2} M_\infty^2\right)^{\gamma/(\gamma - 1)} - 1\right] \tag{7-51}$$

This relation may be simplified to

$$p_0 - p_\infty = \frac{2q}{\gamma M_\infty^2} \left(1 + \frac{M_\infty^2}{4} + \frac{2-\gamma}{24} M_\infty^4 + \cdots \right) \qquad (7\text{-}52)$$

when $M_\infty^2\left(\dfrac{\gamma-1}{2}\right) < 1$.

For very small Mach numbers Eq.(7-52) reduces to the familiar incompressible flow formula

$$p_0 - p_\infty = \tfrac{1}{2}\rho u_\infty^2 \qquad (7\text{-}53)$$

We thus observe that a measurement of the static and stagnation pressures permits a determination of the flow velocity, by either Eq. (7-53) or Eq. (7-51), depending on the fluid medium.

A basic total pressure probe may be constructed in several different ways, as shown in Fig. 7-36. In each instance the opening in the probe is oriented in a direction exactly parallel to the flow when a measurement of the total stream pressure is desired. If the probe is inclined at an angle θ to the free-stream velocity, a somewhat lower pressure will be observed. This reduction in pressure is indicated in Fig. 7-36 according to Ref. [7]. Configuration a represents an open-ended tube placed in the flow. Configuration b is called a shielded probe and consists of a venturi-shaped tube placed in the flow with an open-ended tube at the throat of the section to sense the stagnation pressure. It may be noted that this probe is rather insensitive to flow

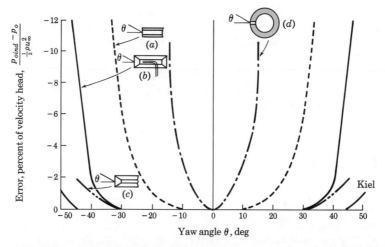

Fig. 7-36 Stagnation pressure response of various probes to changes in yaw angle. (a) Open-ended tube; (b) channel tube; (c) chamfered tube; (d) tube with orifice in side. (*From Ref.* [7].)

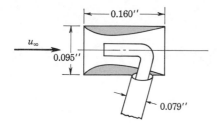

Fig. 7-37 Kiel probe (Model 3696) for measurement of stagnation pressure. (*Courtesy Airflo Instrument Co., Glastonbury, Conn.*)

direction. Configuration c represents an open-ended tube with a chamfered opening. The chamfer is about 15°, and the ratio of OD to ID of the tube is about 5. Configuration d represents a tube having a small hole drilled in its side, which is placed normal to the flow direction. This type of probe, as might be expected, is the most sensitive to changes in yaw angle. Also indicated in Fig. 7-36 is a portion of the curve for a Kiel probe, which is similar in construction to configuration b except that a smoother venturi shape is used as shown in Fig. 7-37. The Kiel probe is the least sensitive to yaw angle.

The measurement of static pressure in a flow stream is considerably more difficult than the measurement of stagnation pressure. A typical probe used for the measurement of both static and stagnation pressures is the *pitot tube* shown in Fig. 7-38. The opening in the front of the probe senses the stagnation pressure, while the small holes around the outer periphery of the tube sense the static pressure. The static-pressure measurement with such a device is strongly dependent on the distance of the peripheral openings from the front opening as well as the yaw angle. Figure 7-39 indicates the dependence of the static-pressure indication on the distance from the leading edge of the probe for both blunt subsonic and conical supersonic configurations. To alleviate this condition, the static-pressure holes are normally placed at least eight diameters downstream from the front of the probe. The dependence of the static and stagnation pressures on yaw angle for a conventional pitot tube is indicated in Fig. 7-40. This device is quite sensitive to flow direction.

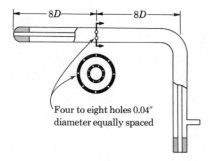

Four to eight holes 0.04″
diameter equally spaced

Fig. 7-38 Schematic drawing of pitot tube.

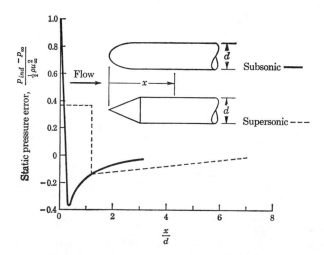

Fig. 7-39 Variation of static pressure along standard subsonic and supersonic probe types. (*From Ref.* [4].)

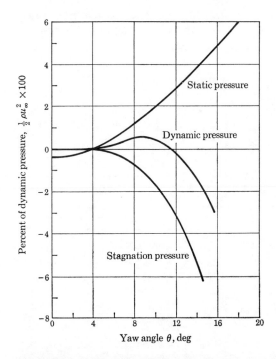

Fig. 7-40 Variation of static, stagnation, and dynamic pressures with yaw angle for pitot tube. (*Courtesy Airflo Instrument Co., Glastonbury, Conn.*)

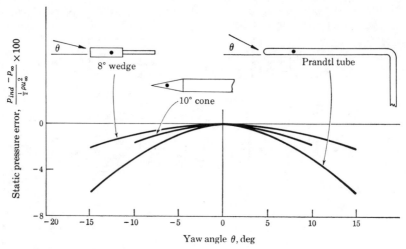

Fig. 7-41 Yaw-angle characteristics of various static-pressure probes. (*From Ref.* [7].)

The static-pressure characteristics of three types of probes are shown in Figs. 7-41 and 7-42 as functions of Mach number and yaw angle.   It may be noted that both the wedge and the Prandtl tube indicate static-pressure

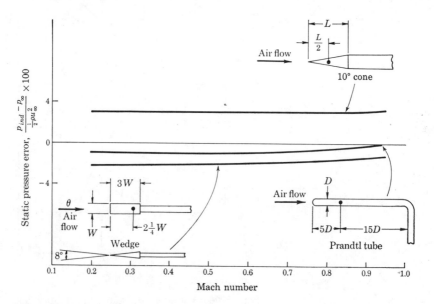

Fig. 7-42 Mach-number characteristics of various static-pressure probes. (*From Ref.* [7].)

values that are too low, while the cone indicates a value that is too high. The wedge is least sensitive to yaw angle. All three probes have two static-pressure holes located 180° apart.

Example 7-6   A pitot tube is inserted in a flowstream of air at 100°F and 14.7 psia. The dynamic pressure is measured as 1.12 in. water when the tube is oriented parallel to the flow. Calculate the flow velocity at that point.

Solution   We use Eq. (7-53) for this calculation. The air density is calculated as

$$\rho = \frac{p_\infty}{RT_\infty} = \frac{(14.7)(144)}{(53.35)(560)} = 0.071 \ \text{lb}_m/\text{ft}^3$$

We also have

$$p_0 - p_\infty = 1.12 \ \text{in. water} = 5.82 \ \text{psf}$$

so that the velocity is

$$u_\infty = \sqrt{\frac{2(p_0 - p_\infty)}{\rho}} = \sqrt{\frac{(2)(32.2)(5.82)}{(0.071)}}$$

$$= 72.6 \ \text{ft/sec}$$

7-16   IMPACT PRESSURE IN SUPERSONIC FLOW

Consider the impact probe shown in Fig. 7-43 which is exposed to a free stream with supersonic flow; that is, $M_1 > 1$. A shock wave will be formed in front of the probe as shown, and the total pressure measured by the probe will not be the free-stream total pressure before the shock wave. It is possible, however, to express the impact pressure at the probe in terms of the free-stream static pressure and the free-stream Mach number. The resulting expression as given in Ref. [10] is

$$\frac{p_\infty}{p_{0_2}} = \frac{\{[2\gamma/(\gamma + 1)]M_1^2 - (\gamma - 1)/(\gamma + 1)\}^{1/(\gamma - 1)}}{\{[(\gamma + 1)/2]M_1^2\}^{\gamma/(\gamma - 1)}} \tag{7-54}$$

where p_∞ is the free-stream static pressure and p_{0_2} is the measured impact pressure behind the normal shock wave. This equation is valid for Reynolds numbers based on probe diameter greater than 400. Equation (7-54) is

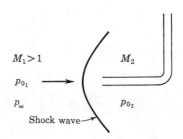

$M_1 > 1$

p_{0_1} →

p_∞

M_2

p_{0_2}

Shock wave →

Fig. 7-43   Impact tube in supersonic flow.

Table 7-1　Operating characteristics of several types of flowmeters according to Ref. [11]

Meter type		Fluid factors					Application factors					Installation factors			
		Fluid range of max. flow liquid, gpm gas, cfm	Useful range-ability	Max. pressure, psig	Temp. range, °F	Max. viscosity, cs	Scale	Indication	Standard accuracy, %	Construction materials	Pressure loss, psi	Line size, in.	Special piping considerations	Typical power requirement	Relative cost
Obstruction meters	Orifice	Liquid 0.2–3,500 Gas 50–100,000	5:1	6,000	−455–2000	4,000	Sq rt rate	Remote diff. press.	1–2 full-scale diff. press.	Most metals	0.1–30	1.5–12 and larger	Straight pipe: 10 diameters upstream, 3 downstream	None	1.0
	Flow nozzle	Liquid 0.5–15,000 Gas 100–500,000	5:1	1,500	−60–1500	4,000	Sq rt rate	Remote diff. press.	1–2 full-scale diff. press.	Bronze, iron, steel, stainless steel	0.1–20	1–24 and larger	Straight pipe: 10 diameters upstream, 3 downstream	None	1.4
	Venturi	Liquid 0.5–15,000 Gas 100–500,000	5:1	1,500	−60–1500	4,000	Sq rt rate	Remote diff. press.	1–2 full-scale diff. press.	Bronze, iron, steel, stainless steel, plastic	0.1–15	1–24 and larger	Straight pipe: 10 diameters upstream, 3 downstream	None	1.5
Drag effects	Glass-tube rotameter	Liquid 5 cc/min–250 Gas 400 cc/min–700	10:1	250	−50–400	100	Linear rate	Local or remote electric or pneumatic	1–2 full scale	Most metals, plastics, and ceramics	0.2" H_2O to 1.0 psi	0.25–4	Vertical only	None	1.0
	Metal-tube rotameter	Liquid 0.5–4,000 Gas 0.6–1,000	10:1	5,000	−300–1600	100	Linear rate	Local or remote electric or pneumatic	1–2 full-scale	Most metals, plastics, and ceramics	1.0" H_2O to 10 psi	0.5–12	Vertical only	None	1.2
	Turbine or propeller	Clean 0.004–50,000 Liquid 0.5–300,000 or Gas	15:1 10:1	15,000	−455–1000	30	Linear total	Remote electric	0.5 reading	Aluminum, stainless steel	2–10	0.125–36	Any position	115 volts 60 Hz 50 watts	5.0
	Positive displacement	Clean 0.25–1,000 Liquid	10:1	150	0–300	2,000	Linear total	Local or remote electric	1 reading	Bronze, iron, steel, stainless steel	1–20	0.5–6	Horizontal recommended	None	1.0
	Magnetic flowmeter	Conductive liquid only 0.002–50,000	20:1	600	−200–360	1,000	Linear rate	Remote electric	0.5–1 full scale	Plastic, stainless steel with nonconductive liner	Very small	0.1–78	Any position	115 volts 60 Hz 200 watts	6.0

called the Rayleigh supersonic pitot formula. We see that in order to determine the value of the Mach number it is necessary to have a measurement of the free-stream static pressure. It is possible to make this measurement with special calibrated probes.

7-17 SUMMARY

Comparisons of the operating range, characteristics, and advantages of several flowmeters have been presented by Miesse and Curth [11]. These comparisons are presented in Table 7-1 and may be taken as an appropriate summary of our discussions in this chapter. The reader should bear in mind that the stated accuracies for the various flowmeters may be improved with suitable calibration. Of course, the overall accuracy of a flow-rate determination is also dependent on the accuracy of the readout equipment. As an example, a venturi might be carefully calibrated within 0.5 percent, but if it were used with a crude instrument for differential pressure measurements, a much poorer precision would result than that for which it was calibrated.

REVIEW QUESTIONS

1. What basic methods are used for calibration of flow-measurement devices?

2. What is meant by positive-displacement flowmeter?

3. What are the relative advantages of the venturi, orifice, and flow-nozzle meters?

4. What type of flow-measurement accuracy would you expect with an orifice; a venturi?

5. What is a sonic nozzle? How is it used? What are its advantages and disadvantages?

6. Why is a rotameter called a drag meter? Could it also be called an area meter?

7. What is the utility of the hot-wire anemometer?

8. Distinguish between the shadowgraph, schlieren, and interferometer flow-visualization techniques. What basic flow variable is measured in each technique?

9. Upon what does the sensitivity of the schlieren depend?

10. What is the primary advantage of the laser anemometer?

11. What is a pitot tube?

12. Why must the tube be tapered in order for a rotameter to indicate flow rate?

13. What particular flow-measurement situations are adapted to the hot-wire anemometer?

14. How would one go about calibrating a venturi for liquid measurement? Could the calibration data be adapted to gas flows? If so, how?

PROBLEMS

7-1. Using Eq. (7-8) as a starting point, obtain Eqs. (7-9) and (7-10). Subsequently, calculate the error in these equations when $\Delta p = p_1/4$ for Eq. (7-9) and $\Delta p = p_1/10$ for Eq. (7-10).

7-2. In the compressible flow equation for a venturi the velocity of approach factor M cancels with a like term in the expansion factor Y_a. Construct a graphical plot which will indicate the error that would result if the approach factor given by Eq. (7-11) were used in conjunction with Eq. (7-8). Use the parameter $\Delta p/\gamma p_1$ as the abscissa for this plot.

7-3. A venturi is to be used for measuring the flow of air at 300 psia and 80°F. The maximum flow rate is 1.0 lb_m/sec. The minimum flow rate is 30 percent of this value. Determine the size of the venturi such that the throat Reynolds number is not less than 10^5. Calculate the differential pressure across the venturi for mass flows of 0.3, 0.5, 0.7, and 1.0 lb_m/sec. Assume $\beta = 0.5$ for the venturi.

7-4. Work Prob. 7-3 for an orifice with pressure taps one diameter upstream and one-half diameter downstream.

7-5. Work Prob. 7-3 for an ASME long-radius flow nozzle.

7-6. A sonic nozzle is to be used to measure a gas flow with an uncertainty of 1 percent. Assuming that the nozzle throat area and discharge coefficients are known exactly, derive an expression for the required relationship between the uncertainties in the stagnation pressure and temperature measurements. Which of the two measurements, pressure or temperature, is more likely to be the controlling factor?

7-7. Calculate the throat area for the sonic nozzle in Example 7-4 if the stated pressure is static pressure upstream, the temperature is total temperature upstream, and the pipe diameter is 1.0 in.

7-8. Show that the parameter that governs the use of a linear relation for a rotameter as in Eq. (7-30) is

$$\frac{ay}{D}$$

provided that $d \approx D$ to the extent that

$$1 - \left(\frac{d}{D}\right)^2 \ll \frac{ay}{D}$$

Under these restrictions plot the error resulting from the linear approximation as a function of the parameter ay/D. Discuss the physical significance of this analysis, and interpret its meaning in terms of specific design recommendations for rotameters.

7-9. Verify that the error in flow rate for a rotameter calculated from Eq. (7-32) is less than 0.2 percent for density variations of ± 5 percent when the float density is designated according to Eq. (7-31).

7-10. The turbine flowmeter whose calibration is shown in Fig. 7-19 is to measure a nominal flow rate of water of 2.5 gpm at 60°F. A single value of the meter constant K is to be used in the data reduction. What deviations from the nominal flow rate are allowable in order that the nominal value of K is accurate within ± 0.25 percent?

7-11. A rotameter is to be designed to measure a maximum flow of 10 gpm of water at 70°F. The bob is 1 in. in diameter and has a total volume of 1 in.³. The bob is constructed so that the density is given in accordance with Eq. (7-31). The total length of the rotameter tube is 13 in., and the diameter of the tube at inlet is 1.0 in. Determine the tube taper for drag coefficients of 0.4, 0.8, and 1.20. Plot the flow rate versus distance from the entrance of the tube for each of these drag coefficients. Determine the meter constant for use in Eq. (7-32), and estimate the error resulting from the use of this relation instead of the exact expression in Eq. (7-28).

7-12. A rotameter is to be used for measurement of the flow of air at 100 psia and 70°F. The maximum flow rate is 0.03 lb_m/sec, the inlet diameter of the meter is 1 in., and the length of the meter is 12 in. The bob is constructed so that its density is five times that of the air and its volume is 1 in.^{3}. Calculate the tube taper for drag coefficients of 0.4, 0.8, and 1.2, and determine the meter constant for use in Eq. (7-30). Plot the error resulting from the use of Eq. (7-30) as a function of flow rate.

7-13. Derive an expression for the product of density and velocity across a hot-wire anemometer in terms of the wire resistance, the current through the wire, and the empirical constants a and b. Subsequently obtain an expression for the uncertainty in this product as a function of the measured quantities.

7-14. The sensitivity of a schlieren system is defined as the fractional deflection obtained at the knife edge per unit angular deflection of a light ray at the test section. Show that this sensitivity may be calculated with

$$S = \frac{f_2}{y_1}$$

Derive an expression for the contrast in terms of the sensitivity, the density gradient, and the test section width.

7-15. Show that the sensitivity of an interferometer, defined as the number of fringe shifts per unit change of density, may be written as

$$S = \frac{\beta L}{\lambda \rho_s}$$

Show that the maximum sensitivity of an interferometer defined as the number of fringe shifts per unit change in Mach number will be about 30 when $L = 6$ in., $\lambda = 5400$ A, $\beta = 0.000293$, and the stagnation density is that of air at standard conditions.

7-16. Calculate the temperatures corresponding to the four fringes nearest to the plate surface in Example 7-5.

7-17. The velocity in turbulent tube flow varies approximately as

$$\frac{u}{u_c} = \left(1 - \frac{r}{r_0}\right)^{\frac{1}{7}}$$

where u_c is the velocity at the center of the tube and r_0 is the tube radius. An experimental setup using air in a 1-ft-diam tube is used to check this relation. ʹ The air temperature is 70°F, and pressure is 15 psia. The maximum flow velocity is 50 ft/sec, and a pitot tube is used to traverse the flow and obtain a measurement of the velocity distribution. Measurements are taken at radii of 0, 2, 4, and 5 in., and the uncertainty in the dynamic-pressure measurement is ± 0.02 in. water. Using the above relation and Eq. (7-53), calculate the nominal velocity and dynamic pressure at each radial location. Then calculate the uncertainty in the velocity measurement at each location based on the uncertainty in the pressure measurement. Assuming that the above relation does represent the true velocity profile, calculate the uncertainty which could result in a mass flow determined from the experimental data. The mass flow would be obtained by performing the integration

$$\dot{m} = \int_0^{r_0} 2\pi r \rho u \, dr$$

7-18. Calculate the dynamic pressure measured by a pitot tube in a flow stream of water at 70°F moving at a velocity of 10 ft/sec.

7-19. Using the supersonic pitot-tube formula given by Eq. (7-54), obtain an expression for the uncertainty in the Mach number as a function of the per cent uncertainty in the pressure ratio (p_∞/p_{0_2}) and the nominal value of the Mach number. Plot the percent uncertainty in the Mach number as a function of Mach number for a constant uncertainty in (p_∞/p_{0_2}) of 1 percent. Assume $\gamma = 1.4$.

7-20. A venturi with throat and upstream diameters of 8 and 16 in. is used to measure the flow of water at 70°F. The flow rate is controlled by a motorized valve downstream from the venturi. The valve is operated so that a constant differential pressure of 12 in. Hg is maintained across the venturi. Suppose someone informs you that this type of control scheme is not very effective because it does not account for possible changes in temperature of the water. Reply to this criticism by plotting the error in the flow rate as a function of water temperature, taking the flow at 70°F as the reference value.

7-21. An obstruction meter is used for the measurement of the flow of moist air at low velocities. Suppose that the flow rate is calculated taking the density as that of dry air at 90°F. Plot the error in this flow rate as a function of relative humidity.

7-22. A small venturi is used for measuring the flow of water in a $\frac{1}{2}$-in.-diam line. The venturi has a throat diameter of $\frac{1}{4}$ in. and is constructed according to ASME specifications. What minimum flow rate at 70°F should be used in order that the discharge coefficient remain in the flat portion of the curve? What pressure reading, in inches of mercury, would be experienced for this flow rate?

7-23. The venturi of Prob. 7-22 is used to measure the flow of air at 100 psia and 120°F (upstream flow conditions). The throat pressure is measured as 80 psia. Calculate the flow rate.

7-24. An orifice is to be used to indicate the flow rate of water in a 1.25-in.-diam line. The orifice diameter is 0.50 in. What pressure reading will be experienced on the orifice for a line-flow velocity of 10 ft/sec? What would be the flow rate for a pressure reading of twice this value?

7-25. A small sonic nozzle having a throat diameter of $\frac{1}{32}$ in. is used to measure and regulate the flow in a 3-in.-diam line. The upstream pressure on the nozzle is varied in accordance with the demand requirements. The downstream pressure is always low enough to insure sonic flow at the throat. What is the flow rate for upstream conditions at 70°F and 150 psia?

7-26. An impact tube is used to measure the Mach number of a certain air flow in a wind tunnel. The static pressure is 3 psia, and the impact pressure at the probe is 16.9 psia. What is the Mach number of the flow ($\gamma = 1.4$ for air)? If the free-stream air temperature is −40°F, what is the flow velocity?

7-27. A pitot tube is used to measure the velocity of an air stream at 70°F and 14.7 psia. If the velocity is 8 ft/sec, what is the dynamic pressure in inches of water? What is the uncertainty of the velocity measurement if the dynamic pressure is measured with a manometer having an uncertainty of 0.02 in. H_2O.

REFERENCES

1. "Fluid Meters, Their Theory and Application," 5th ed., ASME, New York, 1959.
2. "Flowmeter Computation Handbook," ASME, New York, 1961.
3. Arnberg, B. T.: Review of Critical Flowmeters for Gas Flow Measurements, *Trans. ASME*, vol. 84D, p. 447, 1962.
4. Gracey, W.: Measurement of Static Pressure on Aircraft, *NACA Tech. Note* 4184, November, 1957.

5. King, L. V.: On the Convection of Heat from Small Cylinders in a Stream of Fluid, with Applications to Hot-wire Anemometry, *Phil. Trans. Roy. Soc. London*, vol. 214, no. 14, p. 373, 1914.
6. Kovasznay, L. S. G.: Hot-wire Method, in "Physical Measurements in Gas Dynamics and Combustion," p. 219, Princeton University Press, Princeton, N.J., 1954.
7. Krause, L. N., and C. C. Gettelman: Considerations Entering into the Selection of Probes for Pressure Measurement in Jet Engines, *ISA Proc.*, vol. 7, p. 134, 1952.
8. Ladenburg, R. W. (ed.): "Physical Measurements in Gas Dynamics and Combustion," Princeton University Press, Princeton, N.J., 1954.
9. Laurence, J. C., and L. G. Landes: Application of the Constant Temperature Hot-wire Anemometer to the Study of Transient Air Flow Phenomena, *J. Instr. Soc. Am.*, vol. 1, no. 12, p. 128, 1959.
10. Liepmann, H. W., and A. Roshko: "Elements of Gas Dynamics," John Wiley & Sons, Inc., New York, 1957.
11. Miesse, C. C., and O. E. Curth: How to Select a Flowmeter, *Prod. Eng.*, p. 35, May 8, 1961.
12. Purdy, K. R., T. W. Jackson, and C. W. Gorton: Viscous Fluid Flow under the Influence of a Resonant Acoustic Field, *J. Heat Transfer*, vol. 86C, p. 97, February, 1964.
13. Schulze, W. M., G. C. Ashby, Jr., and J. R. Erwin: Several Combination Probes for Surveying Static and Total Pressure and Flow Direction, *NACA Tech. Note* 2830, November, 1952.
14. Smith, E., R. H. Reed, and H. D. Hodges: The Measurement of Low Air Speeds by the Use of Titanium Tetrachloride, *Texas Eng. Exp. Sta. Research Rpt.* 25, May, 1951.
15. Tuve, G. L.: "Mechanical Engineering Experimentation," McGraw-Hill Book Company, New York, 1961.
16. Schraub, F. A., S. J. Kline, et al.: Use of Hydrogen Bubbles for Quantitative Determination of Time-dependent Velocity Fields in Low Speed Water Flows, *ASME Paper no.* 64-WA/FE-20, December, 1964.
17. "Flow Visualization Symposium," ASME, New York, 1960.
18. Dean, R. C. (ed.): "Aerodynamic Measurements," Eagle Press, 1953.
19. Rayle, R. E.: "An Investigation of the Influence of Orifice Geometry on Static Pressure Measurements," *S.M. thesis*, Dept. Mech. Eng., Massachusetts Institute of Technology, 1949. (See also Ref. 18.)
20. Yeh, Y., and H. Z. Cummins: Localized Fluid Flow Measurements with an He-Ne Laser Spectrometer, *Appl. Phys. Lett.*, vol. 4, no. 10, pp. 176–178, May, 1964.
21. Forman, J. W., E. W. George, and R. D. Lewis: Measurement of Localized Flow Velocities in Gases with a Laser Doppler Flowmeter, *Appl. Phys. Lett.*, vol. 7, no. 4, pp. 77–78, August, 1965.
22. Forman, J. W., R. D. Lewis, J. R. Thornton, and H. J. Watson: Laser Doppler Velocimeter for Measurement of Localized Flow Velocities in Liquids, *Proc. IEEE*, vol. 54, pp. 424–425, March, 1966.
23. Berman, N. S., and V. A. Santos: Laminar Velocity Profiles in Developing Flows Using a Laser Doppler Technique, *AIChE J.*, vol. 15, no. 3, pp. 323–327, May, 1969.
24. Goldstein, R. J., and D. K. Kreid: Measurements of Laminar Flow Development in a Square Duct Using a Laser Doppler Flowmeter, *J. Appl. Mech.*, vol. 34, no. 4, pp. 813–818, December, 1967.
25. Goldstein, R. J., and W. F. Hagen: Turbulent Flow Measurements Utilizing the Doppler Shift of Scattered Laser Radiation, *Phys. Fluids*, vol. 10, pp. 1349–1352, June, 1967.

26. Welch, N. E., and W. J. Tomme: The Analysis of Turbulence from Data Obtained with a Laser Velocimeter, *AIAA Paper* 67–179, January, 1967.
27. Welch, N. E., and R. H. Hines: The Practical Application of the Laser Anemometer for Fluid Flow Measurements, presented at the ISA Electro-Optical Systems Conf., New York, N.Y., September, 1969.
28. Hodges, A. E., and T. N. Pound: Further Development of a Smoke Producer Using Vaporized Oil, *Aerodynamics Note* 233, Australian Defense Scientific Service Aeronautical Res. Labs., November, 1964.

8
The Measurement of Temperature

8-1 INTRODUCTION

To most people, temperature is an intuitive concept which tells whether a body is "hot" or "cold." In the exposition of the second principle of thermodynamics, temperature is related to heat, for it is known that heat flows only from a high temperature to a low temperature, in the absence of other effects. In the kinetic theory of gases and statistical thermodynamics it is shown that temperature is related to the average kinetic energy of the molecules of an ideal gas. Further extensions of statistical thermodynamics show the relationship between temperature and the energy levels in liquids and solids. We shall not be able to discuss the many theoretical aspects of the concept of temperature but may only note that it is important in every branch of physical science; hence the experimental engineer should be familiar with the methods employed in temperature measurement. Detailed discussions of the thermodynamic meaning of temperature are given by Obert [8], Sears [10], Tribus [12], and Rossini [9].

Since pressure, volume, electrical resistance, expansion coefficients, etc.,

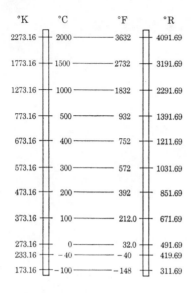

Fig. 8-1 Relationship between Fahrenheit and centigrade temperature scales.

are all related to temperature through the fundamental molecular structure, they change with temperature, and these changes can be used to measure temperature. Calibration may be achieved through comparison with established standards as discussed in Chap. 2. The International Temperature Scale serves to *define* temperature in terms of observable characteristics of materials.

8-2 TEMPERATURE SCALES

The two temperature scales in use are the Fahrenheit and centigrade scales. These scales are based on a specification of the number of increments between the freezing point and boiling point of water at standard atmospheric pressure. The centigrade scale has 100 units between these points, while the Fahrenheit scale has 180 units. The absolute centigrade scale is called the Kelvin scale, while the absolute Fahrenheit scale is termed the Rankine scale. Both absolute scales are so defined that they will correspond as closely as possible with the absolute thermodynamic temperature scale. The zero points on both absolute scales represent the same physical state, and the ratio of two values is the same, regardless of the absolute scale used; i.e.,

$$\left(\frac{T_2}{T_1}\right)_{\text{Rankine}} = \left(\frac{T_2}{T_1}\right)_{\text{Kelvin}} \tag{8-1}$$

The boiling point of water is arbitrarily taken as 100° on the centigrade scale and 212° on the Fahrenheit scale. The relationship between the scales is indicated in Fig. 8-1. It is evident that the following relations apply:

$$°F = 32.0 + \tfrac{9}{5}°C \tag{8-2a}$$

$$°R = \tfrac{9}{5}°K \tag{8-2b}$$

It may be noted that the name Celsius has been officially adopted to replace the name centigrade, but the adoption is not yet in widespread use.

8-3 THE IDEAL-GAS THERMOMETER

The behavior of an ideal gas at low pressures furnishes the basis for a temperature-measurement device that may serve as a secondary experimental standard. The ideal-gas equation of state is

$$pV = mRT \tag{8-3}$$

where V is the volume occupied by the gas, m is the mass, and R is the gas constant for the particular gas given by

$$R = \mathcal{R}/M$$

$\mathcal{R}$ is the universal gas constant having a value of 1,545 ft-lb$_f$/(lb$_m$)(mole)(°R), and M is the molecular weight of the gas. For the gas thermometer a fixed volume is filled with gas and exposed to the temperature to be measured, as shown in Fig. 8-2. At the temperature T the gas-system pressure is measured. Next, the volume is exposed to a standard reference temperature (as discussed in Sec. 2-4), and the pressure is measured under these conditions. According to Eq. (8-3) at constant volume,

$$T = T_{\text{ref}}\left(\frac{p}{p_{\text{ref}}}\right)_{\text{const vol}} \tag{8-4}$$

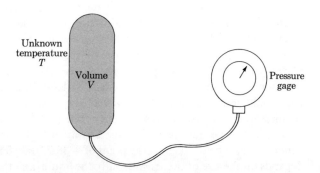

Fig. 8-2 Schematic of ideal-gas thermometer.

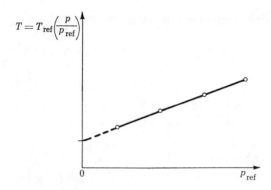

Fig. 8-3 Results of measurements with ideal-gas thermometer.

Now suppose that some of the gas is removed from the volume and the pressure measurements are repeated. In general, there will be a slight difference in the pressure ratio in Eq. (8-4) as the quantity of gas is varied. However, regardless of the gas used, the series of measurements may be repeated and the results plotted as in Fig. 8-3. When the curve is extrapolated to zero pressure, the true temperature as defined by the ideal-gas equation of state will be obtained. A gas thermometer may be used to measure temperatures as low as 1°K by extrapolation.

8-4 TEMPERATURE MEASUREMENT BY MECHANICAL EFFECTS

Several temperature-measurement devices may be classified as mechanically operative. In this sense we shall be concerned with those devices operating on the basis of a change in mechanical dimension with a change in temperature.

The liquid-in-glass thermometer is one of the most common types of temperature-measurement devices. The construction details of such an instrument are shown in Fig. 8-4. A relatively large bulb at the lower portion of the thermometer holds the major portion of the liquid, which expands when heated and rises in the capillary tube upon which are etched appropriate scale markings. At the top of the capillary tube another bulb is placed to provide a safety feature in case the temperature range of the thermometer is inadvertently exceeded. Alcohol and mercury are the most commonly used liquids. Alcohol has the advantage that it has a higher coefficient of expansion than mercury, but it is limited to low-temperature measurements because it tends to boil away at high temperatures. Mercury cannot be used below its freezing point of $-38.78°F$. The size of the capillary depends on the size of the sensing bulb, the liquid, and the desired temperature range for the thermometer.

In operation, the bulb of the liquid-in-glass thermometer is exposed to the environment whose temperature is to be measured.   A rise in temperature causes the liquid to expand in the bulb and rise in the capillary, thereby indicating the temperature.   It is important to note that the expansion registered by the thermometer is the *difference* between the expansion of the liquid and the expansion of the glass.   The difference is a function not only of the heat transfer to the bulb from the environment, but also of the heat conducted into the bulb from the stem; the more the stem conduction relative to the heat transfer from the environment, the larger the error.   To account for such conduction effects the thermometer is usually calibrated for a certain specified depth of immersion.   High-grade mercury-in-glass thermometers have the temperature scale markings engraved on the glass along with a mark which designates the proper depth of immersion.   Very precise mercury-in-glass thermometers may be obtained from the National Bureau of Standards with calibration information for each thermometer.

Mercury-in-glass thermometers are generally applicable up to about 600°F, but their range may be extended to 1000°F by filling the space above the mercury with a gas like nitrogen.   This increases the pressure on the mercury, raises its boiling point, and thereby permits the use of the thermometer at higher temperatures.

A very widely used method of temperature measurement is the

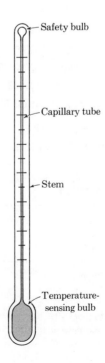

Safety bulb

Capillary tube

Stem

Temperature-sensing bulb

Fig. 8-4   Schematic of a mercury-in-glass thermometer.

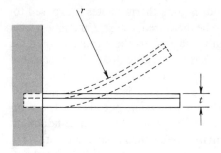

Fig. 8-5 The bimetallic strip.

bimetallic strip. Two pieces of metal with different coefficients of thermal expansion are bonded together to form the device shown in Fig. 8-5. When the strip is subjected to a temperature higher than the bonding temperature, it will bend in one direction; when it is subjected to a temperature lower than the bonding temperature, it will bend in the other direction. Eskin and Fritze [3] have given calculation methods for bimetallic strips. The radius of curvature r may be calculated as

$$r = \frac{t\{3(1 + m)^2 + (1 + mn)[m^2 + (1/mn)]\}}{6(\alpha_2 - \alpha_1)(T + T_0)(1 + m)^2} \tag{8-5}$$

where t = combined thickness of the bonded strip

m = ratio of thicknesses of low- to high-expansion materials

n = ratio of moduli of elasticity of low- to high-expansion materials

α_1 = lower coefficient of expansion

a_2 = higher coefficient of expansion

T = temperature

T_0 = initial bonding temperature

The thermal-expansion coefficients for some commonly used materials are given in Table 8-1.

Table 8-1 Mechanical properties of some commonly used thermal materials

Material	Thermal coefficient of expansion per °C	Modulus of elasticity, psi
Invar	1.7×10^{-6}	21.4×10^6
Yellow brass	2.02×10^{-5}	14.0×10^6
Monel 400	1.35×10^{-5}	26.0×10^6
Inconel 702	1.25×10^{-5}	31.5×10^6
Stainless-steel type 316	1.6×10^{-5}	28×10^6

Example 8-1 A bimetallic strip is constructed of strips of yellow brass and Invar bonded together at 100°F. Each material has a thickness of 0.014 in. Calculate the radius of curvature produced when the strip is subjected to a temperature of 200°F.

Solution We use Eq. 8-5 with properties from Table 8-1.

$$T - T_0 = 100°F = 55.6°C$$
$$m = 1.0$$

$$n = \frac{21.4 \times 10^6}{14.0 \times 10^6} = 1.53$$

$$\alpha_1 = 1.7 \times 10^{-6}\ °C^{-1}$$
$$\alpha_2 = 2.02 \times 10^{-5}\ °C^{-1}$$
$$t = (2)(0.014) = 0.028 \text{ in.}$$

Thus,

$$r = \frac{(0.028)[(3)(1 + 1)^2 + (1 + 1.53)(1 + 1/1.53)]}{6(2.02 - 0.17)(10^{-5})(55.6)(1 + 1)^2}$$

$$= 18.35 \text{ in.}$$

Fluid-expansion thermometers represent one of the most economical, versatile, and widely used devices for industrial temperature-measurement applications. The principle of operation is indicated in Fig. 8-6. A bulb containing a liquid, gas, or vapor is immersed in the environment. The bulb is connected by means of a capillary tube to some type of pressure-measuring device such as the Bourdon gage shown. An increase in temperature causes the liquid or gas to expand, thereby increasing the pressure on the gage; the pressure is thus taken as an indication of the temperature. The entire system consisting of the bulb, capillary, and gage may be calibrated directly. It is clear that the temperature of the capillary tube may influence the reading of the device because some of the volume of fluid is contained therein. If an

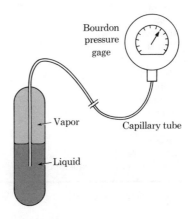

Bourdon pressure gage

Vapor

Capillary tube

Liquid

Fig. 8-6 Fluid-expansion thermometer.

equilibrium mixture of liquid and vapor is used in the bulb, however, this problem may be alleviated, provided that the bulb temperature is always higher than the capillary-tube temperature. In this circumstance the fluid in the capillary will always be in a subcooled liquid state, while the pressure will be uniquely specified for each temperature in the equilibrium mixture contained in the bulb.

Capillary tubes as long as 200 ft may be used with fluid-expansion thermometers. The transient response is primarily dependent on the bulb size and the thermal properties of the enclosed fluid. Highest response may be achieved by using a small bulb connected to some type of electric-pressure transducer through a short capillary.

8-5 TEMPERATURE MEASUREMENT BY ELECTRICAL EFFECTS

Electrical methods of temperature measurement are very convenient because they furnish a signal that is easily detected, amplified, or used for control purposes. In addition, they are usually quite accurate when properly calibrated and compensated.

ELECTRICAL-RESISTANCE THERMOMETER

One quite accurate method of temperature measurement is the electrical-resistance thermometer. It consists of some type of resistive element, which is exposed to the temperature to be measured. The temperature is indicated through a measurement of the change in resistance of the element. Several types of materials may be used as resistive elements, and their characteristics

Table 8-2 Resistance-temperature coefficients α at room temperature, $°C^{-1}$†

Nickel	0.0067
Iron (alloy)	0.002 to 0.006
Tungsten	0.0048
Aluminum	0.0045
Copper	0.0043
Lead	0.0042
Silver	0.0041
Gold	0.004
Platinum	0.00392
Mercury	0.00099
Manganin	± 0.00002
Carbon	-0.0007
Electrolytes	-0.02 to -0.09
Semiconductor (thermistors)	-0.068 to $+0.14$

† According to Lion [6].

are given in Table 8-2.   The linear temperature coefficient of resistance α is defined by

$$\alpha = \frac{R_2 - R_1}{R_1 T_2 - R_2 T_1} \tag{8-6}$$

where R_2 and R_1 are the resistances of the material at temperatures T_2 and T_1, respectively.   The relationship in Eq. (8-6) is usually applied over a narrow temperature range such that the variation of resistance with temperature approximates a linear relation.   For wider temperature ranges the resistance of the material is usually expressed by a quadratic relation

$$R = R_0(1 + aT + bT^2) \tag{8-7}$$

where R = resistance at temperature T
  R_0 = resistance at $0°F$
  a, b = experimentally determined constants

It may be noted that the platinum resistance thermometer is used for the International Temperature Scale between the oxygen point and antimony point, as described in Chap. 2.

Various methods are employed for construction of resistance thermometers, depending on the application.   In all cases care must be taken to ensure that the resistance wire is free of mechanical stresses and so mounted that moisture cannot come in contact with the wire and influence the measurement.

The resistance measurement may be performed with some type of bridge circuit, as described in Chap. 4.   For steady-state measurements a null condition will suffice, while transient measurements will usually require the use of a deflection bridge.   One of the primary sources of error in the electrical-resistance thermometer is the effect of the resistance of the leads which connect the element to the bridge circuit.   Several arrangements may be used to correct for this effect, as shown in Fig. 8-7.   The Siemen's three-lead arrangement is the simplest type of corrective circuit.   At balance conditions the center lead carries no current and the effect of the resistance of the other two leads is canceled out.   The Callender four-lead arrangement solves the problem by inserting two additional lead wires in the adjustable leg of the bridge so that the effect of the lead wires on the resistance thermometer is canceled out.   The floating-potential arrangement in Fig. 8-7c is the same as the Siemen's connection, but an extra lead is inserted.   This extra lead may be used to check the equality of lead resistance.   The thermometer reading may be taken in the position shown, followed by additional readings with the two right and left leads interchanged, respectively.   Through

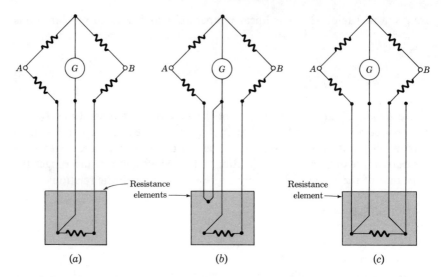

Fig. 8-7 Methods of correcting for lead resistance with electrical-resistance thermometer. (*a*) Siemen's three-lead arrangement; (*b*) Callender four-lead arrangement; (*c*) floating-potential arrangement.   Battery connections made at *A* and *B*.

this interchange procedure the best average reading may be obtained and the lead error minimized.

Example 8-2   A platinum resistance thermometer is used at room temperature. Assuming a linear temperature variation with resistance, calculate the sensitivity of the thermometer in ohms per degrees Fahrenheit.

Solution   The meaning of a linear variation of resistance with temperature is

$$R = R_0[1 + \alpha(T - T_0)]$$

where R_0 is the resistance at the reference temperature T_0.   The sensitivity is thus

$$S = \frac{dR}{dt} = \alpha R_0$$

R_0 depends on the length and size of the resistance wire.   At room temperature, $\alpha = 0.00392°C^{-1} = 0.00218°F^{-1}$ for platinum.

THERMISTORS

The thermistor is a semiconductor device that has a negative-temperature coefficient of resistance in contrast to the positive coefficient displayed by most metals.   Furthermore, the resistance follows an exponential variation with temperature instead of a polynomial relation like Eq. (8-7).   Thus, for a thermistor,

$$R = R_0 \exp\left[\beta\left(\frac{1}{T} - \frac{1}{T_0}\right)\right] \tag{8-8}$$

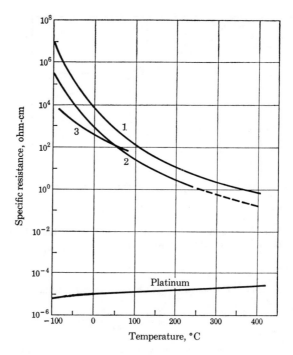

Fig. 8-8 Resistivity of three thermistor materials compared with platinum according to Ref. [1].

where R_0 is the resistance at the reference temperature T_0 and β is an experimentally determined constant. The numerical value of β varies between 3500 and 4600°K, depending on the thermistor material and temperature. The resistivities of three thermistor materials as compared with platinum are given in Fig. 8-8 according to Ref. [1]. A typical static voltage-current curve is shown in Fig. 8-9, while a typical set of transient voltage-current characteristics is illustrated in Fig. 8-10. The numbers on the curve in Fig. 8-9 designate the degrees centigrade rise in temperature above ambient temperature for the particular thermistor.

The thermistor is an extremely sensitive device, and consistent performance within 0.01°C may be anticipated with proper calibration. A rather nice feature of the thermistor is that it may be used for temperature compensation of electric circuits. This is possible because of the negative-temperature characteristic that it exhibits, so that it can be used to counteract the increase in resistance of a circuit with a temperature increase.

Example 8-3 Calculate the temperature sensitivity for thermistor No. 1 in Fig. 8-8 at 100°C. Express the result in ohm-cm °C^{-1}. Take $\beta = 4120$°K at 100°C.

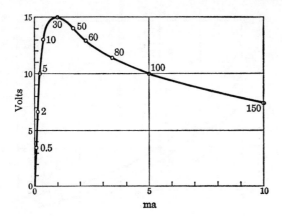

Fig. 8-9 Static voltage-current curve for a typical thermistor according to Ref. [1].

Solution   The sensitivity is obtained by differentiating Eq. (8-8).

$$S = \frac{dR}{dT} = R_0 \exp\left[\beta\left(\frac{1}{T} - \frac{1}{T_0}\right)\right]\frac{-\beta}{T^2}$$

We wish to express the result in resistivity units; thus the resistivity at 100°C is inserted for R_0.   Also,

$$T = T_0 = 100°\text{C} = 373°\text{K}$$

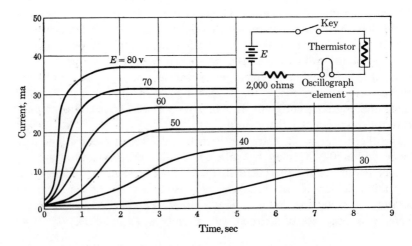

Fig. 8-10 Typical set of transient voltage-current curves for a thermistor according to Ref. [1].   Circuit for measurement is shown in insert.

so that

$$S = -\rho_{100°C} \frac{4120}{(373)^2}$$

$$= -\frac{(110)(4120)}{(373)^2} = -3.26 \text{ ohm-cm } °C^{-1}$$

THERMOELECTRIC EFFECTS

The most common electrical method of temperature measurement uses the thermocouple. When two dissimilar metals are joined together as in Fig. 8-11, an emf will exist between the two points A and B, which is primarily a function of the junction temperature. This phenomenon is called the Seebeck effect. If the two materials are connected to an external circuit such that a current is drawn, the emf may be altered slightly owing to a phenomenon called the Peltier effect. Further, if a temperature gradient exists along either or both of the materials, the junction emf may undergo an additional slight alteration. This is called the Thomson effect. There are, then, three emfs present in a thermoelectric circuit: the Seebeck emf, caused by the junction of dissimilar metals; the Peltier emf, caused by a current flow in the circuit; and the Thomson emf, which results from a temperature gradient in the materials. The Seebeck emf is of prime concern since it is dependent on junction temperature. If the emf generated at the junction of two dissimilar metals is carefully measured as a function of temperature, then such a junction may be utilized for the measurement of temperature. The main problem arises when one attempts to measure the potential. When the two dissimilar materials are connected to a measuring device, there will be another thermal emf generated at the junction of the materials and the connecting wires to the voltage-measuring instrument. This emf will be dependent on the temperature of the connection, and provision must be made to take account of this additional potential.

Two rules are available for analysis of thermoelectric circuits:

1. If a third metal is connected in the circuit as shown in Fig. 8-12, the net emf of the circuit is not affected as long as the new connections are at the same temperature. This statement may be proved with the aid of the second law of thermodynamics and is known as the *law of intermediate metals*.

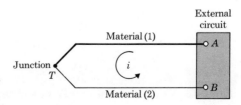

Fig. 8-11 Junction of two dissimilar metals indicating thermoelectric effect.

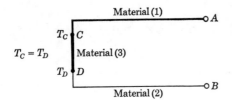

Fig. 8-12 Influence of a third metal in a thermoelectric circuit; law of intermediate metals.

2. Consider the arrangements shown in Fig. 8-13. The simple thermocouple circuits are constructed of the same materials but operate between different temperature limits. The circuit in Fig. 8-13a develops an emf of E_1 between temperatures T_1 and T_2; the circuit in Fig. 8-13b develops an emf of E_2 between temperatures T_2 and T_3. The *law of intermediate temperatures* states that this same circuit will develop an emf of $E_3 = E_1 + E_2$ when operating between temperatures T_1 and T_3, as shown in Fig. 8-13c.

It may be observed that all thermocouple circuits must involve at least two junctions. If the temperature of one junction is known, then the temperature of the other junction may be easily calculated with the thermoelectric properties of the materials. The known temperature is called the *reference temperature*. A common arrangement for establishing the reference temperature is the ice bath shown in Fig. 8-14. An equilibrium mixture of ice and air-saturated distilled water at standard atmospheric pressure produces a known temperature of 32°F. When the mixture is contained in a Dewar flask, it may be maintained for extended periods of time. Note that the arrangement in Fig. 8-14a maintains both thermocouple wires at a reference temperature of 32°F, whereas the arrangement in Fig. 8-14b maintains only one at the reference temperature. The system in Fig. 8-14a would be necessary if the binding posts at the voltage-measuring instrument were at different temperatures, while the connection in Fig. 8-14b would be satisfactory if the binding posts were at the same temperature. To be effective the system in Fig. 8-14a must have copper binding posts; i.e., the binding posts and leads must be of the same material.

It is common to express the thermoelectric emf in terms of the potential generated with a reference junction at 32°F. Standard thermocouple tables

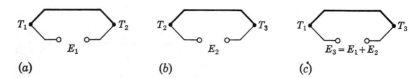

Fig. 8-13 Circuits illustrating the law of intermediate temperatures.

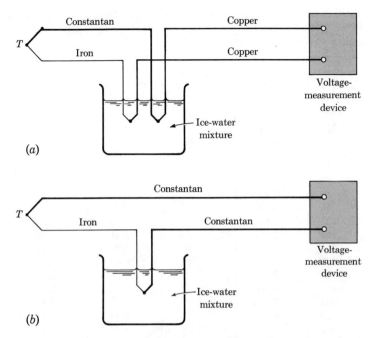

Fig. 8-14 Conventional methods for establishing reference temperature in thermocouple circuit.   Iron-constantan thermocouple illustrated.

have been prepared on this basis, and a summary of the output characteristics of the most common thermocouple combinations is given in Table 8-3. These data are shown graphically in Fig. 8-15, along with the behavior of some of the more exotic thermocouple materials.   The output voltage E of a simple thermocouple circuit is usually written in the form

$$E = AT + \tfrac{1}{2}BT^2 + \tfrac{1}{3}CT^3 \tag{8-9}$$

where T is the temperature in °C and E is based on a reference junction temperature of 0°C.   The constants A, B, and C are dependent on the thermocouple material.

The sensitivity, or thermoelectric power, of a thermocouple is given by

$$S = \frac{dE}{dT} = A + BT + CT^2 \tag{8-10}$$

Table 8-4 gives the approximate values of the sensitivity of various materials relative to platinum at 0°C.   A summary of the operating range and characteristics of the most common thermocouple materials is given in Figs. 8-15 and 8-16.

Table 8-3 Thermal emf in absolute millivolts for commonly used thermocouple combinations according to Ref. [14]

(Based on reference junction at 32°F)

Tempera-ture, °F	Copper-constantan	Chromel-constantan	Iron-constantan	Chromel-alumel	Platinum 10% rhodium
−300	−5.284	−8.30	−7.52	−5.51	
−250	−4.747		−6.71	−4.96	
−200	−4.111	−6.40	−5.76	−4.29	
−150	−3.380		−4.68	−3.52	
−100	−2.559	−3.94	−3.49	−2.65	
−50	−1.654		−2.22	−1.70	
0	−0.670	−1.02	−0.89	−0.68	
50	0.389		0.50	0.40	
100	1.517	2.27	1.94	1.52	0.221
150	2.711		3.41	2.66	0.401
200	3.967	5.87	4.91	3.82	0.595
250	5.280		6.42	4.97	0.800
300	6.647	9.71	7.94	6.09	1.017
350	8.064		9.48	7.20	1.242
400	9.525	13.75	11.03	8.31	1.474
450	11.030		12.57	9.43	1.712
500	12.575	17.95	14.12	10.57	1.956
600	15.773	22.25	17.18	12.86	2.458
700	19.100	26.65	20.26	15.18	2.977
800		31.09	23.32	17.53	3.506
1000		40.06	29.52	22.26	4.596
1200		49.04	36.01	26.98	5.726
1500		62.30		33.93	7.498
1700		70.90		38.43	8.732
2000				44.91	10.662
2500				54.92	13.991
3000					17.292

The output of thermocouples is in the millivolt range and may be measured by the D'Arsonval movement dc millivoltmeter. The voltmeter is basically a current-sensitive device, and hence the meter reading will be dependent on both the emf generated by the thermocouple and the total circuit resistance, including the resistance of the connecting wires. The complete system, consisting of the thermocouple lead wires and millivolt-meter, may be calibrated directly to furnish a reasonably accurate temperature determination. For all precision measurements the thermocouple output is determined with a potentiometer circuit similar to the one described in Sec. 4-4. Potentiometers are available as either manually balanced types like the one shown in Fig. 8-17 or automatically balanced instruments of the indicating or recording type. For very precise laboratory work a microvolt potentiom-

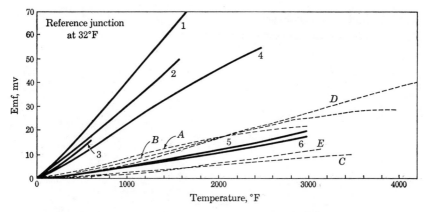

Legend:

1 Chromel-Constantan
2 Iron-Constantan (type J)
3 Copper-Constantan (type T)
4 Chromel-Alumel (type K)
5 Platinum-Platinum rhodium (type R)
6 Platinum-Platinum rhodium (type S)

Tentative curves:

A Rhenium-Molybdenum
B Rhenium-Tungsten
C Iridium-Iridium rhodium
D Tungsten-Tungsten rhenium
E Plat. rhodium-Plat. 10% rhodium

Fig. 8-15 Emf temperature relations for thermocouple materials. (*Courtesy Honeywell, Inc.*)

eter like the Leeds and Northrup K-2 or Rubicon Type B is normally used. These instruments can read potentials within 1 μv. It may be noted that the resistance of the lead wires is of no consequence when a potentiometer is used at balance conditions since the current flow is zero in the thermocouple circuit.

Table 8-4 Thermoelectric sensitivity $S = dE/dT$ **of thermoelement made of materials listed against platinum, μv °C^{-1}†**

(Reference junction kept at a temperature of 0°C)

Bismuth	−72	Silver	6.5
Constantan	−35	Copper	6.5
Nickel	−15	Gold	6.5
Potassium	−9	Tungsten	7.5
Sodium	−2	Cadmium	7.5
Platinum	0	Iron	18.5
Mercury	0.6	Nichrome	25
Carbon	3	Antimony	47
Aluminum	3.5	Germanium	300
Lead	4	Silicon	440
Tantalum	4.5	Tellurium	500
Rhodium	6	Selenium	900

† According to Lion [6].

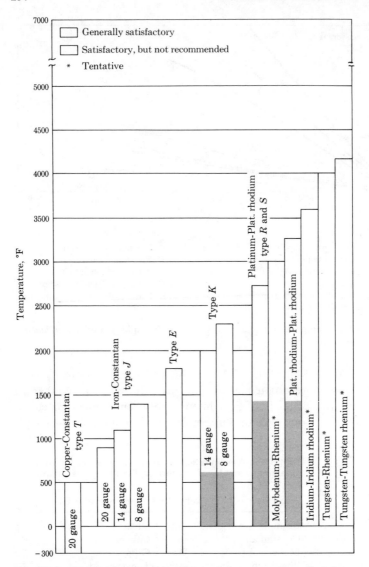

Fig. 8-16 Summary of operating range of thermocouples. See also Fig. 8-15. (*Courtesy Honeywell, Inc.*)

Example 8-4 An iron-constantan thermocouple is connected to a potentiometer whose terminals are at 75°F. The potentiometer reading is 3.59 mv. What is the temperature of the thermocouple junction?

Solution The thermoelectric potential corresponding to 75°F is obtained from Table 8-3 as

$$E_{75} = 1.22 \text{ mv}$$

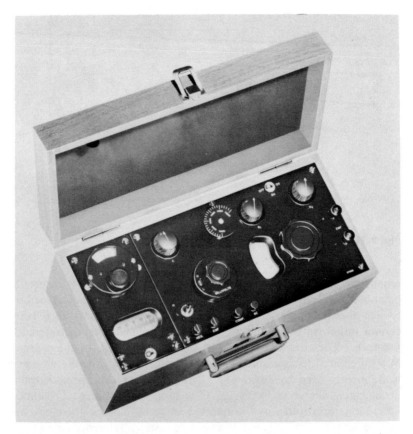

Fig. 8-17 Photograph of a typical manually balanced thermocouple potentiometer. (*Courtesy Honeywell, Inc.*)

The emf of the thermocouple based on a 32°F reference temperature is thus

$E_T = 1.22 + 3.59 = 4.81$ mv

From Table 8-3 the corresponding temperature is 197°F.

In order to provide a more sensitive circuit, thermocouples are occasionally connected in a series arrangement as shown in Fig. 8-18. Such an arrangement is called a *thermopile*, and for the three-junction situation, the output would be three times that of a single thermocouple arrangement *provided that* the temperatures of the hot and cold junctions are uniform. The thermopile arrangement is useful for obtaining a substantial emf for measurement of a small temperature difference between the two junctions. In this way a relatively inexpensive instrument may be used for voltage measurement, whereas a microvolt potentiometer might otherwise be

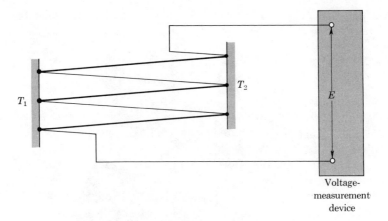

Fig. 8-18 Thermopile connection.

required.   When a thermopile is installed, it is important to ensure that the junctions are electrically insulated from one another.   We have seen that the typical thermocouple measures the difference in temperature between a certain unknown point and another point designated as the reference temperature.   The circuit could just as well be employed for the measurement of a differential temperature.   For small differentials, the thermopile circuit is frequently used to advantage.   Now consider the series thermocouple arrangement shown in Fig. 8-19.   The four junctions are all maintained at different temperatures and connected in series.   Since there are an even number of junctions, it is not necessary to install a reference junction because the same type of metal is connected to both terminals of the potentiometer. If we note that the current will flow from plus to minus and assume that junction A produces a potential drop in this direction, then junctions B and D will produce a potential drop in the opposite direction and junction C will

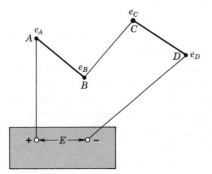

Fig. 8-19 Series connection of thermo-couples.

generate a potential drop in the same direction as junction A. Thus the total emf measured at the potentiometer terminals is

$$E = e_A - e_B + e_C - e_D \qquad (8\text{-}11)$$

The reading will be zero when all the junctions are at the same temperature and will take on some other value at other conditions. Note, however, that the emf of this series connection is *not* indicative of any particular temperature. It is *not* representative of an average of the junction temperatures.

The parallel connection in Fig. 8-20 may be used for obtaining the average temperature of a number of points. Each of the four junctions may be at a different temperature and hence will generate a different emf. The bucking potential furnished by the potentiometer will be the average of the four junction potentials. There can be a small error in this reading, however, because there is a small current flow in the lead wires as a result of the difference in potential between the junctions. Thus, the resistance of the lead wires will influence the reading to some extent.

A more suitable way of obtaining an average temperature is to use the thermopile circuit in Fig. 8-18. Each of the "hot" junctions may be at a different temperature, while all the "cold" junctions may be maintained at a

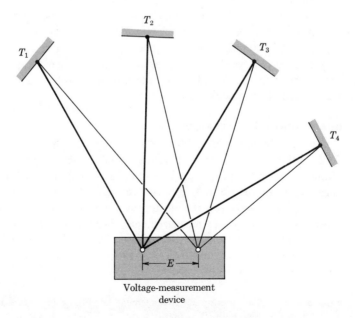

Voltage-measurement
device

Fig. 8-20   Parallel connection of thermocouples.

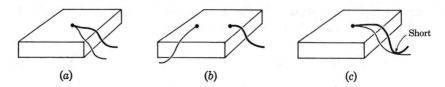

Fig. 8-21 Installation of thermocouples on a metal plate. (*a*) Only junction bead contacts plate; (*b*) contact at two points; (*c*) contact at bead and along wires.

fixed reference value. The average emf is then given by

$$E_{\text{avg}} = \frac{E}{n} \qquad\qquad (8\text{-}12)$$

where n is the number of junction pairs and E is the total reading of the thermopile. The average temperature corresponding to the average emf given in Eq. (8-12) may then be determined.

From the above discussion it is clear that the thermocouple measures the temperature at the last point of electric contact of the two dissimilar materials. Consider the situations shown in Fig. 8-21. The thermocouple installation in Fig. 8-21*a* is made so that only the junction bead makes contact with the metal plate whose temperature is to be measured. The installation in Fig. 8-21*b* allows contact at two points. If there is a temperature gradient in the metal plate, then the emf of the thermocouple will be indicative of the average of the temperatures of these two points. In Fig. 8-21*c* only the junction bead contacts the metal plate, but the two thermocouple wires are in electric contact a short distance away from the plate. The temperature indicated by the thermocouple will be that temperature at the shorted electric contact.

QUARTZ-CRYSTAL THERMOMETER

A novel and highly accurate method of temperature measurement is based on the sensitivity of the resonant frequency of a quartz crystal to temperature change. When the proper angle of cut is used with the crystal, there is a very linear correspondence between the resonant frequency and temperature. Commercial models for the device utilize electronic counters and digital readout for the frequency measurement. For absolute temperature measurements usable sensitivities of 0.001°C are claimed for the device. Since the measurement process relies on a frequency measurement, the device is particularly insensitive to noise pickup in connecting cables.

Example 8-5 The thermopile arrangement of Fig. 8-18 uses copper-constantan thermocouples with $T_1 = 300°F$ and $E = 3.2$ mv for the three-junction pairs. Calculate the value of T_2.

Solution   We first calculate the thermoelectric sensitivity at 300°F using the values in Table 8-3.

$$S_{300} = \left(\frac{dE}{dT}\right)_{300} \approx \frac{E_{350} - E_{250}}{350 - 250} = \frac{8.064 - 5.280}{100} = 0.0278 \text{ mv } °F^{-1}$$

The thermoelectric emf generated by each junction pair is

$$E = \frac{3.2}{3} = 1.067 \text{ mv}$$

so that the temperature difference is

$$T_2 - T_1 = \frac{E}{S} = \frac{1.067}{0.0278} = 38.4°F$$

and

$$T_2 = 300 + 38.4 = 338.4°F$$

Example 8-6   The thermoelectric effect has been used for electric-power generation. Calculate the thermoelectric sensitivity of a device using bismuth and tellurium as the dissimilar materials, and estimate the maximum voltage output for a 100°F temperature difference at approximately room temperature using one junction.

Solution   The sensitivity is calculated from the data of Table 8-4 as

$$S = S_{\text{tellurium}} - S_{\text{bismuth}}$$

$$= 500 - (-72) = 572 \ \mu v \ °C^{-1}$$

The voltage output for a 100°F temperature difference is calculated as

$$E = S \ \Delta T = (572 \times 10^{-6})(\tfrac{5}{9})(100)$$

$$= 3.18 \times 10^{-2} \text{ volt}$$

8-6   TEMPERATURE MEASUREMENT BY RADIATION

In addition to the methods described in the preceding sections, it is possible to determine the temperature of a body through a measurement of the thermal radiation emitted by the body. Two methods are commonly employed for measurement: (1) optical pyrometry and (2) emittance determination. Before we discuss these methods, it is necessary to describe the nature of thermal radiation.

Thermal radiation is electromagnetic radiation emitted by a body as a result of its temperature. This radiation is distinguished from other types of electromagnetic radiation such as radio waves and X rays, which are not propagated as a result of temperature. Thermal radiation lies in the wavelength region from about 0.1 to 100 μ ($1\mu = 10^{-6}$ m). The total thermal radiation emitted by a *blackbody* (ideal radiation) is given as

$$E_b = \sigma T^4 \tag{8-13}$$

where σ = Stefan-Boltzmann constant and is equal to 0.1714×10^{-8} Btu/ (hr)(ft²)(°R⁴)

T = absolute temperature, °R

E_b = emissive power, Btu/(hr)(ft²)

The emissive power of the blackbody varies with wavelength according to the Planck distribution equation

$$E_{b\lambda} = \frac{C_1 \lambda^{-5}}{e^{C_2/\lambda T} - 1} \tag{8-14}$$

where λ = wavelength, μ

T = temperature, °R

$C_1 = 1.187 \times 10^{+8}$ Btu-μ^4/(hr)(ft²)

$C_2 = 2.5896 \times 10^4$ μ-°R

$E_{b\lambda}$ is called the monochromatic blackbody emissive power. A plot of Eq. (8-14) is given in Fig. 8-22 for two temperatures.

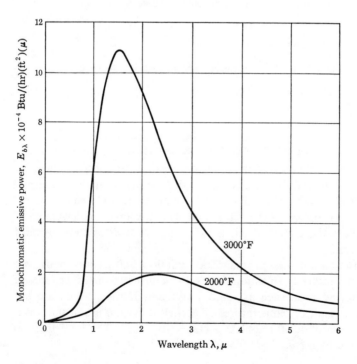

Fig. 8-22 Blackbody emissive power for two temperatures.

When thermal radiation strikes a material surface, the following relation applies:

$$\alpha + \rho + \tau = 1 \qquad (8\text{-}15)$$

where α = absorptivity or the fraction of the incident radiation absorbed

ρ = reflectivity or the fraction reflected

τ = transmissivity or the fraction transmitted

For most solid materials, $\tau = 0$ so that

$$\alpha + \rho = 1 \qquad (8\text{-}16)$$

The emissivity ϵ is defined as

$$\epsilon = \frac{E}{E_b} \qquad (8\text{-}17)$$

where E is the emissive power of an actual surface and E_b is the emissive power of a blackbody at the same temperature. Kirchhoff's identity furnishes the additional relationship

$$\epsilon = \alpha \qquad (8\text{-}18)$$

under conditions of thermal equilibrium. A gray body is one for which the emissivity is constant for all wavelengths; i.e.,

$$\epsilon_\lambda = \frac{E_\lambda}{E_{b\lambda}} = \epsilon \qquad (8\text{-}19)$$

Actual surfaces frequently exhibit highly variable emissivities over the wavelength spectrum. Figure 8-23 illustrates the distinctive features of blackbody and gray-body radiation. For purposes of analysis the real surface is frequently approximated by a gray body having an emissivity equal to the average total emissivity of the real surface as defined by Eq. (8-17).

Let us now consider the measurement of temperature through the use of optical pyrometry. This method refers to the identification of the temperature of a surface with the color of the radiation emitted. As a surface is heated, it becomes dark red, orange, and finally white in color. The maximum points in the blackbody radiation curves shift to shorter wavelengths with increase in temperature according to Wien's law

$$\lambda_{max}T = 5215.6 \; \mu\text{-}°\text{R} \qquad (8\text{-}20)$$

where λ_{max} is the wavelength at which the maximum points in the curves in Fig. 8-22 occur. The shift in these maximum points explains the change in

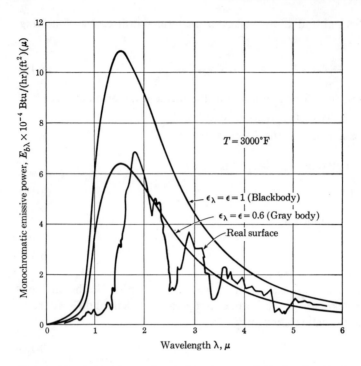

Fig. 8-23 Comparison of emissive powers of blackbody, ideal gray body, and actual surfaces.

color as a body is heated; i.e., higher temperatures result in a concentration of the radiation in the shorter wavelength portion of the spectrum. The temperature-measurement problem consists of a determination of the variation of temperature with color of the object. For this purpose an instrument is constructed as shown schematically in Fig. 8-24. The radiation from the source is viewed through the lens and filter arrangement. An absorption filter at the front of the device reduces the intensity of the incoming radiation so that the standard lamp may be operated at a lower level. The standard lamp is placed in the optical path of the incoming radiation. By an adjustment of the lamp current, the color of the filament may be made to match the color of the incoming radiation. The red filter is installed in the eyepiece to ensure that comparisons are made for essentially monochromatic radiation, thus eliminating some of the uncertainties resulting from variation of radiation properties with wavelength.

Figure 8-25 illustrates the appearance of the lamp filament as viewed from the eyepiece. When balance conditions are achieved, the filament will seem to disappear in the total incoming radiation field. Temperature calibration is made in terms of the lamp heating current.

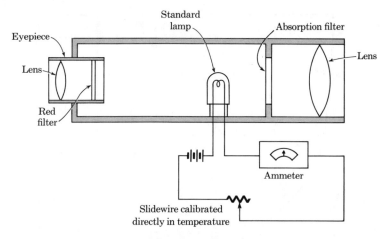

Fig. 8-24 Schematic of optical pyrometer.

The temperature of a body may also be measured by determining the total emitted energy from the body and then calculating the temperature from

$$E = \epsilon \sigma T^4 \tag{8-21}$$

In order to determine the temperature, the emissivity of the material must be known so that

$$T = \left(\frac{E}{\epsilon \sigma}\right)^{\frac{1}{4}} \tag{8-22}$$

The *apparent* blackbody temperature is the value as calculated from Eq. (8-22) with $\epsilon = 1$, or

$$T_a = \left(\frac{E}{\sigma}\right)^{\frac{1}{4}} \tag{8-23}$$

If the apparent temperature is taken as the measured value, the *error* in

Fig. 8-25 Appearance of lamp filament in eyepiece of optical pyrometer.

Filament
too cold

Null
condition

Filament
too hot

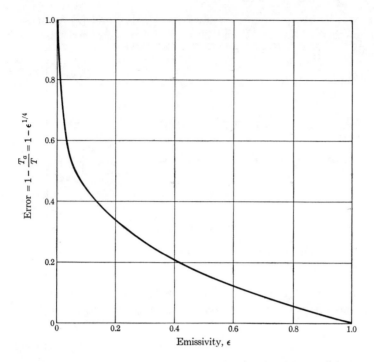

Fig. 8-26   Temperature error due to nonblackbody surface conditions.

temperature due to nonblackbody conditions is thus

$$\text{Error} = \frac{T - T_a}{T} = 1 - \frac{T_a}{T} = 1 - \epsilon^{\frac{1}{4}} \tag{8-24}$$

Figure 8-26 gives this error as a function of emissivity.

Several methods are available to measure the emitted thermal energy from a body, and these methods will be discussed in Chap. 12. For now, it is important to realize that temperature may be calculated from the above equations once this measurement is made.

In practice the optical pyrometer is the more widely used of the two radiation temperature methods since it is relatively inexpensive and portable and the determination does not depend strongly on the surface properties of the material. The measurement of radiant energy from a surface can be quite accurate, however, when suitable instruments are employed. If the surface emissive properties are accurately known, this measurement can result in a very accurate determination of temperature. It will be noted that a total radiation measurement forms the basis for the upper limit of the International Temperature Scale, as discussed in Chap. 2.

The interested reader should consult Refs. [2], [4], and [7] for more information on thermal radiation and temperature measurements by the above methods.

Example 8-7 The emitted energy from a piece of metal is measured, and the temperature is determined to be 1950°F, assuming a surface emissivity of 0.82. It is later found that the true emissivity is 0.75. Calculate the error in the temperature determination.

Solution The emitted energy is given by

$$\frac{q}{A} = \epsilon \sigma T^4$$

When $T = 1950°F = 2410°R$, $\epsilon = 0.82$. We wish to calculate the value of the true temperature T' such that

$$\frac{q}{A} = \epsilon' \sigma (T')^4$$

where $\epsilon' = 0.75$. Thus

$$(0.82)(2410)^4 = (0.75)(T')^4$$

and

$$T' = (2410)\left(\frac{0.82}{0.75}\right)^{0.25} = 2464°R$$

so that the temperature error is

$$\Delta T = 2464 - 2410 = 54°F$$

8-7 EFFECT OF HEAT TRANSFER ON TEMPERATURE MEASUREMENT

A heat-transfer process is associated with all temperature measurements. When a thermometer is exposed to an environment, the temperature is determined in accordance with the total heat-energy exchange with the temperature-sensing element. In some instances the temperature of the thermometer can be substantially different from the temperature which is to be measured. In this section we wish to discuss some of the methods that may be used to correct the temperature readings. It may be noted that the errors involved are classified as fixed errors.

Heat transfer may take place as a result of one or more of three modes: conduction, convection, or radiation. In general, all three modes must be taken into account in analyzing a temperature-measurement problem. Conduction is described by Fourier's law:

$$q = -kA\frac{\partial T}{\partial x} \tag{8-25}$$

where k = thermal conductivity

A = area through which the heat transfer takes place

q = heat-transfer rate in the direction of the decreasing temperature gradient

If a temperature gradient exists along a thermometer, heat may be conducted into or out of the sensing element in accordance with this relation.

Convection heat transfer is described in accordance with Newton's law of cooling:

$$q = hA(T_s - T_\infty) \tag{8-26}$$

where h = convection heat-transfer coefficient

A = surface area exchanging heat with the fluid

T_s = surface temperature

T_∞ = fluid temperature

The radiation heat transfer between two surfaces is proportional to the difference in absolute temperatures to the fourth power according to the Stefan-Boltzmann law of thermal radiation:

$$q_{1-2} = \sigma F_G F_\epsilon(T_1^4 - T_2^4) \tag{8-27}$$

where F_G is a geometric factor and F_ϵ is a factor that describes the radiation properties of the surfaces.

Consider the temperature-measurement problem illustrated in Fig. 8-27. A thermocouple junction is installed in the flat plate whose temperature is to be measured. The plate is exposed to a convection environment on both sides, and the thermocouple wires are exposed to this same environment. The thermocouple wires are covered with insulating material as shown. If

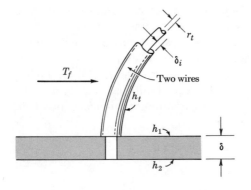

Fig. 8-27 Schematic of general thermocouple installation in a flat plate.

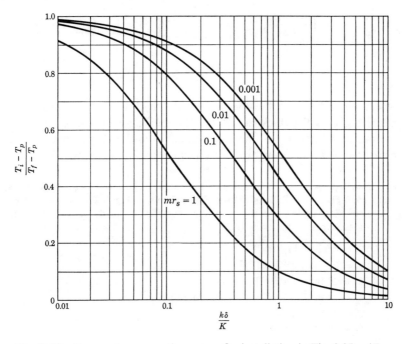

Fig. 8-28 Temperature-correction curves for installation in Fig. 8-27. (*From Ref.* [*11*].)

the plate temperature is higher than the convection environment, heat will be conducted out along the thermocouple wire and the temperature of the junction will be lower than the true plate temperature. An analytical solution to this problem has been presented by Schneider [11], and the results are given in Fig. 8-28. This solution neglects radiation heat-transfer effects.

The nomenclature for the parameters in Fig. 8-28 is as follows:

h_1, h_2 = convection heat-transfer coefficients on the sides of the plate as indicated

h_t = convection heat-transfer coefficient from each of the thermo-couple wires

k = thermal conductivity of the plate material

δ = plate thickness

T_f = fluid temperature surrounding the thermocouple wire

T_i = temperature indicated by the thermocouple

T_p = true plate temperature (temperature a large distance away from the thermocouple junction)

r_s = $\sqrt{2} r_t$, where r_t is the radius of each of the thermocouple wires

$$m = \left(\frac{h_1 + h_2}{k\delta}\right)^{\frac{1}{2}} \tag{8-28}$$

$$K = \sqrt{2}\pi(k_A^{\frac{1}{2}} + k_B^{\frac{1}{2}})r_t^{\frac{3}{2}}\left(\frac{1}{h_t} + \frac{\delta_i}{k_i}\right)^{-\frac{1}{2}} \tag{8-29}$$

where k_A and k_B are the thermal conductivities of the two thermocouple materials, δ_i is the thickness of the thermocouple insulation, and k_i is the thermal conductivity of the insulation. Additional information on temperature corrections is provided by Hennecke and Sparrow [16].

Example 8-8   A copper-constantan thermocouple is attached to a $\frac{1}{16}$-in.-thick stainless-steel plate [$k = 35$ Btu/(hr)(ft)(°F)] as shown in Fig. 8-27.   The diameter of the wires is 0.04 in., and the following estimates are made of the significant heat-transfer parameters:

$h_1 = 4.0$ Btu/(hr)(ft²)(°F)
$h_2 = 2.5$ Btu/(hr)(ft²)(°F)
$h_t = 20$ Btu/(hr)(ft²)(°F)
$k_A = 14$ Btu/(hr)(ft)(°F)   (constantan)
$k_B = 220$ Btu/(hr)(ft)(°F)   (copper)
$T_f = 500°F$

The thermocouple wires are coated with an electrically insulating lacquer, which may be considered negligibly thin insofar as heat transfer is concerned.   The plate is cooled on the side opposite the thermocouple installation, and the thermocouple indicates a temperature of 300°F.   Calculate the true plate temperature.

Solution   We observe that

$\delta = \frac{1}{16}$ in. $= 0.0071$ ft
$r_t = 0.02$ in. $= 0.00167$ ft
$\delta_i \approx 0$
$T_i = 300°F$

From Eq. (8-28),

$$m = \left[\frac{4.0 + 2.5}{(35)(0.0142)}\right]^{\frac{1}{2}} = 3.62$$

From Eq. (8-29),

$$K = \sqrt{2}\pi(14^{\frac{1}{2}} + 220^{\frac{1}{2}})(0.00167)^{\frac{3}{2}}(\tfrac{1}{20} + 0)^{-\frac{1}{2}}$$
$$= 0.0251$$

The parameters for use with Fig. 8-28 are thus

$$mr_s = \sqrt{2}(0.00167)(3.62) = 0.00856$$

$$\frac{k\delta}{K} = \frac{(35)(0.0071)}{(0.0251)} = 9.9$$

Using these values, we obtain from Fig. 8-28

$$\frac{T_i - T_p}{T_f - T_p} = 0.08$$

or

$$300 - T_p = (0.08)(500 - T_p) \quad \text{and} \quad T_p = 282.6°F$$

The thermocouple conduction error is $T_i - T_p = 17.4°F$. This error could be reduced by using smaller thermocouple wires or by insulating the wires.

Now let us consider the general problem of the temperature measurement of a gas stream and the influence of radiation on this measurement. The situation is illustrated in Fig. 8-29. The temperature of the thermometer is designated as T_t, the true temperature of the gas is T_g, and the effective radiation temperature of the surroundings is T_s. If it is assumed that the surroundings are very large, then the following energy balance may be made:

$$hA(T_g - T_t) = \sigma A \epsilon (T_t^4 - T_s^4) \tag{8-30}$$

where h = convection heat-transfer coefficient from the gas to the thermometer
A = surface area of the thermometer
ϵ = surface emissivity of the thermometer

Equation (8-30) may be used to determine the true gas temperature.

In practice, the radiation error in a temperature measurement is reduced by placing a radiation shield around the thermometer, which reflects most of the radiant energy back to the thermometer. A simple radiation-shield arrangement is shown in Fig. 8-30. The environment is assumed to be very large, and we wish to find the true gas temperature knowing the indicated temperature T_t and the other heat-transfer parameters, such as the heat-transfer coefficients and emissivities. The thermometer size may vary considerably, from the rather substantial dimensions of a mercury-in-glass thermometer to a tiny thermistor bead embedded in the tip of a hypodermic needle. An energy balance that neglects conduction effects may be made for both the thermometer and the shield using a radiation-network analysis to

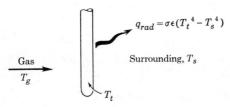

Gas
T_g

$q_{rad} = \sigma \epsilon (T_t^4 - T_s^4)$

Surrounding, T_s

T_t

Fig. 8-29 Schematic illustrating influence of radiation on temperature of thermometer.

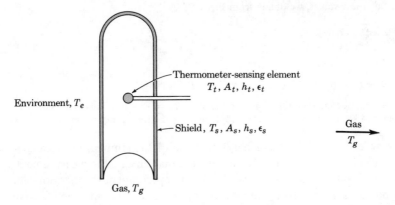

Fig. 8-30 A simple radiation-shield arrangement for a thermometer.

obtain expressions for the radiation heat transfer.† The results of the analysis are

Thermometer

$$h_t(T_g - T_t) = \frac{\epsilon_t}{1 - \epsilon_t} (E_{bt} - J_t) \tag{8-31}$$

Shield

$$2h_s(T_g - T_s) = \frac{E_{bs} - J_t}{(1/F_{ts})(A_s/A_t) + 1/\epsilon_s - 1} + \frac{E_{bs} - E_{be}}{1/\epsilon_s - 1 + 1/F_{se}} \tag{8-32}$$

where

$$J_t = \frac{\{E_{bt}[\epsilon_t/(1 - \epsilon_t)] + F_{te}E_{be}\}A_t/A_s + E_{bs}/[(1/F_{ts})(A_s/A_t) + 1/\epsilon_s - 1]}{(A_t/A_s)[F_{te} + \epsilon_t/(1 - \epsilon_t)] + 1/[(1/F_{ts})(A_s/A_t) + 1/\epsilon_s - 1]} \tag{8-33}$$

ϵ_t and ϵ_s are the emissivities of the thermometer and shield, respectively, h_t and h_s are the convection heat-transfer coefficients from the gas to the thermometer and shield, respectively, A_t is the surface area of the thermometer for convection and radiation, A_s is the surface area of the radiation shield *on each side*, and the blackbody emissive powers E_b are given by

$$E_{bt} = \sigma T_t^4 \tag{8-34}$$
$$E_{bs} = \sigma T_s^4 \tag{8-35}$$
$$E_{be} = \sigma T_e^4 \tag{8-36}$$

† An explanation of the radiation-network method is given in Ref. [5].

The radiation-shape factors F are defined as

F_{ts} = fraction of radiation that leaves the thermometer and arrives at the shield

F_{se} = fraction of radiation that leaves the outside surface of shield and gets to the environment (this fraction is 1.0)

F_{te} = fraction of radiation that leaves the thermometer and gets to the environment

The analysis that arrives at Eqs. (8-31) and (8-32) has been simplified somewhat by assuming that all the radiation exchange between the shield and environment may be taken into account by adjusting the value of F_{se} to include radiation from the inside surface of the shield. Some type of convection heat-transfer analysis must be used to determine the values of h_t and h_s, depending on whether natural or forced convection is involved, etc. An iterative solution must usually be performed on Eqs. (8-31) and (8-32) except in special cases. Some useful charts are given in Ref. [15] for correcting temperature measurements for the effects of conduction and radiation.

In general, the determination of convection heat-transfer coefficients requires very detailed consideration of the fluid dynamics of the problem, and empirical relations must often be used for the calculation. For more information the reader should consult Refs. [2] and [7]. Some frequently used formulas for calculating convection heat transfer are summarized in Table 8-5.

Example 8-9 A mercury-in-glass thermometer is placed inside a cold room in a frozen-food warehouse to measure the change in air temperature when the door is left open for extended periods of time. In one instance the thermometer reads 34°F, while the automatic temperature-control system for the room indicates that the *wall temperature* of the room is 15°F. The convection heat-transfer coefficient for the thermometer is estimated at 2 Btu/(hr)(ft²)(°F), and $\epsilon = 0.9$ for glass. Estimate the true air temperature.

Solution We use Eq. (8-30) to calculate the true air temperature T_g:

$$h(T_g - T_t) = \epsilon\sigma(T_t^4 - T_s^4)$$

$$(2)(T_g - 494) = (0.1714 \times 10^{-8})(0.9)(494^4 - 475^4)$$

Thus,

$$T_g = 500.7°R = 40.7°F$$

We may note, of course, the rather substantial difference between the indicated air temperature of 34°F and the actual temperature of 40.7°F.

Table 8-5 Frequently used convection heat-transfer formulas

Physical situation	Type of fluid	Range of validity	Heat-transfer relation	Fluid properties evaluated at
Forced convection over flat plate, plate heated over entire length	Gas or liquid	Laminar: $\mathrm{Re}_x < 5 \times 10^5$ Turbulent: $\mathrm{Re}_x > 5 \times 10^5$ Laminar: $\mathrm{Re}_L < 5 \times 10^5$ Turbulent: $\mathrm{Re}_L > 5 \times 10^5$	$\overline{\mathrm{Nu}}_x = 0.332\ \mathrm{Re}_x^{\frac{1}{2}}\ \mathrm{Pr}^{\frac{1}{3}}$ $\mathrm{St}_x \mathrm{Pr}^{\frac{2}{3}} = 0.0288\ \mathrm{Re}_x^{-\frac{1}{5}}$ $\overline{\mathrm{Nu}}_L = 0.664\ \mathrm{Re}_L^{\frac{1}{2}}\ \mathrm{Pr}^{\frac{1}{3}}$ $\overline{\mathrm{Nu}}_L = (0.036\ \mathrm{Re}_L^{0.8} - 836)\ \mathrm{Pr}^{\frac{1}{3}}$	Film temperature T_f
Forced convection in smooth circular tube	Gas or liquid	$\mathrm{Re}_d > 3{,}000$ $\mathrm{Re}_d < 2{,}100$ and $\mathrm{Re}_d \left(\dfrac{d}{L}\right) > 10$	$\overline{\mathrm{Nu}}_d = 0.023\ \mathrm{Re}_d^{0.8}\ \mathrm{Pr}^n$ $n = 0.4$ for heating; $n = 0.3$ for cooling $\overline{\mathrm{Nu}}_d = 1.85(\mathrm{Re}_d\ \mathrm{Pr})^{\frac{1}{3}} \left(\dfrac{d}{L}\right)^{\frac{1}{3}} \left(\dfrac{\mu}{\mu_w}\right)^{0.14}$	Average bulk temperature of fluid
Forced-convection crossflow over cylinder	Gas Liquids	$40 < \mathrm{Re}_d < 4{,}000$ $40 < \mathrm{Re}_d < 4{,}000$	$\overline{\mathrm{Nu}}_d = 0.615\ \mathrm{Re}_d^{0.466}$ $\overline{\mathrm{Nu}}_d\ 0.685\ \mathrm{Re}_d^{0.466}\ \mathrm{Pr}^{\frac{1}{3}}$	Film temperature T_f Film temperature T_f
Forced convection over spheres	Gas Liquids	$17 < \mathrm{Re}_d < 70{,}000$ $1 < \mathrm{Re}_d < 200{,}000$	$\overline{\mathrm{Nu}}_d = 0.37\ \mathrm{Re}_d^{0.6}$ $\overline{\mathrm{Nu}}_d \mathrm{Pr}^{-0.3}\left(\dfrac{\mu}{\mu_w}\right)^{0.25} = 1.2 + 0.5\ \mathrm{Re}_d^{0.54}$	Film temperature T_f Free-stream temperature T_∞
Free convection from vertical flat plate	Gas or liquid	$10^4 < \mathrm{Gr}_L\ \mathrm{Pr} < 10^9$ $10^9 < \mathrm{Gr}_L\ \mathrm{Pr} < 10^{12}$	$\overline{\mathrm{Nu}}_L = 0.59(\mathrm{Gr}_L\ \mathrm{Pr})^{\frac{1}{4}}$ $\overline{\mathrm{Nu}}_L = 0.13(\mathrm{Gr}_L\ \mathrm{Pr})^{\frac{1}{3}}$	Film temperature T_f Film temperature T_f
Free convection from horizontal cylinders	Gas or liquid	$10^4 < \mathrm{Gr}_d\ \mathrm{Pr} < 10^9$ $10^9 < \mathrm{Gr}_d\ \mathrm{Pr} < 10^{12}$	$\overline{\mathrm{Nu}}_d = 0.53(\mathrm{Gr}_d\ \mathrm{Pr})^{\frac{1}{4}}$ $\overline{\mathrm{Nu}}_d = 0.13(\mathrm{Gr}_d\ \mathrm{Pr})^{\frac{1}{3}}$	Film temperature T_f Film temperature T_f

Definition of symbols: All quantities in consistent set of units so that Nu, Re, Pr, Gr, and St are dimensionless.

$\overline{\mathrm{Nu}}_d = \dfrac{\bar{h}d}{k}$; $\overline{\mathrm{Nu}}_L = \dfrac{\bar{h}L}{k}$; $\overline{\mathrm{Nu}}_x = \dfrac{h_x x}{k}$; $\mathrm{Re}_d = \dfrac{\rho u_\infty d}{\mu}$ for flow over cylinders or spheres; $\mathrm{Re}_d = \dfrac{\rho u_m d}{\mu}$ for flow in tubes; $\mathrm{Re}_x = \dfrac{\rho u_\infty x}{\mu}$; $\mathrm{Re}_L = \dfrac{\rho u_\infty L}{\mu}$; $\mathrm{Pr} = \dfrac{c_p \mu}{k}$;

$\mathrm{Gr}_L = \dfrac{\rho^2 g \beta (T_w - T_\infty) L^3}{\mu^2}$ $\mathrm{Gr}_d = \dfrac{\rho^2 g \beta (T_w - T_\infty) d^3}{\mu^2}$; $\mathrm{St}_x = \dfrac{h_x}{\rho u_\infty c_p}$

where

- d = diameter of tube
- g = acceleration of gravity
- h = average heat-transfer coefficient over entire surface
- h_x = local heat-transfer coefficient on flat plate
- k = thermal conductivity of fluid
- L = total length or height of flat plate or length of tube
- T_f = film temperature $(T_w + T_\infty)/2$

- T_w = wall or surface temperature
- T_∞ = free-stream temperature
- u_m = mean fluid velocity in tube
- u_∞ = free-stream velocity past flat plate, cylinder, or sphere
- x = distance from leading edge of flat plate
- β = volume coefficient of expansion of fluid
- μ = dynamic viscosity of fluid
- μ_w = viscosity evaluated at wall temperature
- ρ = fluid density

8-8 TRANSIENT RESPONSE OF THERMAL SYSTEMS

When a temperature measurement is to be made under nonsteady-state conditions, it is important that the transient response characteristics of the thermal system be taken into account. Consider the simple system shown in Fig. 8-31. The thermometer is represented by the mass m and specific heat c and is suddenly exposed to a convection environment of temperature T_∞. The convection heat-transfer coefficient between the thermometer and fluid is h, and radiation heat transfer is assumed to be negligible. It is also assumed that the thermometer is substantially uniform in temperature at any instant of time; i.e., the thermal conductivity of the thermometer material is sufficiently large in comparison with the surface conductance because of the convection heat-transfer coefficient. The energy balance for the transient process may be written as

$$hA(T_\infty - T) = mc \frac{dT}{d\tau} \qquad (8\text{-}37)$$

The solution of Eq. (8-37) gives the temperature of the thermometer as a function of time:

$$\frac{T - T_\infty}{T_0 - T_\infty} = e^{(-hA/mc)\tau} \qquad (8\text{-}38)$$

where T_0 is the thermometer temperature at time zero. Equation (8-38) represents the familiar exponential decay behavior described in Sec. 2-7. The time constant for the system in Fig. 8-31 is

$$\tau = \frac{mc}{hA} \qquad (8\text{-}39)$$

When the heat-transfer coefficient is sufficiently large, there may be substantial temperature gradients within the thermometer itself and a different type of analysis must be used to determine the temperature variation with time. The interested reader should consult Ref. [2] for more information on this subject.

8-9 THERMOCOUPLE COMPENSATION

Suppose a thermocouple is used to measure a transient temperature variation. The response of the thermocouple will be dependent on several factors, as

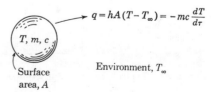

Fig. 8-31 Simple thermal system subjected to a sudden change in ambient temperature.

$$q = hA(T - T_\infty) = -mc \frac{dT}{d\tau}$$

T, m, c

Surface area, A

Environment, T_∞

outlined above, and will follow a variation like that of Eq. (8-38) when subjected to a step change in environment temperature. If a compensating electric network is applied to the system, it is possible to increase the frequency response of the thermocouple. The disadvantage of the compensating network is that it reduces the thermocouple output; however, if the measuring instrument is sufficiently sensitive, this problem is not too critical.

A typical thermocouple-compensation network is shown in Fig. 8-32. The thermocouple input voltage is represented by E_i, and the output voltage is represented by E_0. This particular network will attenuate low frequencies more than high frequencies and has a frequency response to a step input that is approximately opposite to that of a thermocouple. The network and thermocouple may thus be used in combination to produce a flat response over a wider frequency range. The output response of the network is given by

$$\frac{E_0}{E_i} = \alpha \frac{1 + j\omega\tau_c}{1 + \alpha j\omega\tau_c} \tag{8-40}$$

where $\alpha = \dfrac{R}{R + R_c}$

$\tau_c = R_c C$

ω = frequency of the input signal

$j = \sqrt{-1}$

The amplitude response is given by the absolute value of the function in Eq. (8-40). Thus,

$$\left(\frac{E_0}{E_i}\right)_{\text{amplitude}} = \alpha\sqrt{\frac{1 + \omega^2\tau_c^2}{1 + \alpha^2\omega^2\tau_c^2}} \tag{8-41}$$

The high-frequency compensation of the network is improved as the value of α is decreased, but this brings about a decreased output since the steady-state output is

$$\left(\frac{E_0}{E_i}\right)_{\text{steady state}} = \frac{R}{R + R_c} = \alpha \tag{8-42}$$

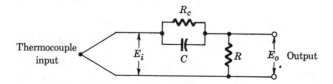

Fig. 8-32 Typical thermocouple-compensation network.

In practice, the compensating circuit is usually designed so that τ_c is equal to the time constant for the thermocouple. Variable resistors may be used to change the compensation with a change in thermal environment conditions.

The overall problem we have described above is one of compensating a thermocouple output so that higher-frequency response may be obtained. The step response of the thermocouple given by Eq. (8-38) will only hold, of course, for the conditions described. Therefore, the compensating network of Fig. 8-32 will achieve the desired objectives in varying degrees, depending on the exact thermal environment to which the thermocouple is exposed. Let us examine the thermocouple and compensating network response in more detail. Figure 8-33 shows the thermocouple response to a step change in temperature of the environment T_∞. We assume that the voltage output of the thermocouple is directly proportional to temperature over this range so that the voltage output would also follow a similar curve. In Fig. 8-34 we have plotted the output voltage of the compensating network for a step-voltage input and for $\alpha = 0.1$. The thermocouple response is obtained by solving the electric-network equivalent of the thermocouple. In that network the thermal capacitance is analogous to electric capacitance, and convection heat-transfer coefficient is analogous to thermal or electric resistance. If Figs. 8-33 and 8-34 are added together, we obtain the plot shown in Fig. 8-35, where the response is shown for different values of the ratio τ_c/τ. For $\tau_c/\tau > 1$, the response curve "overshoots" the steady-state voltage ratio of 0.1. It is clear that the curve $\tau = \tau_c$ offers a broader frequency response than the uncompensated curve. Careful inspection will also show that some overshoot can be desirable. The curve for $\tau_c/\tau = 1.1$ indicates a substantial

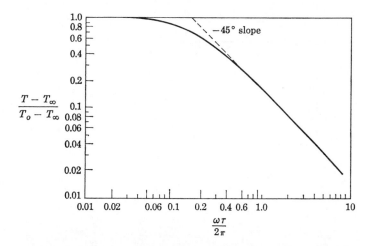

Fig. 8-33 Thermocouple response to step change in environment temperature.

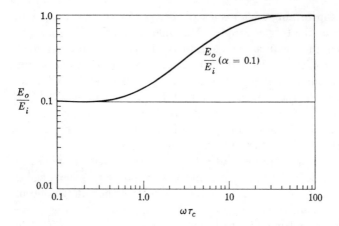

Fig. 8-34 Response of the network in Fig. 8-32 for a step-voltage input with $\alpha = 0.1$.

extension of the frequency response within a plus or minus range of about 5 percent. Excessive overshoot, of course, gives a decided nonlinear response. Optimal dynamic response of the network will be obtained when $\tau_c/\tau \approx 1.1$, provided the thermocouple input behaves nicely. In practice,

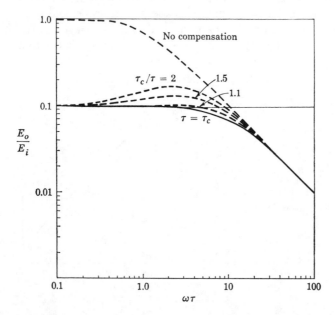

Fig. 8-35 Combined response of thermocouple and compensating network to step change in environment temperature.

variable resistors will be provided in the circuits to adjust the compensation in accordance with the actual variations in the thermal environment. Further information on transient response of thermocouples is given by Shepard and Warshawsky [17].

Example 8-10   A thermocouple bead has the approximate shape of a sphere $\frac{1}{16}$ in. in diameter. The properties may be taken as those of iron ($\rho = 490$ lb$_m$/ft^3, $c = 0.11$ Btu/lb$_m$-°F). Suppose such a bead is exposed to a convection environment where $h = 15$ Btu/(hr)(ft^2)(°F). Estimate the time constant for the thermocouple.

Solution   We use Eq. (8-39) to calculate the time constant with

$$m = \rho V = \rho \tfrac{4}{3}\pi r^3$$
$$A = 4\pi r^2$$

so that

$$\tau = \frac{(490)(0.11)}{(3)(15)(32)(12)} = 3.12 \times 10^{-3}\ \text{hr}$$

$$= 11.2\ \text{sec}$$

Example 8-11   Design a thermocouple-compensation network like that shown in Fig. 8-32 having a steady-state attenuation of a factor of 10. Let the network have the same time constant as the thermocouple bead in Example 8-10.

Solution   We take

$$\alpha = \frac{R}{R + R_c} = 0.1 \quad \text{and} \quad \tau_c = R_c C = 11.2$$

We must arbitrarily choose a value for one of the resistances. Let us choose a value of R_c which will make C a reasonable number. Take

$$R_c = 1\ \text{megohm} = 10^6\ \text{ohms}$$

Then $C = \dfrac{11.2}{10^6} = 1.12 \times 10^{-5} = 11.2\ \mu\text{f}$. Using the above value of R_c we obtain

$$R = 1.11 \times 10^5\ \text{ohms}$$

A more practical set of values would be

$$C = 10\ \mu\text{f}$$
$$R_c = 1.12\ \text{megohms}$$
$$R = 0.124\ \text{megohm}$$

8-10 HIGH-SPEED TEMPERATURE MEASUREMENTS

The measurements of temperatures in high-speed gas-flow streams are sometimes difficult because of the fact that a stationary probe must measure the temperature of the gas that is brought to rest at the probe surface. As the gas velocity is reduced to zero, the kinetic energy is converted to thermal energy and evidenced as an increase in temperature.   If the gas is brought to rest adiabatically, the resulting stagnation temperature is given by

$$T_0 = T_\infty + \frac{u_\infty{}^2}{2c_p g_c} \tag{8-43}$$

where T_0 = stagnation temperature

T_∞ = free-stream or *static* temperature

u_∞ = flow velocity

c_p = constant-pressure specific heat of the gas

The stagnation temperature of the gas may also be expressed in terms of the Mach number as

$$\frac{T_0}{T_\infty} = 1 + \frac{\gamma - 1}{2} M^2 \tag{8-44}$$

where $\gamma = c_p/c_v$ and has a value of 1.4 for air.   An inspection of Eqs. (8-43) and (8-44) shows that the stagnation temperature is substantially the same as the static temperature for low-speed flow.   The difference in these temperatures is only 4°F for a velocity of 226 ft/sec in air.

For the actual case of a probe inserted in a high-speed flow stream, the temperature indicated by the probe will *not*, in general, be equal to the stagnation temperature given by Eq. (8-43). The actual temperature is called the *recovery temperature* and is strongly dependent on the probe configuration.   The *recovery factor* is defined by

$$r = \frac{T_r - T_\infty}{T_0 - T_\infty} \tag{8-45}$$

where T_r is the recovery temperature.   In practice, the recovery factor must be determined from a calibration of the probe at different flow conditions. It is usually in the range $0.75 < r < 0.99$.   One of the important objectives of a probe design is to achieve a configuration which produces a nearly constant recovery factor over a wide range of flow velocities.

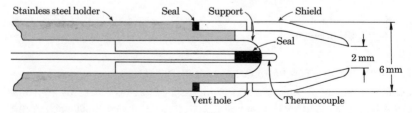

Fig. 8-36 High-speed temperature probe according to Winkler [13].

Table 8-6 Summary of characteristics of temperature-measurement devices

Device	Applicable temperature range, °F	Approximate accuracy, °F	Transient response	Cost	Remarks
Liquid-in-glass thermometer:					
a. Alcohol	−90 to 150	±1	Poor	Low	Used as inexpensive low-temperature thermometers
b. Mercury	−35 to 600	±0.5	Poor	Variable	Accuracy of ±0.1°F may be obtained with special calibrated thermometers
c. Gas-filled mercury	−35 to 1000	±0.5	Poor	Variable	
Fluid-expansion thermometer:					
a. Liquid or gas	−150 to 1000	±2	Poor	Low	Widely used for industrial temperature measurements
b. Vapor-pressure	20 to 400	±2	Poor	Low	
Bimetallic strip	−100 to 1000	±0.5	Poor	Low	Widely used in simple temperature-control devices
Electrical-resistance thermometer	−300 to 1800	±0.005†	Fair to good, depending on size of element	Readout equipment can be rather expensive for high-precision work	Most accurate of all methods
Thermistor	−100 to 500	±0.02†	Very good	Low but readout equipment may be expensive for high-precision work	Useful for temperature-compensation circuits; thermistor beads may be obtained in very small sizes
Copper-constantan thermocouple	−300 to 650	±0.5	Good, depends on wire size	Low	Linear sensitivity
Iron-constantan thermocouple	−300 to 1200	±0.5	Good, depends on wire size	Low	Superior in reducing atmospheres
Chromel-alumel thermocouple	−300 to 2200	±0.5	Good, depends on wire size	Low	Resistant to oxidation at high temperatures
Platinum-platinum 10% rhodium thermocouple	0 to 3000	±0.5	Good, depends on wire size	High	Low output; most resistant to oxidation at high temperatures; accuracy of ±0.15°F may be obtained in carefully controlled conditions
Optical pyrometer	1200 up	±20	Poor	Medium	Widely used for measurement of industrial furnace temperatures
Radiation pyrometer	0 up	±1 at low ranges, ±5 to 20 at high temperatures; depends on blackbody conditions and type of pyrometer. See Chap. 12.	Good, depending on type of pyrometer	Medium to high	Increased applications resulting from new high-precision devices being developed

† Accuracy is that which *may* be achieved. Inexpensive versions of the devices may not give the optimum accuracy.

A typical high-speed temperature probe is shown in Fig. 8-36 according to Winkler [13]. The gas stream enters the front of the probe and is diffused to a lower velocity in the enclosing shield arrangement. The flow subsequently leaves the probe through the side ventholes. The shield also serves to reduce radiation losses from the thermocouple-sensing element. The recovery factor for this probe is essentially 1.0.

8-11 SUMMARY

A summary of the range and application of the various temperature-measurement devices is given in Table 8-6. It will be noted that many of the devices overlap in their range of application. Thus, the selection of a device for use in a particular application is often a matter of preference. In many cases, the selection is based on a consideration of the type of readout equipment that is already available at the installation where the device is to be used.

REVIEW QUESTIONS

1. Distinguish between the Fahrenheit and centigrade temperature scales.
2. Upon what principle does the mercury-in-glass thermometer operate?
3. Describe the resistance characteristics of thermistors.
4. State the law of intermediate metals for thermocouples.
5. What is the Seebeck effect?
6. State the law of intermediate temperatures for thermocouples.
7. Why is a reference temperature necessary when using thermocouples?
8. Why does the resistance of thermocouple wires not influence a temperature reading when a potentiometer is used?
9. In general, why does heat transfer influence the accuracy of a temperature measurement?
10. In general, would one classify the response of thermal systems as fast, slow, or intermediate?

PROBLEMS

8-1. Calculate the temperature at which the Fahrenheit and centigrade scales coincide.

8-2. A certain mercury-in-glass thermometer has been calibrated for a prescribed immersion depth. The thermometer is immersed too much, such that the extra depth is equal to a distance of 10° on the scale. The true temperature reading may be calculated with

$$T_{true} = T_{ind} - 0.000088(T_{ind} - T_{amb})D$$

where T_{ind} = indicated temperature

T_{amb} = ambient temperature of the exposed stem

D = extra immersion depth of the thermometer past the correct mark

Calculate the thermometer error for an indicated temperature of 210°F and an ambient temperature of 70°F.

8-3. The bimetallic strip material of Example 8-1 is to be used in an on-off–temperature-control device, which will operate at a nominal temperature of 200°F. Calculate the deflection at the end of a 4-in. strip for deviations of ±1 and 2°F from the nominal temperature.

8-4. Suppose the bimetallic strip in Prob. 8-3 is used to indicate temperatures of 150, 200, and 300°F. The tip of the strip is connected to an appropriate mechanism, which amplifies the deflection so that it may be read accurately. The uncertainty in the length of the strip is ±0.01 in., and the uncertainty in the thickness of each material is ±0.0002 in. The perpendicular deflection from the 100°F position is taken as an indication of the temperature. Calculate the uncertainty in each of the above temperatures, assuming zero uncertainty in all the material properties; i.e., calculate the uncertainty in the deflection at each of these temperatures.

8-5. A bimetallic strip of yellow brass and Monel 400 is bonded at 120°F. The thickness of the yellow brass is 0.014 ± 0.0002 in., and the thickness of the Monel is 0.010 ± 0.0001 in. The length of the strip is 5 in. Calculate the deflection sensitivity defined as the deflection per degree Fahrenheit temperature difference. Estimate the uncertainty in this deflection sensitivity.

8-6. The specific volume of mercury is given by the relation

$$v = v_0(1 + aT + bT^2)$$

where T is in degrees centigrade and

$$a = 0.1818 \times 10^{-3}$$
$$b = 0.0078 \times 10^{-6}$$

A high-temperature thermometer is constructed of a Monel-400 tube having an inside diameter of 0.032 ± 0.0002 in. After the inside is evacuated, mercury is placed on the inside such that a column height of 4.00 ± 0.01 in. is achieved when the thermometer temperature is 500°F. If the tube is to be used for a temperature measurement, calculate the uncertainty at 500°F if the uncertainty in the height measurement is ±0.01 in.

8-7. A fluid-expansion thermometer uses liquid Freon ($C = 0.22$ Btu/lb$_m$-°F, $\rho = 85.2$ lb$_m$/ft^3) enclosed in a 0.25 in.-ID copper ($C = 0.092$, $\rho = 560$) cylinder with a wall thickness of 0.032 in. The thermometer is exposed to a crossflow of air at 400°F, 15 psia, and 20 ft/sec. Estimate the time constant for such a thermometer. Neglect all effects of the capillary tube. Repeat for a crossflow of liquid water at 180°F and 20 ft/sec. Recalculate the time constant for both of these situations for a stainless-steel bulb ($C = 0.11$, $\rho = 490$).

8-8. It is desired to measure a temperature differential of 5°F using copper-constantan thermocouples at a temperature level of 200°F. A millivolt recorder is available for the emf measurement which has an uncertainty of 0.004 mv. Precision thermocouple wire is available, which may be assumed to match the characteristics in Table 8-3 exactly. How many junction pairs in a thermopile must be used in order that the uncertainty in the temperature differential measurement does not exceed 0.05°F? Neglect all errors due to heat transfer.

8-9. A thermopile of chromel-alumel with four junction pairs is used to measure a differential temperature of 4.0°F at a temperature level of 400°F. The sensitivity of the thermocouple wire is found to match that in Table 8-3 within ±0.5 percent, and the millivolt potentiometer has an uncertainty of 0.002 mv. Calculate the uncertainty in the differential temperature measurement.

8-10. For a certain thermistor, $\beta = 3420$°K, and the resistance at 200°F is known to be 1,010 ± 3 ohms. The thermistor is used for a temperature measurement, and the resistance is measured as 2.315 ± 4 ohms. Calculate the temperature and the uncertainty.

8-11. Calculate the thermoelectric sensitivity for iron-constantan and copper-constantan at $0°C$. Plot the error which would result if these values were assumed constant over the temperature range from -50 to $+600°F$.

8-12. A chromel-alumel thermocouple is exposed to a temperature of $1560°F$. The potentiometer is used as the cold junction, and its temperature is estimated to be $83°F$. Calculate the emf indicated by the potentiometer.

8-13. Four iron-constantan thermocouple junctions are connected in series. The temperatures of the four junctions are 200, 300, 100, and $32°F$. Calculate the emf indicated by a potentiometer.

8-14. A millivolt recorder is available with a total range of 0 to 10 mv and an accuracy of ± 0.25 percent of full scale. Calculate the corresponding temperature ranges for use with iron-constantan, copper-constantan, and chromel-alumel thermocouples, and indicate the temperature accuracies for each of these thermocouples.

8-15. When a material with spectral emissivity less than unity is viewed with an optical pyrometer, the apparent temperature will be somewhat less than the true temperature and will depend on the wavelength at which the measurements are made. The error is given by

$$\frac{1}{T} - \frac{1}{T_a} = \frac{\lambda}{C_2} \ln \epsilon_\lambda$$

where T = true temperature, $°R$
T_a = apparent temperature, $°R$
ϵ_λ = spectral emissivity at the particular value of λ
C_2 = constant from Eq. (8-14)

For a measurement at 0.665μ and an apparent temperature of $2400°F$, plot the error as a function of ϵ_λ.

8-16. A radiant energy measurement is made to determine the temperature of a hot block of metal. The emitted energy from the surface of the metal is measured as 5010 ± 80 Btu/(hr)(ft²) and the surface emissivity is estimated as $\epsilon = 0.90 \pm 0.05$. Calculate the surface temperature of the metal, and estimate the uncertainty.

8-17. A radiometer is used to measure the radiant energy flux from a material having a temperature of $542 \pm 1.0°F$. The emissivity of the material is 0.95 ± 0.03. The radiometer is then used to view a second material having exactly the same geometric shape and orientation as the first material and an emissivity of 0.72 ± 0.05. The uncertainty in the radiant energy flux measurement is ± 40 Btu/(hr)(ft²) for both measurements, and the second measurement is 2.23 times as large as the first measurement. Calculate the temperature of the second material and the uncertainty.

8-18. Consider the thermometer-shield arrangement shown in Fig. 8-30. Assuming that the shield area is very large compared with the thermometer area and that all radiation leaving the thermometer is intercepted by the shield, calculate the true gas temperature using the following data:

$h_t = 2.0$ Btu/(hr)(ft²)(°F)
$h_s = 1.5$ Btu/(hr)(ft²)(°F)
$\epsilon_t = 0.8$
$\epsilon_s = 0.2$
$T_t = 600°F$
$T_s = 100°F$

(It is assumed from the statement of the problem that $F_{ts} = 1.0$, $F_{se} = 1.0$, and $F_{te} = 0$.)

What temperature would the thermometer indicate if the shield were removed? Calculate the error reduction owing to the use of the radiation shield.

8-19. Rework Prob. 8-18 assuming that $A_s/A_t = 100$ and that the radiation shape factors are given as $F_{ts} = 0.9$, $F_{te} = 0.1$, and $F_{se} = 1.1$ (recall the assumption pertaining to F_{se}).

8-20. A platinum resistance thermometer is placed in a duct to measure the temperature of an airflow stream. The thermometer is placed inside a cylindrical shell 0.25 in. in diameter, which has a polished outside surface with $\epsilon = 0.08 \pm 0.02$. The airstream velocity is known to be 10 ft/sec, and the pressure is 14.7 psia. The thermometer indicates a temperature of 240°F. The duct wall temperature is measured at 380°F. Calculate the true air temperature, and estimate the uncertainty in this temperature. Assume that the uncertainty in the convection heat-transfer coefficient calculated with the appropriate formula from Table 8-5 is ± 15 percent. Assume that the uncertainty in the resistance-thermometer measurement is $\pm 0.05°$F.

8-21. Repeat Prob. 8-20, but assume that the cylinder is a chrome-plated copper rod with a thermocouple embedded in the surface. The uncertainty in the thermocouple measurement is assumed to be $\pm 0.5°$F.

8-22. A small copper sphere is constructed with a thermocouple embedded in the center and is used to measure the air temperature in an oven. The walls of the oven are at $1200 \pm 20°$F. The emissivity of the copper surface is 0.57 ± 0.04, and the temperature indicated by the thermocouple is $1015 \pm 1.0°$F. The convection heat-transfer coefficient is 5.0 Btu/(hr)(ft²)(°F) ± 15 percent. Calculate the true air temperature and the uncertainty. Calculate the true air temperature if the surface of the sphere had been chromeplated with $\epsilon = 0.06 \pm 0.02$. Estimate the uncertainty in this circumstance.

8-23. The thermocouple wires in Example 8-8 are coated with an insulation material having a thickness of 0.003 in. and a thermal conductivity of 0.1 Btu/(hr)(ft)(°F). Calculate the true plate temperature, assuming the other conditions are the same as in the example.

8-24. A $\frac{1}{32}$-in. constantan strip is heated by the gas stream on one side and cooled on the other side as shown. Two bare 0.004-in.-diam copper wires are attached to the strip to indicate the temperature difference in the transverse direction. A potentiometer indicates 0.15 ± 0.001 mv, and the sensitivity of copper constantan may be taken as 0.0279 ± 0.0001 mv °F⁻¹. Estimates of the significant heat-transfer parameters are as follows:

$$h_1 = 3.5 \text{ Btu/(hr)(ft}^2)(°F)$$
$$h_2 = 2.5 \text{ Btu/(hr)(ft}^2)(°F)$$
$$h_{t_1} = 15 \text{ Btu/(hr)(ft}^2)(°F)$$
$$h_{t_2} = 10 \text{ Btu/(hr)(ft}^2)(°F)$$
$$k_{cu} = 220 \text{ Btu/(hr)(ft)(°F)}$$
$$k_{con} = 14 \text{ Btu/(hr)(ft)(°F)}$$

Calculate the true temperature difference, and estimate the uncertainty.

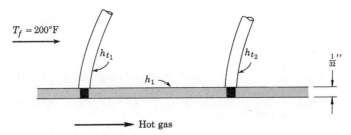

8-25. Calculate the time constant for the copper-rod thermometer of Prob. 8-21. Consider only the convection heat transfer from the gas in calculating this value.

8-26. A thermocouple is placed inside a $\frac{1}{8}$-in.-OD, $\frac{1}{16}$-in.-ID copper tube. The tube is then placed inside a furnace whose walls are at 1500°F. The air temperature in the furnace is 1200°F. Calculate the temperature indicated by the thermocouple, and estimate the time τ to obtain

$$\frac{T - T_i}{T_t - T_i} = 0.5$$

where T_i = initial temperature of the thermocouple before it is placed in the furnace

T_t = indicated temperature after a long time

T = temperature at time τ

Assume $\epsilon = 0.78$ for copper.

8-27. A certain thermocouple has a time constant of 1.2 sec. Design a compensation network like that shown in Fig. 8-32 having a steady-state attenuation factor of 8. Select the value of the resistance so that the capacitor has an even multiple value.

8-28. A certain high-speed temperature probe having a recovery factor of 0.98 ± 0.01 is used to measure the temperature of air at Mach 3.00. The thermocouple installed in the probe is accurate within $\pm 2.0°F$ and indicates a temperature of 715°F. Calculate the free-stream temperature and the uncertainty.

8-29. A 4-in.-square copper plate is suspended vertically in a room and a thermocouple attached to measure the plate temperature. The walls of the room are at 110°F, and the thermocouple indicates a temperature of 95°F. Assuming the copper has an emissivity of 0.7, calculate the air temperature in the room. Determine the convection heat-transfer coefficient from Table 8-5. If the uncertainty in the surface emissivity is ± 0.05, what is the resulting uncertainty in the calculated air temperature?

8-30. Repeat Prob. 8-29 for polished copper having an emissivity of $\epsilon = 0.05 \pm 0.01$.

REFERENCES

1. Becker, J. A., C. B. Green, and G. L. Pearson: Properties and Uses of Thermistors, *Trans. AIEE*, vol. 65, pp. 711–725, November, 1946.
2. Eckert, E. R. G., and R. M. Drake: "Heat and Mass Transfer," 2d ed., McGraw-Hill Book Company, New York, 1959.
3. Eskin, S. G., and J. R. Fritze: Thermostatic Bimetals, *Trans. ASME*, vol. 62, pp. 433–442, July, 1940.
4. Hackforth, H. L.: "Infrared Radiation," McGraw-Hill Book Company, New York, 1960.
5. Holman, J. P.: "Heat Transfer," 2d ed., McGraw-Hill Book Company, New York, 1968.
6. Lion, K. S.: "Instrumentation in Scientific Research," McGraw-Hill Book Company, New York, 1959.
7. McAdams, W. H.: "Heat Transmission," 3d ed., McGraw-Hill Book Company, New York, 1954.
8. Obert, E. F.: "Concepts of Thermodynamics," McGraw-Hill Book Company, New York, 1960.
9. Rossini, F. D. (ed.): "Thermodynamics and Physics of Matter," Princeton University Press, Princeton, N.J., 1955.
10. Sears, F. W.: "Thermodynamics," Addison-Wesley Publishing Company, Inc., Reading, Mass., 1953.

11. Schneider, P. J.: "Conduction Heat Transfer," Addison-Wesley Publishing Company, Inc., Reading, Mass., 1955.

12. Tribus, M.: "Thermostatics and Thermodynamics," D. Van Nostrand Company, Inc., Princeton, N.J., 1961.

13. Winkler, E. M.: Design and Calibration of Stagnation Temperature Probes for Use at High Supersonic Speeds and Elevated Temperatures, *J. Appl. Phys.*, vol. 25, p. 231, 1954.

14. ———, Reference Tables for Thermocouples, *Natl. Bur. Std. (U.S.), Cir.* 561, April, 1955.

15. West, W. E., and J. W. Westwater: Radiation-conduction Correction for Temperature Measurements in Hot Gases, *Ind. Eng. Chem.*, vol. 45, pp. 2, 152, 1953.

16. Hennecke, D. K., and E. M. Sparrow: Local Heat Sink on a Convectively Cooled Surface, Application to Temperature Measurement Error, *Int. J. Heat and Mass Trans.*, vol. 13, p. 287, February, 1970.

17. Shepard, C. E., and I. Warshawsky: Electrical Techniques for Compensation of Thermal Time Lag of Thermocouples and Resistance Thermometer Elements, *NACA Tech. Note* 2703, May, 1952.

9
Thermal- and Transport-property Measurements

9-1 INTRODUCTION

Several types of thermal properties are essential for energy-balance calculations in heat-transfer applications. Values of these properties for a variety of substances and materials are already available in tabular form in various handbooks; however, with the new materials, which appear regularly, it is important that the engineer be familiar with some basic methods of measuring these properties.

Most thermal-property measurements involve a determination of heat flow and temperature. We have already discussed several types of temperature-measuring devices in Chap. 8 and shall have occasion to refer to them from time to time in discussing thermal-property measurements. Heat flow is usually measured by making an energy balance on the device under consideration. For example, a metal plate might be heated with an electric heater and the plate immersed in a tank of water during this heating process. The convection heat loss from the plate could thus be determined by making a measurement of the electric power dissipated in the heater. As another example, consider the heating of water by passing it through a heated pipe.

The convection heat transfer from the pipe wall to the water may be determined by measuring the mass flow rate of water and the inlet and exit water temperatures to the heated section of pipe.   The energy gained by the water is therefore the heat transfer from the pipe, provided that the outside surface of the pipe is insulated so that no losses are incurred.   The techniques for measurement of heat transfer by thermal radiation are discussed in Chap. 12.

Thermal conductivity may be classified as a transport property since it is indicative of the energy transport in a fluid or solid.   In gases and liquids the transport of energy takes place by molecular motion, while in solids transport of energy by free electrons and lattice vibration is important. Fluid viscosity is also classified as a transport property because it is dependent on the momentum transport that results from molecular motion in the fluid. Mass diffusion is similarly classified as a transport process because it results from molecular movement.   The diffusion coefficient is the transport property in this case.   In this chapter we shall consider some simple methods for measurement of these transport properties.

The measurement of heat flow falls under the general subject of calorimetry.   In this chapter we shall discuss some simple calorimetric determinations that may be performed.   The broad subject of thermodynamic-property measurement is beyond the scope of our discussion.

9-2 THERMAL-CONDUCTIVITY MEASUREMENTS

Thermal conductivity is defined by the Fourier equation

$$q_x = -kA \frac{\partial T}{\partial x} \tag{9-1}$$

where q_x = heat-transfer rate

A = area through which the heat is transferred

$\partial T/\partial x$ = temperature gradient in the direction of the heat transfer

Experimental determinations of thermal conductivity are based on this relationship.   Consider the thin slab of material shown in Fig. 9-1.   If the heat-transfer rate through the material, the material thickness, and the difference in temperature are measured, then the thermal conductivity may be calculated from

$$k = \frac{q \, \Delta x}{A(T_1 - T_2)} \tag{9-2}$$

In an experimental setup the heat may be supplied to one side of the slab by an electric heater and removed from the other side by a cooled plate.   The

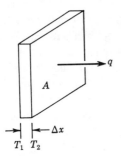

Fig. 9-1 Simple thermal-conductivity measurement.

temperatures on each side of the slab may be measured with thermocouples or thermistors, whichever is more appropriate.

The main problem of the above method for determining thermal conductivity is that heat may escape from the edges of the slab, or if the edges are covered with insulation, a two-dimensional temperature profile may result, which can cause an error in the determination. This problem may be alleviated by the installation of guard heaters, as shown in Fig. 9-2. In this arrangement the heater is placed in the center and a slab of the specimen is placed on each side of the heater plate. A coolant is circulated through the device to remove the energy, and thermocouples are installed at appropriate places to measure the temperatures. A guard heater surrounds the main heater, and its temperature is maintained at that of the main heater. This prevents heat transfer out from the edges of the specimen and thus maintains a one-dimensional heat transfer through the material whose thermal conductivity is to be determined. The guarded hot plate, as it is called, is widely used for determining the thermal conductivity of nonmetals, i.e., solids of rather low thermal conductivity. For materials of high thermal conductivity a small temperature difference would be encountered which would require much more precise temperature measurement.

A very simple method for the measurement of thermal conductivities

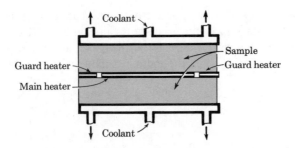

Fig. 9-2 Schematic of guarded hot-plate apparatus for measurement of thermal conductivity.

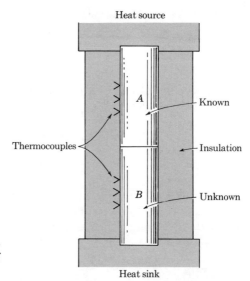

Fig. 9-3 Apparatus for measurement of thermal conductivity of metals.

of metals is depicted in Fig. 9-3. A metal rod A of known thermal conductivity is connected to a rod of the metal B whose thermal conductivity is to be measured. A heat source and heat sink are connected to the ends of the composite rod, and the assembly is surrounded by insulating material to minimize heat loss to the surroundings and to ensure one-dimensional heat flow through the rod. Thermocouples are attached, or embedded, in both the known and unknown materials as shown. If the temperature gradient through the known material is measured, the heat flow may be determined. This heat flow is subsequently used to calculate the thermal conductivity of the unknown material. Thus,

$$q = -k_A A \left(\frac{dT}{dx}\right)_A = -k_B A \left(\frac{dT}{dx}\right)_B \qquad (9-3)$$

The temperatures may be measured at various locations along the unknown and the variation of thermal conductivity with temperature determined from these measurements. Van Dusen and Shelton [6] have used this method for the determination of thermal conductivities of metals up to 600°C.

Example 9-1 A guarded hot-plate apparatus is used to measure the thermal conductivity of a metal having $k = 50$ Btu/(hr)(ft)(°F). The thickness of the specimen is 0.125 ± 0.002 in., and the heat flux may be measured within 1 percent. Calculate the accuracy necessary on the ΔT measurement in order to ensure an overall uncertainty in the measurement of k of 5 percent. If one of the plate temperatures is nominally 300°F, calculate the nominal value of the other plate temperature and the tolerable uncertainty in each temperature measurement, assuming a nominal heat flux of 20,000 Btu/(hr)(ft²).

Solution   We use Eq. (9-2) to estimate the uncertainty in conjunction with Eq. (3-2).

$$k = \frac{(q/A)\,\Delta x}{\Delta T} \tag{a}$$

Equation (3-2) may be written as

$$\frac{w_R}{R} = \left[\left(\frac{\partial R}{\partial x_1}\right)^2 \left(\frac{w_1}{R}\right)^2 + \left(\frac{\partial R}{\partial x_2}\right)^2 \left(\frac{w_2}{R}\right)^2 + \left(\frac{\partial R}{\partial x_3}\right)^2 \left(\frac{w_3}{R}\right)^2 \right]^{\frac{1}{2}} \tag{b}$$

We have

$$\frac{w_k}{k} = 0.05$$

$$\frac{\partial k}{\partial (q/A)} = \frac{\Delta x}{\Delta T} \qquad\qquad \frac{w_{q/A}}{k} = \frac{0.01\,q/A(\Delta T)}{(q/A)\,\Delta x} = \frac{0.01\,\Delta T}{\Delta x}$$

$$\frac{\partial k}{\partial (\Delta x)} = \frac{q/A}{\Delta T} \qquad\qquad \frac{w_{\Delta x}}{k} = \frac{(0.002)\,\Delta T}{(q/A)(0.125)}$$

$$\frac{\partial k}{\partial (\Delta T)} = -\frac{(q/A)\,\Delta x}{(\Delta T)^2} \qquad\qquad \frac{w_{\Delta T}}{k} = \frac{w_{\Delta T}\,\Delta T}{(q/A)\,\Delta x}$$

Inserting these expressions in Eq. (b) gives

$$0.05 = \left[(0.01)^2 + (0.002/0.125)^2 + \left(\frac{w_{\Delta T}}{\Delta T}\right)^2 \right]^{\frac{1}{2}}$$

or $w_{\Delta T}/\Delta T = 0.0146 = 1.46$ percent.   We now calculate the nominal value of ΔT as

$$\Delta T = \frac{(q/A)\,\Delta x}{k} = \frac{(2 \times 10^4)(0.125)}{(12)(50)} = 4.1667°\text{F}$$

Since $\Delta T = T_1 - T_2$, the nominal value of T_1 is

$$T_1 = 300 + 4.1667 = 304.1667°\text{F}$$

The tolerable uncertainty in each temperature measurement is calculated from

$$\frac{w_{\Delta T}}{\Delta T} = \left\{ \left[\frac{\partial (\Delta T)}{\partial T_1} \frac{w_{T_1}}{\Delta T} \right]^2 + \left[\frac{\partial (\Delta T)}{\partial T_2} \frac{w_{T_2}}{\Delta T} \right]^2 \right\}^{\frac{1}{2}}$$

$$w_{\Delta T} = (w_{T_1}^2 + w_{T_2}^2)^{\frac{1}{2}}$$

and we assume the uncertainties are the same for both temperatures so that

$$w_T = \frac{1}{\sqrt{2}}\,w_{\Delta T} = 0.707\,w_{\Delta T}$$

$$w_T = (0.707)(0.0146)(4.1667) = \pm 0.043°\text{F}$$

We thus see that a very accurate measurement of temperature is necessary in order to give only a 5 percent accuracy in the determination of k.

Example 9-2   An apparatus like that in Fig. 9-3 is used to measure the thermal conductivity of the metal of Example 9-1.   The same heat flux is used with a bar 3.0 ± 0.005 in. long.   Determine the accuracy necessary for the determination of

ΔT and the tolerable uncertainty in the temperature measurements.   Assume the same conditions as in Example 9-1.

Solution   For a 5 percent uncertainty in k the uncertainty in ΔT is given as

$$0.05 = \left[(0.01)^2 + \left(\frac{0.005}{3.0}\right)^2 + \left(\frac{w_{\Delta T}}{\Delta T}\right)^2\right]^{\frac{1}{2}}$$

$$\frac{w_{\Delta T}}{\Delta T} = 0.049 = 4.9\%$$

The nominal value of ΔT is given as

$$\Delta T = \frac{(2 \times 10^4)(3)}{(12)(50)} = 100°F$$

The allowable uncertainty in the determination of the temperatures is

$$w_T = \frac{1}{\sqrt{2}} w_{\Delta T} = (0.707)(0.049)(100) = \pm 3.46°F$$

This value is very easy to obtain.   Even if the heat flux were reduced to 2×10^3 Btu/(hr)(ft²), the nominal temperature difference would be 10°F and the allowable uncertainty would be $\pm 0.346°F$.   This is still a value that may be attained with careful experimental techniques.

9-3 THERMAL CONDUCTIVITY OF LIQUIDS AND GASES

Kaye and Higgins [3] have used a guarded hot-plate method for determining thermal conductivity of liquids.   Their apparatus is shown in Fig. 9-4.   The

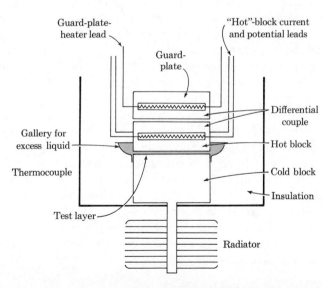

Fig. 9-4 Guarded hot-plate apparatus for measurement of thermal conductivity of liquids according to Ref. [3].

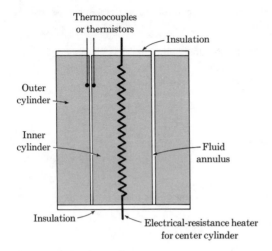

Fig. 9-5 Concentric-cylinder method for measurement of thermal conductivity of liquids.

diameter of the plates is 5 cm, and the thickness of the liquid layer is approximately 0.05 cm. This layer must be sufficiently thin so that convection currents are minimized. An annular arrangement, as shown in Fig. 9-5, may also be used for the determination of liquid thermal conductivities. Again, the thickness of the liquid layer must be thin enough to minimize thermal-convection currents.

A concentric-cylinder arrangement may also be used for the measurement of the thermal conductivity of gases. Keyes and Sandell [4] have used such a device for the measurement of thermal conductivity of water vapor, oxygen, nitrogen, and other gases. The inner and outer cylinders were both constructed of silver with a length of 5 in. and an outside diameter of $1\frac{1}{2}$ in. The gap space for the gas was 0.025 in. Vines [7] has utilized such a device for the measurement of high-temperature-gas thermal conductivities. A schematic of his apparatus is shown in Fig. 9-6. The emitter serves as the heat source, while the heat stations on either end act as guard heaters. The emitter has an outside diameter of 6 mm and a length of 50 mm, while the receiver has an inside diameter of 10 mm and a length of 125 mm with a wall thickness of 1 mm. For most of the tests it was possible to maintain a temperature difference of 5 to 10°C between the emitter and the receiver. The heat-transfer rate is measured by determining the electric-power input to the emitter, while thermocouples installed on the surface of the emitter and receiver determine the temperature difference.

For the concentric-cylinder device the relation that is used to calculate the thermal conductivity is

$$k = \frac{q \ln (r_2/r_1)}{2\pi L(T_1 - T_2)} \tag{9-4}$$

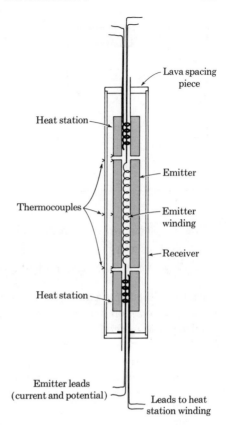

Fig. 9-6 Apparatus for determining thermal conductivity of gases at high temperatures according to Vines [7].

where q = heat-transfer rate

r_2, r_1 = outside and inside radii of the annular space containing the fluid, respectively

T_2, T_1 = temperatures measured at these radii

An investigation by Leidenfrost [8] proposes an apparatus for measuring thermal conductivities of gases and liquids from -180 to $+500°C$. Complete experimental data are not available, but accuracies of 0.1 percent are claimed for the device.

Thermal conductivities of several materials are given in the Appendix.

9-4 MEASUREMENT OF VISCOSITY

The defining equation for *dynamic* or *absolute viscosity* is

$$\tau = \mu \frac{du}{dy} \tag{9-5}$$

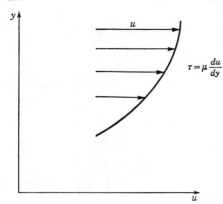

Fig. 9-7 Diagram indicating relation of viscosity to velocity gradient and fluid shear.

where τ = shear stress between fluid layers in laminar flow

$\quad \mu$ = viscosity

du/dy = normal velocity gradient as indicated in Fig. 9-7

Various methods are employed for measurement of the viscosity. The two most common methods are the rotating-concentric-cylinder method and the capillary-flow method. We shall discuss the rotating-cylinder method first.

Consider the parallel plates as shown in Fig. 9-8. One plate is stationary, and the other moves with constant velocity u. The velocity profile for the fluid between the two plates is a straight line, and the velocity gradient is

$$\frac{du}{dy} = \frac{u}{b} \tag{9-6}$$

The system could be used to measure the viscosity by measuring the force required to maintain the moving plate at the constant velocity u. The system is impractical from a construction standpoint, however, and the conventional approach is to approximate the parallel flat-plate situation with the rotating concentric cylinders shown in Fig. 9-9. The inner cylinder is stationary and attached to an appropriate torque-measuring device, while the outer cylinder is driven at a constant angular velocity ω. If the annular space b is sufficiently small in comparison with the radius of the inner cylinder,

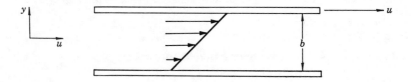

Fig. 9-8 Velocity distribution between large parallel plates.

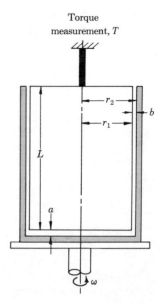

Fig. 9-9 Rotating concentric-cylinder apparatus for measurement of viscosity.

then the rotating-cylinder arrangement approximates the parallel-plate situation and the velocity profile in the gap space may be assumed to be linear. Then,

$$\frac{du}{dy} = \frac{r_2\omega}{b} \tag{9-7}$$

where, now, the y distance is taken in the radial direction and it is assumed that $b \ll r_1$. Now, if the torque T is measured, the fluid shear stress is expressed by

$$\tau = \frac{T}{2\pi r_1{}^2 L} \tag{9-8}$$

where L is the length of the cylinder. The viscosity is determined by combination of Eqs. (9-5), (9-7), and (9-8) to give

$$\mu = \frac{Tb}{2\pi r_1{}^2 r_2 L\omega} \tag{9-9}$$

If the concentric-cylinder arrangement is constructed such that the gap space a is small, then the bottom disk will also contribute to the torque and influence the calculation of the viscosity. The torque on the bottom disk is

$$T_d = \frac{\mu\pi\omega}{2a} r_1{}^4 \tag{9-10}$$

where a is the gap spacing, as shown in Fig. 9-9. Combining the torques due to the bottom disk and the annular space gives

$$T = \mu\pi\omega r_1{}^2\!\left(\frac{r_1{}^2}{2a} + \frac{2Lr_2}{b}\right) \tag{9-11}$$

Once the torque, angular velocity, and dimensions are measured, the viscosity may be calculated from Eq. (9-11).

Perhaps the most common method of viscosity measurement consists of a measurement of the pressure drop in laminar flow through a capillary tube. Consider the tube cross section shown in Fig. 9-10. If the Reynolds number defined by

$$\mathrm{Re}_d = \frac{\rho u_m d}{\mu} \tag{9-12}$$

is less than 1,000, laminar flow will exist in the tube and the familiar parabolic-velocity profile will be experienced as shown. If the fluid is incompressible and the flow is steady, it can be shown that the volume rate of flow Q can be written as

$$Q = \frac{\pi r^4 (p_1 - p_2)}{8\,\mu L} \tag{9-13}$$

A viscosity determination may be made by measuring the volume rate of flow and pressure drop for flow in such a tube. To ensure that laminar flow will exist, a small-diameter capillary tube is used; the small diameter reduces the Reynolds number as calculated from Eq. (9-12). In Eq. (9-12) the

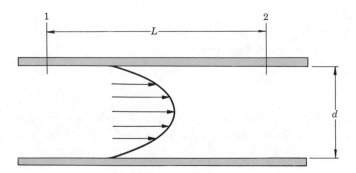

Fig. 9-10 Laminar flow through a capillary tube.

product ρu_m may be calculated from

$$\rho u_m = \frac{\dot{m}}{\pi r^2} \tag{9-14}$$

where $\dot{m}$ is the mass rate of flow.

When a viscosity measurement is made on a gas, the compressibility of the gas must be taken into account. The resulting expression for the mass flow of the gas under laminar flow conditions in the capillary is

$$\dot{m} = \frac{\pi r^4}{16 \, \mu R T} (p_1{}^2 - p_2{}^2) \tag{9-15}$$

where R is the gas constant for the particular gas.

Care must be taken to ensure that the flow in the capillary is *fully developed*, i.e., the parabolic-velocity profile has been established. This means that the pressure measurements should be taken far enough downstream from the entrance of the tube to ensure that developed flow conditions persist. It may be expected that the flow will be fully developed when

$$\frac{L}{d} > \frac{\text{Re}_d}{8} \tag{9-16}$$

where L is the distance from the entrance of the tube.

The Saybolt viscosimeter is an industrial device that uses the capillary-tube principle for measurement of viscosities of liquids. A schematic of the device is shown in Fig. 9-11. A cylinder is filled to the top with the liquid and enclosed in a constant-temperature bath to ensure uniformity of temperature during the measurements. The liquid is then allowed to drain from the bottom through the short capillary tube. The time necessary for 60 cc to drain is recorded, and this time is taken as indicative of the viscosity of the liquid. Since the capillary tube is short, a fully developed laminar-velocity profile is not established and it is necessary to apply a correction to account for the actual profile. If the velocity profile were fully developed, the kinematic viscosity would vary directly with the time for drainage; i.e.,

$$\nu = \frac{\mu}{\rho} = c_1 t$$

To correct for the nonuniform velocity profile, another term is added to give

$$\nu = c_1 t + \frac{c_2}{t}$$

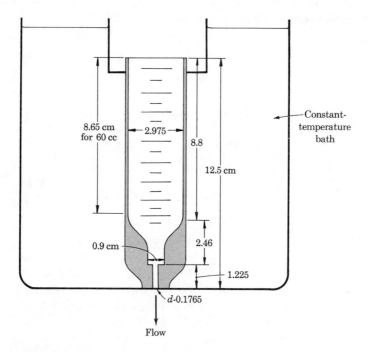

Fig. 9-11 Schematic of the Saybolt viscosimeter.

With the constants inserted, the relation is

$$\nu = \left(0.00237t - \frac{1.93}{t}\right) \times 10^{-3} \text{ ft}^2/\text{sec} \tag{9-17}$$

The symbol t designates the drainage time in seconds for 60 cc of liquid, and it is common to express the viscosity in units of *Saybolt seconds* when this method of measurement is used.   Equation (9-17) provides a method of converting to more explicit units for kinematic viscosity.   Viscosities for several fluids are given in the Appendix.

Example 9-3   A Saybolt viscosimeter is used to measure the viscosity of a certain motor oil.   The time recorded for drainage of 60 cc is estimated at 183 ± 0.5 sec. Calculate the percentage uncertainty in the viscosity.

Solution   Equation (3-2) is used in conjunction with Eq. (9-17) to estimate the uncertainty. We have

$$\frac{\partial \nu}{\partial t} = c_1 - \frac{c_2}{t^2} = \left(0.00237 + \frac{1.93}{t^2}\right) \times 10^{-3}$$

so that the uncertainty in the viscosity is given as

$$w_\nu = \frac{\partial \nu}{\partial t} w_t$$

$$= \left(0.00237 + \frac{1.93}{t^2}\right)(10^{-3})(0.5) \text{ ft}^2/\text{sec}$$

The nominal value of the viscosity is

$$\nu = \left(0.00237t - \frac{1.93}{t}\right)10^{-3} \text{ ft}^2/\text{sec}$$

At $t = 183$ sec

$$w_\nu = 1.21 \times 10^{-6} \text{ ft}^2/\text{sec}$$
$$\nu = 4.22 \times 10^{-4} \text{ ft}^2/\text{sec}$$

so that $w_\nu/\nu = 2.87 \times 10^{-3} = 0.287$ percent.

9-5 GAS DIFFUSION

Consider a container of a certain gas (2) as shown in Fig. 9-12. At one end of the container another gas (1) is introduced and allowed to diffuse into gas 2. The diffusion rate of gas 1 at any instant of time is given by Fick's law of diffusion as

$$n_1 = -D_{12}A \frac{\partial \overline{N}_1}{\partial x} \tag{9-18}$$

where n_1 is the molal rate of diffusion, A is the cross-sectional area for diffusion, $\overline{N}_1$ is the molal concentration of component 1 at any point, and D_{12} is defined as the diffusion coefficient. It may be noted that Eq. (9-18) is similar to the Fourier law for heat conduction. When a mass balance is made on an element, the one-dimensional diffusion equation results:

$$\frac{\partial \overline{N}_1}{\partial t} = D_{12} \frac{\partial^2 \overline{N}_1}{\partial x^2} \tag{9-19}$$

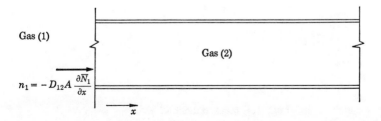

Fig. 9-12 Schematic of gas-diffusion process.

A solution to this equation may be used as the basis for an experimental determination of the diffusion coefficient D_{12}.

A simple experimental setup for the measurement of gas-diffusion

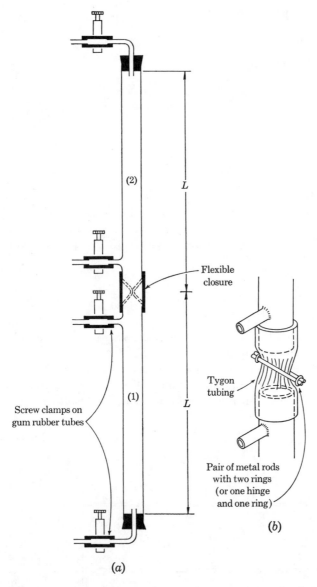

Fig. 9-13 Loschmidt apparatus for measurement of diffusion coefficients in gases. (a) Apparatus; (b) detail of flexible connection.

coefficients is shown in Fig. 9-13. Two sections of glass tubing are connected by a flexible Tygon tubing connection, which may be clamped as shown. The sections of tubing are charged with the two gases whose diffusion characteristics are to be investigated. After the initial charging process, the flexible connection is opened and the two gases are allowed to diffuse into one another. After a period of time the Tygon connection is closed again, the exact time recorded, and the two sections allowed to reach equilibrium concentrations. These equilibrium concentrations are then measured with appropriate analytical equipment. It is only necessary to measure the concentration of one of the gases in both sections since the concentration of the other gas will be obtained from a mole balance. However, measuring both gives a check and information about the accuracy of the experiment. Some check of this kind is always very desirable, although not always feasible.

A solution of Eq. (9-19) gives these concentrations in terms of the diffusion coefficient and dimensions of the tubes. Assuming that the lower portion is filled with gas (1) and the upper portion with gas (2) initially, the solution is

$$F = \frac{N_{1A} - N_{1B}}{N_{1A} + N_{1B}} = \frac{8}{\pi^2} \sum_{k=0}^{\infty} \frac{1}{(2k+1)^2} \exp\left[\frac{-\pi^2 D_{12} t (2k+1)^2}{4L^2}\right] \qquad (9\text{-}20)$$

where the subscripts A and B refer to the lower and upper sections, respectively, t is the time the two gases are allowed to diffuse, and L is the length of *each* tube. N_{1A} and N_{1B} represent the number of moles of component 1 in the lower and upper parts of the apparatus, respectively, after time t. For an experimental determination of D_{12} the optimum time to be allowed for the diffusion process is

$$t_{\text{opt}} = \frac{4L^2}{\pi^2 D_{12}} \qquad (9\text{-}21)$$

assuming that the only experimental errors are those involved in the determination of F. If the diffusion process is allowed to run for t_{opt} or longer, it is found that the first term in the series of Eq. (9-20) is the primary term and that the higher-order terms are negligible. Then,

$$F = 0.2982 \qquad \text{at } t = t_{\text{opt}} \qquad (9\text{-}22)$$

If the second- and higher-order terms are neglected, it is easy to calculate the value of D_{12} once a determination of F and t have been made. Neglecting these higher-order terms gives

$$F = \frac{8}{\pi^2} \exp \frac{-\pi^2 D_{12} t}{4L^2} \qquad (9\text{-}23)$$

$$\text{or } D_{12} = \frac{-4L^2}{\pi^2 t} \ln \frac{\pi^2 F}{8} \tag{9-24}$$

It is necessary to know only an approximate value for D_{12} in order to calculate the value of t_{opt} from Eq. (9-21). Then, the experiment is conducted using a diffusion time greater than this value. The exact value of D_{12} is then calculated from the experimental data using Eq. (9-24).

The simple experimental setup shown in Fig. 9-13 is called a *Loschmidt apparatus*. In the apparatus the heavier gas is usually contained in the lower section, while the lighter gas is placed in the upper section. Extreme care must be exerted to ensure that the system is free of leaks and that each section contains a pure component at the start of the experiment. Care must also be exerted to ensure that both gases are initially at the same temperature and pressure. The apparatus should be constructed so that the volume of the inlet tubing is negligible in comparison with the volume of the main apparatus. While the diffusion process is in progress, the temperature of the apparatus should be maintained constant. Diffusion coefficients for several gases are given in the Appendix.

Example 9-4   An apparatus like that shown in Fig. 9-13 is constructed so that $L = 50$ cm. CO_2 is placed in the lower tube and air in the upper tube. Both gases are at standard atmospheric pressure and $77°F$. The diffusion coefficient for CO_2 in air is $0.164 \text{ cm}^2/\text{sec}$. Calculate the mole fraction of CO_2 in each tube for (1) 5 min and (2) 2 hr after the connection between the tubes is opened.

Solution   The mole fractions are given by

$$x_{1A} = \frac{N_{1A}}{N_{1_0}} \qquad x_{1B} = \frac{N_{1B}}{N_{1_0}}$$

where N_{1_0} is the initial number of moles in each tube and equal to $N_{1A} + N_{1B}$. Gas 1 is the CO_2 in this case. From Eq. (9-20),

$$F = x_{1A} - x_{1B} \tag{a}$$

We calculate the optimum time as

$$t_{opt} = \frac{(4)(50)^2}{\pi^2(0.164)} = 6,180 \text{ sec}$$

The 5-min condition is far short of this so that the series solution of Eq. (9-20) must be used to find F. Thus, with $t = 300$ sec,

$$F = \frac{8}{\pi^2} [e^{-300/6,180} + \tfrac{1}{9}e^{-(9)(300)/6,180} + \tfrac{1}{25}e^{-(25)(300)/6,180} + \cdots]$$

$$= \frac{8}{\pi^2}(0.9526 + 0.0718 + 0.0119 + \cdots)$$

$$= 0.8400$$

Using Eq. (a) and

$$x_{1A} + x_{1B} = 1.0 \tag{b}$$

gives $x_{1A} = 0.92$ and $x_{1B} = 0.08$.

For the 2-hr time $t = (2)(3,600) = 7,200$ sec, and only the first term of the series need be used. Thus, from Eq. (9-23),

$$F = \frac{8}{\pi^2} e^{-7,200/6,180} = 0.2525$$

Again, using Eqs. (a) and (b) gives

$$x_{1A} = 0.6262 \quad \text{and} \quad x_{1B} = 0.3738$$

9-6 CALORIMETRY

The subject of calorimetry is concerned with a determination of energy quantities. These quantities may be classified in terms of the thermodynamic properties of a system, such as enthalpy, internal energy, specific heat, and heating value, or in terms of energy flow that results from a transport of mass across the boundaries of the thermodynamic system. Calorimetry is a very broad subject covering a majority of the experimental determinations that are used for measurement of thermodynamic properties. We shall give an example of calorimetry which shows the overall features of energy and mass balances and their importance; this is the flow calorimeter which is used for the measurement of heating values of gaseous or liquid fuels.

The actual device that is usually used for the experiment is called the Junkers calorimeter. A schematic of the calorimeter is given in Fig. 9-14. The gaseous fuel is burned inside the calorimeter where it gives up heat to the cooling water. The water flow rate is determined by weighing, and the inlet and outlet water temperatures are measured with precision mercury-in-glass thermometers as shown. The products of combustion are cooled to a sufficiently low temperature so that water vapor is condensed. This condensate is collected in the graduated flask as shown. The gas flow rate is usually measured with a positive-displacement flowmeter. A detailed description of the calorimeter is given by Shoop and Tuve [5], and the interested reader should consult their discussion for procedural information. We shall discuss the system in a general way and indicate an analysis which may be performed to determine the heating value of the fuel.

The flow schematic for the flow calorimeter is shown in Fig. 9-15. For convenience, all streams entering the device are designated with the subscript 1, while all streams leaving the device are designated with the subscript 2. The fuel and air are burned inside the calorimeter, and the major portion of the heat of combustion is removed by the cooling water.

The following experimental measurements are made: the inlet and exit cooling water temperatures T_{w_1} and T_{w_2}, the mass flow rate of fuel $\dot{m}_f$, the mass flow rate of cooling water $\dot{m}_w$, the condensate temperature T_{c_2}, the entering fuel and air temperatures T_{f_1} and T_{a_1}, and the relative humidity of the inlet air ϕ_1. In addition, an analysis of the products of combustion is made to determine the oxygen, carbon dioxide, and carbon monoxide content. From all these data, mass and energy balances may be made to determine the heating value of the fuel. Let us consider the simple case of the combustion of methane CH_4. We write the chemical balance equation as

$$a CH_4 + x O_2 + 3.76x N_2 + b H_2O$$
$$\rightarrow d CO_2 + e CO + f O_2 + g H_2O + 3.76x N_2 \quad (9\text{-}25)$$

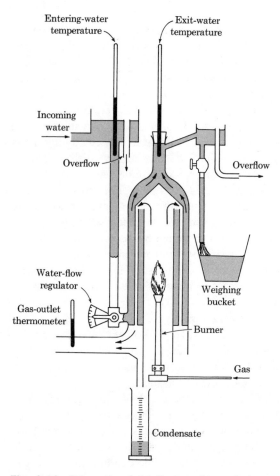

Fig. 9-14 Schematic of the Junkers flow calorimeter.

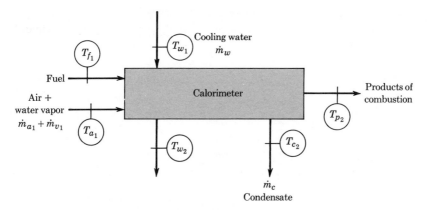

Fig. 9-15   Flow schematic for calorimeter of Fig. 9-14.

An analysis of the products gives the values of d, e, and f for 100 moles of dry products.   The mass balance then becomes

Carbon

$$a = d + e \tag{9-26}$$

Hydrogen

$$4a + 2b = 2g \tag{9-27}$$

Oxygen

$$2x + b = 2d + e + 2f + g \tag{9-28}$$

Equation (9-26) permits an immediate determination of a.   The relative humidity of the incoming air ϕ_1 gives a relationship for b as

$$\frac{b}{4.76x} = \frac{p_{g_1}\phi_1}{p_1 - \phi_1 p_{g_1}} \tag{9-29}$$

where p_{g_1} is the saturation pressure of water vapor at the inlet air temperature and p_1 is the total pressure of the incoming air–water-vapor mixture.   We thus have three unknowns (b, g, and x) and three equations [(9-27) to (9-29)] which may be solved for the unknowns.   With the mass balance now complete an energy balance may be made to determine the chemical energy content of the incoming fuel.   We observe that

$$\dot{m}_{f_1} = 16a \qquad \dot{m}_{a_1} = 32x + (28)(3.76x)$$
$$\dot{m}_{v_1} = 18b$$

The water vapor in the products of combustion is calculated from

$$\dot{m}_{v_2} = 18g - \dot{m}_c \tag{9-30}$$

The energy balance becomes

$$\dot{m}_{f_1}(h_{f_1} + E_{f_1}) + \dot{m}_{a_1}h_{a_1} + \dot{m}_{v_1}h_{v_1} - \dot{m}_{p_2}h_{p_2} - \dot{m}_{c_2}h_{c_2}$$
$$= \dot{m}_w(h_{w_2} - h_{w_1}) \tag{9-31}$$

where E_{f_1} is designated as the chemical energy of the fuel. The energy term for the products $\dot{m}_p h_{p_2}$ is calculated from the energies of the individual constituents. Thus,

$$\dot{m}_p h_{p_2} = \dot{m}_{CO_2}h_{CO_2} + \dot{m}_{CO}h_{CO} + \dot{m}_{O_2}h_{O_2} + \dot{m}_{N_2}h_{N_2} + \dot{m}_{v_2}h_{v_2} \tag{9-32}$$

The mass flow rate of these constituents is determined from the balance conditions in Eqs. (9-27) to (9-29).

The purpose of the above discussion is to illustrate the number of experimental measurements that must be made in even a simple calorimetric determination like the one shown and the fact that all these measurements influence the accuracy of the final result. Of course, the calculated value of heating value (or chemical energy content) will be more sensitive to some measurements than others. For details on the measurement techniques the reader should consult Ref. [5]. It may be noted that standardized procedures are available for this particular test, which make it unnecessary to perform all of the analysis indicated above; nevertheless, this analysis indicates the theory behind the measurement process.

The *bomb calorimeter* is another device that is frequently used for heating-value determinations in solid and liquid fuels. In contrast to the flow calorimeter described above, the measurements are made under constant-volume, nonflow conditions as shown in Fig. 9-16. A measured sample of the fuel is enclosed in a metal container, which is subsequently charged with oxygen at high pressure. The bomb is then placed in a container of water and the fuel ignited through external electric connections. The temperature of the water is measured as a function of time after the ignition process; and from a knowledge of the mass of water in the system, mass and specific heat of the container, and transient heating and cooling curves, the energy release during combustion may be determined. A motor-driven stirrer ensures uniformity in temperature of the water surrounding the bomb. In some circumstances external heating may be supplied to the jacket water to maintain a uniform temperature, while in other instances the jacket may be left empty in order to maintain nearly adiabatic conditions on the inner water container. A compensation for the heat lost to the environment may be

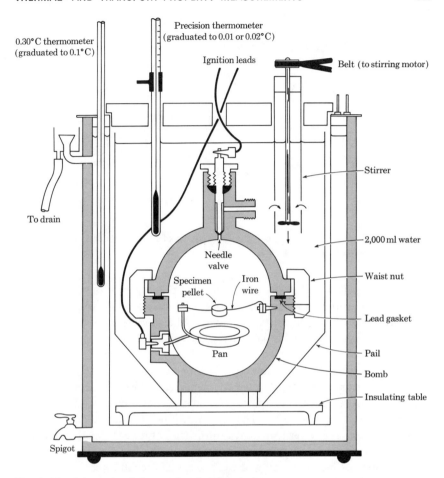

Fig. 9-16 Schematic of the nonflow bomb calorimeter.

made through an analysis of the transient heating and cooling curve. Doolittle [1] gives detailed procedures for use of the bomb calorimeter, and the interested reader should consult this reference for more information.

9-7 CONVECTION HEAT-TRANSFER MEASUREMENTS

Determination of convection heat-transfer coefficients cover a very broad range of experimental activities. In this section we shall illustrate only two simple experimental setups: one for a forced-convection system and one for a free-convection system. The interested reader should consult Ref. [2] for more information.

Consider the experimental setup shown in Fig. 9-17. It is desired to

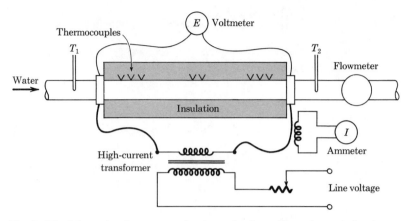

Fig. 9-17 Schematic of apparatus for determination of forced-convection heat-transfer coefficients in smooth tubes.

obtain convection heat-transfer coefficients for the flow of water in a smooth tube. Heat is supplied to the tube by electric heating as shown. The tube is usually made of a high-resistance material like stainless steel to reduce the electric current necessary for heating. Thermocouples are spot-welded or cemented to the outside surface of the tube to measure the wall temperature. Either thermocouples or thermometers are inserted in the flow to measure the water temperature at inlet and outlet to the heated section. A voltmeter and ammeter measure the power input to the tube, while some type of flowmeter measures the water-flow rate. The electrically heated tube delivers a constant heat flux to the water [constant Btu/(hr)(ft²) of tube surface] so that it is reasonable to assume a straight-line variation of water bulk temperature from inlet to outlet. Thus the wall and bulk temperatures are known along the length of the tube, and the heat-transfer coefficient may be calculated at any axial location from

$$q = hA(T_w - T_B) \tag{9-33}$$

where A is the total *inside-heated* surface area of the tube, T_w and T_B are the wall and bulk temperatures at the particular location, and q is the total heat-transfer rate given by

$$q = EI \tag{9-34}$$

E and I are the voltage and current impressed on the test section. The heated surface area is

$$A = \pi \, d_i L \tag{9-35}$$

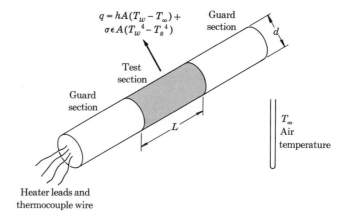

Fig. 9-18 Schematic of heated-cylinder arrangement for determination of free-convection heat-transfer coefficients.

The outside surface of the tube must be covered with insulation to ensure that all the electric heating is dissipated in the water.

A simple experimental setup for determination of natural-convection heat-transfer coefficients on a horizontal cylinder is shown in Fig. 9-18. The heated horizontal cylinder consists of three sections. The center section is the test element, while the two end sections are guard heaters that eliminate end losses from the center section. All three sections are electrically heated and instrumented with thermocouples, which measure their respective surface temperatures. The electric current to the guard heaters is adjusted so that their temperature is just equal to the temperature of the center section. At this balance condition all the electric heating of the center section is dissipated to the surrounding air or enclosure. The heat loss from the test element is given by

$$q = hA(T_w - T_\infty) + \sigma\epsilon A(T_w{}^4 - T_s{}^4) \tag{9-36}$$

where h is the free-convection heat-transfer coefficient, T_w is the test section temperature, ϵ is the surface emissivity, T_∞ is the surrounding air temperature, T_s is the temperature of the surrounding room or enclosure, and A is the surface area given by

$$A = \pi \, dL \tag{9-37}$$

The heat-transfer rate q is determined from measurement of the electric-power input to the test section. To determine the value of h from Eq. (9-36), the surface emissivity must be known. In practice, the test-section surface is usually nickel- or chrome-plated so that the radiation loss is small compared with the convection loss.

Example 9-5 A 1×1 ft square metal plate is used for a determination of free-convection heat-transfer coefficients. The plate is placed in a vertical position and exposed to room air at 70°F. The plate is electrically heated to a uniform surface temperature of 120°F, and the heating rate is measured as 15.0 ± 0.2 watts. The emissivity of the surface is estimated as 0.07 ± 0.02. Determine the nominal value and uncertainty of the heat-transfer coefficient, assuming that the temperature measurements are exact. Assume that the effective radiation temperature of the surroundings is 70°F.

Solution The nominal value of the heat-transfer coefficient is obtained from Eq. (9-36):

$$\frac{q}{A} = h(T_w - T_\infty) + \sigma\epsilon(T_w{}^4 - T_s{}^4) \tag{a}$$

where $T_w = 120°F \qquad T_\infty = T_s = 70°F$

$$\epsilon = 0.07 \qquad \frac{q}{A} = (15)(3.413) \text{ Btu/(hr)(ft}^2)$$

The resulting value of h is

$$h = 0.942 \text{ Btu/(hr)(ft}^2)(°F)$$

From Eq. (a) the explicit relation for h is

$$h = \frac{1}{T_w - T_\infty}\left[\frac{q}{A} - \sigma\epsilon(T_w{}^4 - T_s{}^4)\right] \tag{b}$$

The uncertainty may be obtained from this expression using Eq. (3-2). We have

$$\frac{\partial h}{\partial(q/A)} = \frac{1}{T_w - T_\infty} \qquad w_{q/A} = 0.2 \text{ watt/ft}^2$$

$$= 0.6826 \text{ Btu/(hr)(ft}^2)$$

$$\frac{\partial h}{\partial\epsilon} = \frac{-\sigma(T_w{}^4 - T_s{}^4)}{T_w - T_\infty} \qquad w_\epsilon = 0.02$$

The uncertainty in h is thus

$$w_h = \left\{\left(\frac{1}{50}\right)^2(0.6826)^2 + \left[\frac{(0.1714 \times 10^{-8})(580^4 - 530^4)}{50}\right]^2(0.02)^2\right\}^{\frac{1}{2}}$$

$$= 0.0271 \text{ Btu/(hr)(ft}^2)(°F)$$

or

$$\frac{w_h}{h} = 0.0288 = 2.88\%$$

9-8 HUMIDITY MEASUREMENTS

The water-vapor content of air is an important parameter in many processes. Proper control of critical operations with fabrics, papers, and chemicals frequently hinges upon a suitable control of the humidity of the surrounding environment. We shall discuss four basic techniques for measurement of the water-vapor content of air. First, let us consider some definitions. The *specific humidity*, or *humidity ratio*, is the mass of water vapor per unit mass of dry air. The *dry-bulb temperature* is the temperature of the air–water-vapor mixture as measured by a thermometer exposed to the mixture. The *wet-bulb temperature* is the temperature indicated by a thermometer covered with a wicklike material saturated with liquid after the arrangement has been allowed to reach evaporation equilibrium with the mixture, as indicated in Fig. 9-19. The *dew point* of the mixture is the temperature at which vapor starts to condense when the mixture is cooled at constant pressure.

The *relative humidity* ϕ is defined as the ratio of the actual mass of vapor to the mass of vapor required to produce a saturated mixture at the same temperature. If the vapor behaves like an ideal gas,

$$\phi = \frac{m_v}{m_{\text{sat}}} = \frac{p_v V / R_v T}{p_g V / R_v T} = \frac{p_v}{p_g} \tag{9-38}$$

where p_v is the actual partial pressure of the vapor and p_g is the saturation pressure of the vapor at the temperature of the mixture. The specific humidity is

$$\omega = \frac{m_v}{m_a} \tag{9-39}$$

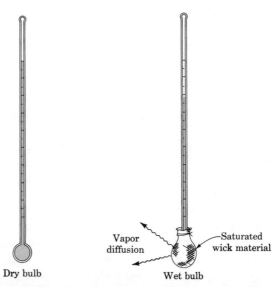

Vapor
diffusion

Saturated
wick material

Dry bulb

Wet bulb

Fig. 9-19 Measurement of dry- and wet-bulb temperatures.

which, for ideal-gas behavior, becomes

$$\omega = 0.622 \frac{p_v}{p_a} \tag{9-40}$$

where p_a is the partial pressure of the air.

The most fundamental method for measuring humidity is a gravimetric procedure, which is used by the National Bureau of Standards for calibration purposes. A sample of the air–water-vapor mixture is exposed to suitable chemicals until the water is absorbed. The chemicals are then weighed to determine the amount of vapor absorbed. Uncertainties as low as 0.1 percent can be obtained with this method.

There is a definite analytical relationship between the dry-bulb, wet-bulb, and dew-point temperatures of a mixture and its humidity. A determination of any two of these temperatures may be used to calculate the humidity. The classic method used for humidity determination in large open spaces is a measurement of the dry- and wet-bulb temperatures with a sling psychrometer. Both thermometers are rotated with a speed of about 1,000 ft/min and the temperatures recorded. The vapor pressure of the mixture may then be calculated with Carrier's equation:

$$p_v = p_{g_w} - \frac{(p - p_{g_w})(T_{DB} - T_{WB})}{2{,}800 - T_{WB}} \tag{9-41}$$

where p_v = actual vapor pressure, psia

p_{g_w} = saturation pressure corresponding to wet-bulb temperature, psia

p = total pressure of mixture, psia

T_{DB} = dry-bulk temperature

T_{WB} = wet-bulk temperature

The specific or relative humidities are then calculated from Eq. (9-38) or (9-40).

The sling psychrometer is not particularly suitable to automation or continuous-recording requirements. For such applications a Dunmore-type electrical humidity transducer is used. This transducer consists of a resistance element, which is constructed by winding dual noble-metal elements on a plastic form with a carefully controlled spacing between them. The windings are then coated with a lithium-chloride solution, which forms a conducting path between the wires. The resistance of this coating is a very strong function of relative humidity and thus may be used to sense the humidity. An ac bridge circuit, as discussed in Chap. 4, is normally used for sensing the resistance of the transducer. Because of wide nonlinear variation of the resistance of the coating with humidity, a single transducer is normally suitable for use only over a narrow range of humidities. If measurements over a wide range of humidities are desired, multiple sensors can be employed, each with a coating suitable for some segment of the humidity spectrum. In general, a single transducer is not used over much more than a 10 percent relative-humidity range. Accuracies for the device are about 1.5 percent relative humidity and may be used over an operating range of -40 to $+150°F$.

The humidity of a mixture may also be determined with a measurement of dew-point temperature. For this purpose the conventional technique is to cool a highly polished metal surface (mirror) until the first traces of fogging or condensation appear when the surface is in contact with the mixture. The temperature of the surface for the onset of fogging is then the dew point. The sensitivity of the measurement process may be improved by using a photocell to detect a beam of light that is reflected from the surface. The intensity of the reflected beam will decrease at the onset of fogging, and this point may be used to mark the temperature signal from the mirror as the dew point. For continuous monitoring of dew point, the mirror-photocell technique may be modified by connecting the mirror to a low-temperature source like a dry-ice–acetone bath and simultaneously to an electric heating element. An appropriate electronic circuit senses the output of the photocell and causes the electric heater to be actuated when the output drops as a result of fogging. The heater then remains on until the mirror surface is clear and the photocell output rises, thereby reducing the heater current. Basically, the system just described is a control system to maintain the mirror at the dew-point temperature regardless of changes in environment. A thermocouple attached to the mirror may then be used to continuously monitor the dew point. A summary of humidity-measurement techniques is given in Refs. [9] to [11]. The thermodynamics of air–water-vapor mixtures is discussed in Ref. [12].

9-9 HEAT-FLUX METERS

There are many applications where a direct measurement of heat flux is desired. In this section we shall discuss a few devices for such measurements. The first type of heat-flux meter we shall consider is the slug sensor shown in Fig. 9-20. A slug of high thermal conductivity is installed in the wall as shown with a surrounding layer of insulation. When subjected to a heat flux at the surface, the temperature of the slug will rise and give an indication of the magnitude of the heat flux. Assuming that the temperature of the slug remains uniform at any instant of time, the transient-energy balance becomes

$$\frac{q}{A} = \frac{mc}{A}\frac{dT_s}{d\tau} + U(T_s - T_w) \tag{9-42}$$

where q/A is the imposed heat flux, m is the mass of the slug, c is the specific

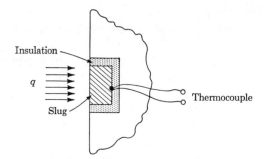

Fig. 9-20 Slug-type heat-flux meter.

heat of the slug, U is a coefficient expressing the conduction loss to the surrounding wall, T_s is the temperature of the slug, and T_w is the surrounding wall temperature. It is clear that a solution of Eq. (9-42) will give the slug temperature as a function of the heat flux and the properties of the slug. Thus, a measurement of T_s may be taken as an indication of the heat flux.

Some problems are encountered in using slug sensors. First, for very high heat fluxes the temperature of the slug may not remain uniform and a more cumbersome analysis must be employed to calculate the heat flux from the temperature measurements. Second, the presence of the sensor alters the temperature profile in the wall; in particular, the temperature at the sensor location is higher than it might have been otherwise (for slug conductivity greater than wall conductivity), thereby influencing the net heat flux at the point of measurement, whether it occurs by convection, or radiation, or a combination of both. Third, the slug sensor is clearly adapted only to transient measurements. As soon as the wall reaches equilibrium, the temperature of the slug will no longer be indicative of the heat flux. Additional information on the response of slug-type sensors is available in Refs. [13] and [14].

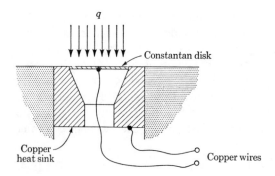

Fig. 9-21 Gardon foil-type heat-flux meter according to Ref. [15].

The Gardon [15] heat-flux meter is depicted in Fig. 9-21. A thin constantan disk is mounted to a copper heat sink, as shown, with the heat sink installed in the wall material. A small copper wire is then attached to the center of the disk and, along with a wire connected to the copper sink, forms the two wires of a copper-constantan thermocouple to measure the temperature difference between the center of the disk and the sides. A heat-flux incident on the disk is absorbed and conducted radially outward, thereby creating the temperature difference that is sensed by the thermocouple. The only losses in the gage are the radiation losses from the back side of the disk to the copper heat sink, but by careful calibration this may be taken into account with little difficulty. Heat fluxes over a range of 5×10^4 to 10^6 may be measured with the device. Time constants of the order of 0.1 sec are experienced. Increased sensitivity may be obtained by using a copper disk along with a positively doped bismuth-telluride center connection [17].

A very versatile heat-flux meter has been described by Hager [16] and is marketed by the RdF Corp., Hudson, N.H. The meter operates on the principle of measuring the temperature drop across a thin insulating material of about 0.0003 in. thickness. The temperature drop is directly proportional to the heat conducted through the material in accordance with Fourier's law [Eq. (9-1)]. The unique feature of these meters is the butt-bonding process, which allows a distortion-free joining of copper-constantan on both sides of the material. The detailed construction is indicated in Fig. 9-22 with a number of junctions connected in series to produce a thermopile. As many as 40 junctions may be produced in this manner within a space of about 1 in.^{2}.

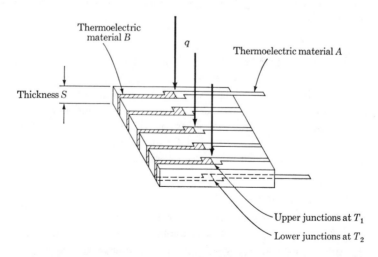

Fig. 9-22 Detail of heat-flux meter operating on conduction principle. (*Courtesy RdF Corp., Hudson, N.H.*)

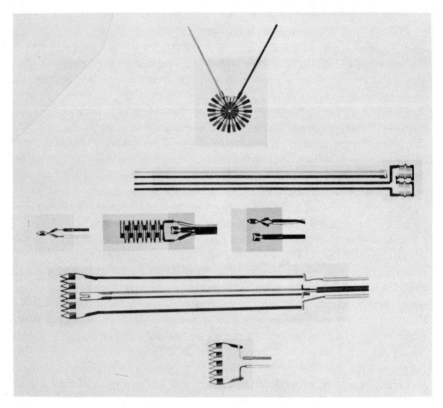

Fig. 9-23 Photographs of assembled microfoil heat-flux meters. (*Courtesy RdF Corp., Hudson, N.H.*)

The junctions are covered with a protective coating, but the total thickness of the heat-flux meter is never more than 0.012 in. The maximum heat flux that may be accommodated with the meter is about 2×10^5 Btu/(hr)(ft²) at a maximum operating temperature of 500°F. Sensitivity ranges from 0.07 to 27 (mv)(ft²)sec/Btu, depending on the number of junctions employed. Time constants range between 0.02 and 0.4 sec. Figure 9-23 shows some photographs of typical meter configurations.

REVIEW QUESTIONS

1. Write the defining equation for thermal conductivity.

2. What is a guarded hot-plate apparatus? Under what conditions is it applied?

3. How are thermal conductivities of metals measured?

4. What are some of the considerations that must be applied for gas and liquid thermal conductivity that are not important in thermal-conductivity measurements for solids?

5. Write the defining equation for viscosity.

6. What is the operating principle of the Saybolt viscosimeter?

7. Write the defining equation for diffusion coefficient.

8. How does a Loschmidt apparatus function?

9. Describe the flow calorimeter.

10. Describe the bomb calorimeter.

11. Write the defining equation for the heat-transfer coefficient.

12. How does a sling psychrometer measure relative humidity?

13. Describe a technique for measuring dew point.

14. What is the operating principle of the slug-type heat-flux meter?

15. What is a Gardon heat-flux meter?

PROBLEMS

9-1. The following data are taken from Ref. [6] on an apparatus like that shown in Fig. 9-3. The measurements were made on a lead sample whose thermal conductivity is taken to be 0.352 watt/cm-°C at 0°C.

Thermocouple positions	Temperature, °C	Distance between adjacent thermocouples, cm
1	277.3	3.14
2	231.6	3.14
3	186.5	3.14
4	143.0	3.15
5	100.3	
6	78.6	3.16
7	37.3	

Using these data construct a graph of the thermal conductivity of lead versus temperature.

9-2. The apparatus of Fig. 9-3 is used for measurement of the thermal conductivity of an unknown metal. The lengths of the unknown and known bars are the same, and five thermocouples are equally spaced on each bar. From the temperature and heat flux measurement the thermal conductivity of the unknown is to be established as a function of temperature. Design the apparatus to measure the thermal conductivity of a material, with $k \sim 100$ Btu/(hr)(ft)(°F). Consider the influence of thermocouple spacing, dimensions of the bars, heat flux measurement, and temperature measurement on accuracy. Take the thermal conductivity of the known material as exactly 20 Btu/(hr)(ft)(°F).

9-3. A concentric-cylinder device like that in Fig. 9-5 is used for a measurement of the thermal conductivity of water at 100°F. Using the dimensions given in the text, discuss the influence of the heat flux and temperature measurements on the accuracy of the thermal-conductivity determination.

9-4. The viscosity of water is 1.65 lb_m/hr-ft at 100°F.   A small capillary tube 50 $\pm$ 0.01 ft long is to be used to check this value.   Calculate the maximum allowable flow rate for an inside tube diameter of 0.100 $\pm$ 0.001 in.   Calculate the pressure drop for this flow rate, and estimate the allowable uncertainty in the pressure drop in order that the uncertainty in the viscosity does not exceed 5 percent.   Assume that the flow rate is measured within $\pm$ 0.01 lb_m/hr.

9-5. The rotating concentric-cylinder apparatus of Fig. 9-9 is used to measure the viscosity of water at 100°F.   The inner cylinder radius is 1.50 $\pm$ 0.001 in., and its length is 4.00 $\pm$ 0.002 in.   The outer cylinder is rotated at a speed of 1,800 $\pm$ 2 rpm. The outer cylinder diameter is 1.60 $\pm$ 0.001 in., and the gap spacing at the bottom of the cylinder is 0.500 $\pm$ 0.005 in.   Calculate the nominal torque which will be measured, and estimate the allowable uncertainty in this measurement in order that the overall uncertainty in the viscosity does not exceed 5 percent.   Repeat the calculation for glycerine at 100°F.

9-6. The tolerance limits on the radii of the inner and outer cylinders of Fig. 9-9 exert a strong influence on the accuracy of the viscosity measurement.   Establish a relation expressing the percent uncertainty in the viscosity measurement as a function of the uncertainty in these radius measurements.   Assume that the torque, angular velocity, and bottom-spacing measurements are exact for this calculation.   Assume that the uncertainty (tolerance) is the same for both cylinders.

9-7. The viscosity of a gas is to be determined by measuring the flow rate and pressure drop through a capillary tube.   The experimental variables which are measured are:

Inlet pressure p_1
Pressure differential $\Delta p = p_1 - p_2$
Temperature T
Mass flow rate $\dot{m}$
Tube diameter $d = 2r$

Derive an expression for the percent uncertainty in the viscosity measurement in terms of the uncertainties in these five primary measurements.   In view of the information in Chaps. 5 to 8, discuss the relative influence of each of the measurements, i.e., which primary measurement will probably have the most effect on the final estimated uncertainty in the viscosity.

9-8. The viscosity of an oil having a density of 51.9 lb_m/ft^3 is measured as 200 Saybolt sec.   Determine the dynamic viscosity in units of centipoise and lb_m/hr-ft.

9-9. A Loschmidt apparatus is used to measure the diffusion coefficient for benzene in air.   At 25°C the diffusion coefficient is 0.088 cm^2/sec for a pressure of 1 standard atmosphere.   Both tubes have a length of 60 cm.   Calculate the mole fraction of benzene in each tube 10 min and 3 hr after the connection between the tubes is opened.

9-10. A Loschmidt apparatus with $L = 50$ cm is used to measure the diffusion coefficient for CO_2 in air.   After a time of 6.52 min the mole fraction of CO_2 in the upper tube is 0.0912.   Calculate the value of the diffusion coefficient.

9-11. For $t = t_{opt}$, calculate the relative magnitudes of the first four terms in the series expansion of Eq. (9-20).   Repeat for $t = \frac{1}{2}t_{opt}$ and $t = 2t_{opt}$.

9-12. Obtain a simplified expression for the heating value of methane as measured with a flow calorimeter with the following conditions:

$$T_{a_1} = T_{f_1} = T_{p_2} = T_{c_2} \qquad \phi_1 = 0$$

The incoming gaseous fuel is saturated with water vapor.

9-13. Design an apparatus like that shown in Fig. 9-17 to measure forced-convection heat-transfer coefficients for water at about 200°F in a range of Reynolds numbers from 50,000 to 100,000 based on tube diameter; i.e.,

$$\text{Re}_d = \frac{\rho u_m d}{\mu}$$

where u_m is the mean flow velocity in the tube. Be sure to specify the accuracy and range of all instruments required and estimate the uncertainty in the calculated values of the heat-transfer coefficients.

9-14. Design an apparatus like that shown in Fig. 9-18 to measure free-convection heat-transfer coefficients for air at atmospheric pressure and ambient temperature of 70°F. The apparatus is to produce data over a range of Grashof numbers from 10^5 to 10^7 where

$$\text{Grashof number} = \text{Gr} = \frac{\rho^2 g \beta (T_w - T_\infty) d^3}{\mu^2}$$

where β = volume coefficient of expansion for air
d = tube diameter
T_w = tube surface temperature
T_∞ = ambient air temperature

The properties of the air are evaluated at the film temperature defined by (see Table 8-5)

$$T_f = \frac{T_w + T_\infty}{2}$$

The Grashof number is dimensionless when a consistent set of units is used. Be sure to specify the accuracy and range of all instruments used and estimate the uncertainty in the calculated values of the heat-transfer coefficient.

9-15. A copper sphere 1 in. in diameter is used as a slug-type heat-flux meter. A thermocouple is attached to the interior and the transient temperature recorded when the sphere is suddenly exposed to a convection environment. Plot the time response of the system for sudden exposure to an environment with $h = 100$ Btu/(hr)(ft²)(°F). Assume the sphere is initially at 200°F and the environment temperature remains constant at 100°F. What is the time constant of this system?

9-16. A heat-flux meter is to be constructed of a thin layer of glasslike substance having a thermal conductivity of 0.4 Btu/(hr)(ft)(°F). The temperature drop across the layer will be measured with a thermopile consisting of 10 copper-constantan junctions. The glass thickness is 0.0008 in. Calculate the voltage output of the thermopile when a heat flux of 10^5 Btu/(hr)(ft²)(°F) is imposed on the layer.

9-17. The dry- and wet-bulb temperatures for a quantity of air at 14.7 psia are measured as 95 and 75°F, respectively. Calculate the relative humidity and the dew point of the air–water-vapor mixture.

9-18. An air–water-vapor mixture is contained at a total pressure of 30 psia. The dry-bulb temperature is 90°F. What will be the wet-bulb temperature if the relative humidity is 50 percent?

9-19. The thermal conductivity of an insulating material is to be determined using a guarded hot-plate apparatus. A 1-ft² sample is used, and the approximate thermal conductivity is about 1.0 Btu/(hr)(ft)(°F). A differential-thermocouple arrangement is used to measure the temperature drop across the sample with an uncertainty of ±0.5°F.

A heat input of 5 kw $\pm$ 1 percent is supplied with a resulting temperature drop of 102°F. The sample thickness is 0.08 in. What is the thermal conductivity and its uncertainty, assuming no heat losses and that the sample dimensions are known exactly?

REFERENCES

1. Doolittle, J. S.: "Mechanical Engineering Laboratory," McGraw-Hill Book Company, New York, 1957.
2. Jakob, M.: "Heat Transfer," vol. 1, John Wiley & Sons, Inc., New York, 1949.
3. Kaye, G. W. C., and W. F. Higgins: The Thermal Conductivities of Certain Liquids, *Proc. Roy. Soc. London*, vol. 117, no. 459, 1928.
4. Keyes, F. G., and D. J. Sandell: New Measurements of Heat Conductivity of Steam and Nitrogen, *Trans. ASME*, vol. 72, p. 768, 1950.
5. Shoop, C. F., and G. L. Tuve: "Mechanical Engineering Practice," 5th ed., McGraw-Hill Book Company, New York, 1956.
6. Van Dusen, M. S., and S. M. Shelton: Apparatus for Measuring Thermal Conductivity of Metals up to 600°C, *J. Res. Natl. Bur. Std.*, vol. 12, no. 429, 1934.
7. Vines, R. G.: Measurements of Thermal Conductivities of Gases at High Temperatures, *Trans. ASME*, vol. 82C, p. 48, 1960.
8. Leidenfrost, W.: An Attempt to Measure the Thermal Conductivity of Liquids, Gases and Vapors with a High Degree of Accuracy over Wide Ranges of Temperature and Pressure, *Intern. J. Heat and Mass Transfer*, vol. 7, no. 4, pp. 447–476, 1964.
9. Fraade, D. J.: Measuring Moisture in Gases, *Instr. Control Systems*, p. 100, April, 1963.
10. Moisture and Humidity, *Instr. Control Systems*, p. 121, October, 1964.
11. Amdur, E. J.: Humidity Sensors, *Instr. Control Systems*, p. 93, June, 1963.
12. Holman, J. P.: "Thermodynamics," McGraw-Hill Book Company, New York, 1969.
13. Kirchhoff, R. H.: Calorimetric Heating-rate Probe for Maximum Response Time Interval, *AIAA J.*, p. 966, May, 1964.
14. Hornbaker, D. R., and D. L. Rall: Thermal Perturbations Caused by Heat Flux Transducers and Their Effect on the Accuracy of Heating Rate Measurements, *ISA Trans.*, p. 100, April, 1963.
15. Gardon, R.: An Instrument for the Direct Measurement of Intense Thermal Radiation, *Rev. Sci. Instr.*, p. 366, May, 1953.
16. Hager, N. E., Jr.: Thin Foil Heat Meter, *Rev. Sci. Instr.*, November, 1965.
17. Carey, R. M.: "Experimental Test of a Heat Flow Meter," M.S. thesis, Southern Methodist University, Dallas, Tex., July, 1963.

10
Force, Torque, and Strain Measurements

10-1 INTRODUCTION

Some types of force measurements applicable to pressure-sensing devices have been discussed in Chap. 6. We now wish to consider other methods for measuring forces and torques and relate these methods to basic strain measurements and experimental stress analysis.

Force is represented mathematically as a vector with a point of application. Physically, it is a directed push or pull. According to Newton's second law of motion, as written for a particle of constant mass, force is proportional to the product of mass and acceleration. Thus,

$$F = \frac{1}{g_c}\, ma \tag{10-1}$$

where the $1/g_c$ term is the proportionality constant. When force is expressed in lb_f, mass in lb_m, and acceleration in ft/sec^2, g_c has the value of 32.1739 lb_m-ft/lb_f-sec^2 and is numerically equal to the acceleration of gravity at sea level. The weight of a body is the force exerted on the body by the

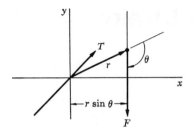

Fig. 10-1   Vector representation for torque.

acceleration of gravity at sea level. The weight of a body is the force exerted on the body by the acceleration of gravity so that

$$F = W = \frac{mg}{g_c} \tag{10-2}$$

At sea level, the weight in lb_f units is numerically equal to the mass in lb_m units.

Torque is represented as a moment vector formed by the cross product of a force and radius vector. In Fig. 10-1 the torque that the force F exerts about the point 0 is given mathematically as

$$\mathbf{T} = \mathbf{r} \times \mathbf{F} = |r|\,|F| \sin \theta \tag{10-3}$$

Mass represents a quantity of matter and is an inertial property related to force through Eq. (10-1). As such, mass is not a physical property that is measured directly; it is determined through a force measurement or by comparison with mass standards. Even with a comparison of two masses, as with a simple balance, a force equivalency is still involved.

In this chapter we shall be concerned with force measurements as related to mechanical systems. Forces arising from electrical and magnetic effects are not considered.

10-2 MASS BALANCE MEASUREMENTS

Consider the schematic of an analytical balance as shown in Fig. 10-2. The balance arm rotates about the fulcrum at point 0 (usually a knife edge) and is shown in an unbalanced position as indicated by the angle ϕ. Point G represents the center of gravity of the arm, and d_G is the distance from 0 to this point. W_B is the weight of the balance arm and pointer. When $W_1 = W_2$, ϕ will be zero and the weight of the balance arms will not influence the measurements. The sensitivity of the balance is a measure of the angular displacement ϕ per unit unbalance in the two weights W_1 and W_2.

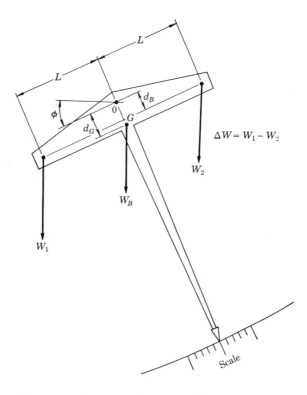

Fig. 10-2 Schematic of analytical balance.

Expressed analytically this relation is

$$S = \frac{\phi}{W_1 - W_2} = \frac{\phi}{\Delta W} \tag{10-4}$$

We now wish to determine the functional dependence of this sensitivity on the physical size and mass of the balance. According to Fig. 10-2 the moment equation for equilibrium is

$$W_1(L \cos \phi - d_B \sin \phi) = W_2(L \cos \phi + d_B \sin \phi) + W_B d_G \sin \phi \tag{10-5}$$

For small deflection angles $\sin \phi \approx \phi$ and $\cos \phi \sim 1.0$ so that Eq. (10-5) becomes

$$W_1(L - d_B\phi) = W_2(L + d_B\phi) + W_B \, d_G\phi$$

or

$$\frac{\phi}{\Delta W} = \frac{L}{(W_1 + W_2)d_B + W_B d_G} \tag{10-6}$$

Near equilibrium, $W_1 \approx W_2$ and S becomes

$$S = \frac{\phi}{\Delta W} = \frac{L}{W_B d_G + 2W d_B} \tag{10-7}$$

where, now, we have used the single symbol W to designate the loading on the balance. If the balance arm is constructed so that $d_B = 0$, we obtain

$$S = \frac{L}{W_B d_G} \tag{10-8}$$

an expression that is independent of the loading of the balance. Precision balances are available that have an accuracy of about 1 part in 10^8. For precision instruments optical methods are usually employed to sense the deflection angle and determine the equilibrium position.

Example 10-1   A balance is constructed with the following dimensions:

$W_B = 50$ g
$d_G = 0.3$ cm
$L = 21.2$ cm
$d_B = 0.01$ cm

The pointer scale is graduated so that readings may be taken within one-quarter of a degree. Estimate the uncertainty due to sensitivity in determining the weight of a mass of 1,000 g. It may be noted that other uncertainties exist in the measurement, such as frictional effects, errors in reading, etc.

Solution   From Eq. (10-7) we have

$$S = \frac{\phi}{\Delta W} = \frac{L}{2W d_B + W_B d_G}$$

$$= \frac{21.2}{(2)(1000)(0.01) + (50)(0.3)} = 0.606 \text{ rad/g}$$

For an uncertainty in ϕ of 0.25° the uncertainty in W is then given by

$$\Delta W = \frac{\phi}{S} = \frac{0.25\pi}{(180)(0.606)} = 0.0072 \text{ g}$$

or about 1 part in 100,000.

When a balance is used for mass determinations, considerable errors may result if corrections are not made for buoyancy forces of the air surrounding the samples. Typically, an unknown mass is balanced with a group of standard brass weights. The forces that the instrument senses are not the weight forces of the unknown mass and brass weights but the weight forces *less* the buoyancy force on each mass. If the measurement is conducted in a vacuum or the unknown brass and mass weights have equal volume, the buoyancy forces will cancel out and there will be no error. Otherwise, the error may be corrected with the following analysis. The two forces exerted on the balance arms will be

$$W_1 = (\rho_u - \rho_a)V_u \qquad (10\text{-}9)$$

$$W_2 = (\rho_s - \rho_a)V_s \qquad (10\text{-}10)$$

where ρ_u = density of the unknown

ρ_s = density of the standard weights

ρ_a = density of the surrounding air

V_u, V_s = volumes of the unknown and standard weights, respectively

The true weights of the unknown and standard weights are

$$W_u = \rho_u V_u \qquad (10\text{-}11)$$

$$W_s = \rho_s V_s \qquad (10\text{-}12)$$

At balance $W_1 = W_2$ and there results

$$W_u = W_s\left(1 + \frac{\rho_a}{\rho_s}\frac{\rho_s - \rho_u}{\rho_u - \rho_a}\right) \qquad (10\text{-}13)$$

Example 10-2 A quantity of a plastic material having a density of about 80 lb_m/ft^3 is weighed on a standard equal-arm balance. Balance conditions are achieved with brass weights totaling 152 g in a room where the ambient air is at 70°F and 14.7 psia. The density of the brass weights may be taken as 530 lb_m/ft^3. Calculate the true weight of the plastic material and the percent error that would result if the balance reading were taken without correction.

Solution The true weight of the brass weights is 152 g. The density of the air may be calculated as

$$\rho_a = \frac{p}{RT} = \frac{(14.7)(144)}{(53.35)(530)} = 0.0749 \ lb_m/ft^3$$

The true weight of the plastic is then calculated from Eq. (10-13) as

$$W_u = 152\left(1 + \frac{0.0749}{530}\frac{530 - 80}{80 - 0.0749}\right)$$

$$= 153.21$$

The error is $1.21/152 \times 100$ or 0.795 percent.

10-3 ELASTIC ELEMENTS FOR FORCE MEASUREMENTS

Elastic elements are frequently employed to furnish an indication of the magnitude of an applied force through a displacement measurement. The simple spring is an example of this type of force-displacement transducer. In this case the force is given by

$$F = ky \qquad\qquad (10\text{-}14)$$

where k is the spring constant and y is the displacement from the equilibrium position. For the simple bar shown in Fig. 10-3 the force is given by

$$F = \frac{AE}{L} y \qquad\qquad (10\text{-}15)$$

where A = cross-sectional area
$\quad L$ = length
$\quad E$ = Young's modulus for the bar material

The deflection of the cantilever beam shown in Fig. 10-4 is related to the loading force by

$$F = \frac{3EI}{L^3} y \qquad\qquad (10\text{-}16)$$

where I is the moment of inertia of the beam. Any one of the three devices mentioned above is suitable for use as a force transducer provided that accurate means are available for indicating the displacements. The differential transformer (Sec. 4-13), for example, may be useful for measurement of these displacements, as well as capacitance and piezoelectric transducers (Secs. 4-14 and 4-15).

Another elastic device frequently employed for force measurements is

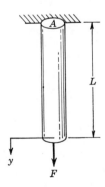

Fig. 10-3 Simple elastic element.

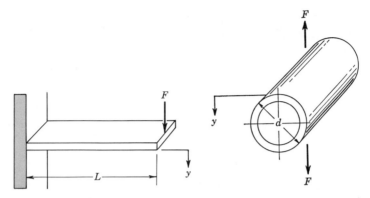

Fig. 10-4 Cantilever elastic element.

Fig. 10-5 Thin-ring elastic element.

the thin ring shown in Fig. 10-5. The force deflection relation for this type of elastic element is

$$F = \frac{16}{\pi/2 - 4/\pi} \frac{EI}{d^3} y \tag{10-17}$$

where d is the outside ring diameter and I is the moment of inertia about the centroidal axis of the ring section. The *proving ring* is a ring transducer that employs a sensitive micrometer for the deflection measurement, as shown in Fig. 10-6. To obtain a precise measurement, one edge of the micrometer is mounted on a vibrating-reed device R, which is plucked to obtain a vibratory motion. The micrometer contact is then moved forward until a noticeable damping of the vibration is observed. Deflection measurements may be made within ± 0.00002 in. with this method. The proving ring is widely used as a calibration standard for large tensile-testing machines.

Example 10-3 A small cantilever beam is constructed of spring steel having $E = 28.3 \times 10^6$ psi. The beam is 0.186 in. wide and 0.035 in. thick, with a length of 1.00 ± 0.001 in. A LVDT is used for the displacement-sensing device, and it is estimated that the uncertainty in the displacement measurement is ± 0.001 in. The uncertainties in the bar dimensions are ± 0.0003 in. Calculate the indicated force and uncertainty due to dimension tolerances when $y = 0.100$ in.

Solution The moment of inertia is calculated with

$$I = \frac{bh^3}{12} \tag{a}$$

where b is the width and h is the thickness. Thus,

$$I = \frac{(0.186)(0.035)^3}{12} = 6.64 \times 10^{-7} \text{ in.}^4$$

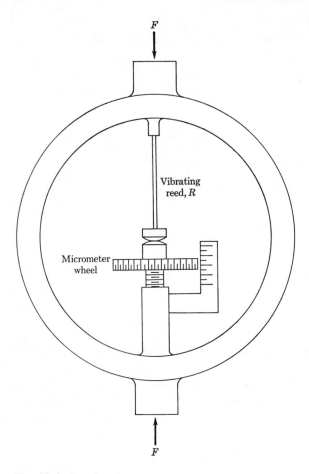

Fig. 10-6   Proving ring.

Equation (10-16) may be rewritten as

$$F = \frac{3Ebh^3}{12L^3} y \tag{b}$$

Using Eq. (3-2) the uncertainty in the force measurement can be written

$$\frac{w_F}{F} = \left[\left(\frac{w_b}{b}\right)^2 + 9\left(\frac{w_h}{h}\right)^2 + 9\left(\frac{w_L}{L}\right)^2 + \left(\frac{w_y}{y}\right)^2 \right]^{\frac{1}{2}} \tag{c}$$

The nominal value of the force becomes

$$F = \frac{(3)(28.3 \times 10^6)(6.64 \times 10^{-7})}{(1.00)^3} 0.100$$

$$= 5.83 \text{ lb}_f$$

From Eq. (c) the uncertainty is

$$\frac{w_F}{F} = \left[\left(\frac{0.0003}{0.186}\right)^2 + 9\left(\frac{0.0003}{0.035}\right)^2 + 9\left(\frac{0.001}{1.00}\right)^2 + \left(\frac{0.001}{0.100}\right)^2\right]^{\frac{1}{2}}$$

$$= 0.0278 \text{ or } 2.78\%$$

In this instance 1.0 percent uncertainty would be present even if the dimensions of the beam were known exactly.

The surface strain (deformation) in elastic elements like those discussed above is, of course, a measure of the deflection from the no-load condition. This surface strain may be measured very readily by the electrical-resistance strain gage to be discussed in subsequent paragraphs. The output of the strain gage may thus be taken as an indication of the impressed force. The main problem with the use of these gages for the force-measurement applications is that a moment may be impressed on the elastic element because of eccentric loading. This would result in an alteration of the basic strain distribution as measured by the strain gage. There are means for compensating for this effect through installation of multiple gages, which are properly interconnected to cancel out the deformation resulting from the impressed moment. The interested reader should consult Refs. [4] and [5] for a summary of these methods.

10-4 TORQUE MEASUREMENTS

Torque, or moment, may be measured by observing the angular deformation of a bar or hollow cylinder, as shown in Fig. 10-7. The moment is given by

$$M = \frac{\pi G(r_0^4 - r_i^4)}{2L} \phi \tag{10-18}$$

where G = shear modulus of elasticity

$\quad r_i$ = inside radius

$\quad r_0$ = outside radius

$\quad L$ = length of the cylinder

$\quad \phi$ = angular deflection

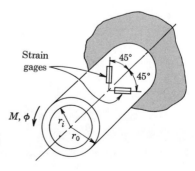

Fig. 10-7 Hollow cylinder as an elastic element for torque measurement.

Strain gages attached at 45° angles as shown will indicate strains of

$$\epsilon_{45°} = \pm \frac{Mr_0}{\pi G(r_0{}^4 - r_i{}^4)} \tag{10-19}$$

Either the deflection or the strain measurement may be taken as an indication of the applied moment. Multiple strain gages may be installed and connected so that any deformation due to axial or transverse load is cancelled out in the final readout circuit.

A rather old device for the measurement of torque and dissipation of power from machines is the Prony brake.† A schematic diagram is shown in Fig. 10-8. Wooden blocks are mounted on a flexible band or rope, which is connected to the arm. Some arrangement is provided to tighten the rope to increase the frictional resistance between the blocks and rotating flywheel of the machine. The torque exerted on the Prony brake is given by

$$T = FL \tag{10-20}$$

The force F may be measured by conventional platform scales or other methods discussed in the previous paragraphs.

The power dissipated in the brake is calculated from

$$P = \frac{2\pi TN}{33,000} \text{ hp} \tag{10-21}$$

where the torque is in ft-lb$_f$ and N is the rotational speed in revolutions per minute.

Various other types of brakes are employed for power measurements on mechanical equipment. The *water brake*, for example, dissipates the output energy through fluid friction between a paddle wheel mounted inside a stationary chamber filled with water. The chamber is freely mounted on

† Named for G. C. F. M. Riche, Baron de Prony (1755–1839), French hydraulic engineer.

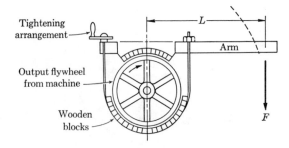

Fig. 10-8 Schematic of a Prony brake.

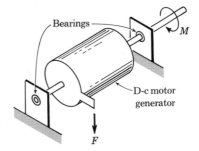

Fig. 10-9 Schematic of cradled dynamometer.

bearings so that the torque transmitted to it can be measured through an appropriate moment arm similar to that used with the Prony brake.

The dc cradled dynamometer is perhaps the most widely used device for power and torque measurements on internal-combustion engines, pumps, small steam turbines, and other mechanical equipment. The basic arrangement of this device is shown in Fig. 10-9. A dc motor generator is mounted on bearings as shown, with a moment arm extending from the body of the motor to a force-measurement device, which is usually a pendulum scale. When the device is connected to a power-producing machine, it acts as a dc generator whose output may be varied by dissipating the power in resistance racks. The torque impressed on the dynamometer is measured with the moment arm and the output power calculated with Eq. (10-21). The dynamometer may also be used as an electric motor to drive some power-absorbing device like a pump. In this case, it furnishes a means for measurement of torque and power input to the machine. Commercial dynamometers are equipped with controls for precise variation of the load and speed of the machine and are available with power ratings as high as 5,000 hp.

10-5 STRESS AND STRAIN

Stress analysis involves a determination of the stress distribution in materials of various shapes and under different loading conditions. Experimental stress analysis is performed by measuring the deformation of the piece under load and inferring from this measurement the local stress which prevails. The measurement of deformation is only one facet of the overall problem, and the analytical work that must be applied to the experimental data in order to determine the local stresses is of equal importance. Our concern in the following sections is with the methods that may be employed for deformation measurements. Some simple analyses of these measurements will be given to illustrate the reasoning necessary to obtain local stress values. For more detailed considerations the reader should consult Refs. [2] and [5] and the periodical publication [8].

Consider the bar shown in Fig. 10-10 subjected to the axial load T. Under no-load conditions the length of the bar is L and the diameter is D.

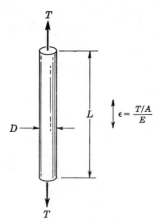

Fig. 10-10   Simple bar in axial strain.

The cross-sectional area of the bar is designated by A.   If the load is applied such that the stress does not exceed the elastic limit of the material, the axial strain is given by

$$\epsilon = \frac{T/A}{E} = \frac{\sigma_a}{E} \tag{10-22}$$

where σ_a is the axial stress and E is Young's modulus for the material.   The unit axial strain ϵ_a is defined by the relation

$$\epsilon_a = \frac{dL}{L} \tag{10-23}$$

i.e., it is the axial deformation per unit length.   Resulting from the deformation in the axial direction is a corresponding deformation in the cross-sectional area of the bar.   The change in area is evidenced by a change in the diameter or, more specifically, by a change in the transverse dimension.   The ratio of the unit strain in the transverse direction to the unit strain in the axial direction is defined as Poisson's ratio and must be determined experimentally for various materials.

$$\mu = -\frac{\epsilon_t}{\epsilon_a} = -\frac{dD/D}{dL/L} \tag{10-24}$$

A typical value for Poisson's ratio for many materials is 0.3.   If the material is in the plastic state, the volume remains constant with the change in strain so that

$$dV = L\,dA + A\,dL = 0$$

or

$$\frac{dA}{A} = -\frac{dL}{L}$$

Expressed in terms of the diameter this relation is

$$2\frac{dD}{D} = -\frac{dL}{L}$$

so that $\mu = 0.5$ under these conditions.

10-6 STRAIN MEASUREMENTS

Let us first consider some basic definitions. Any strain measurement must be made over a finite length of the workpiece. The smaller this length, the more nearly the measurement will approximate the unit strain at a point. The length over which the average strain measurement is taken is called the *base length*. The *deformation sensitivity* is defined as the minimum deformation that can be indicated by the appropriate gage. *Strain sensitivity* is the minimum deformation that can be indicated by the gage per unit base length.

A simple method of strain measurement is to place some type of grid marking on the surface of the workpiece under zero-load conditions and then measure the deformation of this grid when the specimen is subjected to a load. The grid may be scribed on the surface, drawn with a fine ink pen, or photoetched. Rubber threads have also been used to mark the grid. The sensitivity of the grid method depends on the accuracy with which the displacement of the grid lines may be measured. A micrometer microscope is frequently employed for such measurements. An alternate method is to photograph the grid before and after the deformation and make the measurements on the developed photograph. Photographic paper can have appreciable shrinkage, so that glass photographic plates are preferred for such measurements. The grid may also be drawn on a rubber model of the specimen and the local strain for the model related to that which would be present in the actual workpiece. Grid methods are usually applicable to materials and processes having appreciable deformation under load. These methods might be applicable to a study of the strains encountered in sheet-metalforming processes. The grid could be installed on a flat sheet of metal before it is formed. The deformation of the grid after forming gives the designer an indication of the local stresses induced in the material during the forming process.

Brittle coatings offer a convenient means for measuring the local stress in a material. The specimen or workpiece is coated with a special substance having very brittle properties. When the specimen is subjected to a load, small cracks appear in the coating. These cracks appear when the state of tensile stress in the workpiece reaches a certain value and thus may be taken as a direct indication of this local stress. The brittle coatings are valuable for obtaining an overall picture of the stress distribution over the surface of the specimen. They are particularly useful for determination of stresses at stress concentration points that are too small or inconveniently located for installation of electrical resistance or other types of strain gages. In some instances, stress data obtained from brittle-coating tests may be used to plan more precise strain measurements with resistance strain gages. A popular brittle coating is manufactured by the Magnaflux Corporation under the trademark of Stresscoat. Strain sensitivities of this coating range from 400 to 2,000 μin./in. The theory and application of brittle coatings are discussed very completely by Durelli et al. [3], and the interested reader should consult this reference for additional information. We may mention, however, that residual stresses in the coating, differential thermal expansion between the coating and the workpiece, and curing procedure can influence the stress properties of the coating significantly.

10-7 ELECTRICAL-RESISTANCE STRAIN GAGES

The electrical-resistance strain gage is the most widely used device for strain measurement. Its operation is based on the principle that the electrical resistance of a conductor changes when it is subjected to a mechanical deformation. Typically, an electric conductor is bonded to the specimen with an insulating cement under no-load conditions. A load is then applied, which produces a deformation in both the specimen and the resistance element. This deformation is indicated through a measurement of the change in resistance of the element and a calculation procedure which is described below.

Let us now develop the basic relations for the resistance strain gage. The resistance of the conductor is

$$R = \rho \frac{L}{A} \tag{10-25}$$

where L = length
$\quad A$ = cross-sectional area
$\quad \rho$ = resistivity of the material

Differentiating Eq. (10-25), we have

$$\frac{dR}{R} = \frac{d\rho}{\rho} + \frac{dL}{L} - \frac{dA}{A} \tag{10-26}$$

The area may be related to the square of some transverse dimension, such as the diameter of the resistance wire. Designating this dimension by D, we have

$$\frac{dA}{A} = 2\frac{dD}{D} \tag{10-27}$$

Introducing the definition of the axial strain and Poisson's ratio from Eq. (10-23) and (10-24), we have

$$\frac{dR}{R} = \epsilon_a(1 + 2\mu) + \frac{d\rho}{\rho} \tag{10-28}$$

The gage factor F is defined by

$$F = \frac{dR/R}{\epsilon_a} \tag{10-29}$$

so that

$$F = 1 + 2\mu + \frac{1}{\epsilon_a}\frac{d\rho}{\rho} \tag{10-30}$$

We may thus express the local strain in terms of the gage factor, the resistance of the gage, and the change in resistance with the strain:

$$\epsilon = \frac{1}{F}\frac{\Delta R}{R} \tag{10-31}$$

The value of the gage factor and the resistance are usually specified by the manufacturer so that the user need only measure the value of ΔR in order to determine the local strain. For most gages the value of F is constant over a rather wide range of strains. It is worthwhile, however, to examine the influence of various physical properties of the resistance material on the value of F. If the resistivity of the material does not vary with the strain, we have, from Eq. (10-30),

$$F = 1 + 2\mu \tag{10-32}$$

Taking a typical value of μ as 0.3 we would obtain $F = 1.6$. In this case, the change in resistance of the material results solely from the change in physical dimensions. If the resistivity decreases with strain, the value of F will be less than 1.6. When the resistivity increases with strain, F will be greater than 1.6. Gage factors for various materials have been observed from -140 to $+175$. If the resistance material is strained to the point that it is operating in the plastic region, $\mu = 0.5$ and the resistivity remains essentially constant. Under these conditions the gage factor approaches a value of 2. For most commercial strain gages the gage factor is the same for both compressive and tensile strains. A high gage factor is desirable in practice because a larger change in resistance ΔR is produced for a given strain input, thereby necessitating less sensitive readout circuitry.

Three common types of resistance strain gages are shown in Fig. 10-11. The bonded wire gage is perhaps the most common, with the wire size varying between about 0.0005 and 0.001 in. The foil gage usually employs a foil less than 0.001 in. thick. The semiconductor gage employs a silicon base material that is strain-sensitive and has the advantage that very large values of F may be obtained ($F \sim 100$). The material is usually produced in brittle wafers having a thickness of about 0.01 in. Semiconductor gages also have very high temperature coefficients of resistance. Table 10-1 presents a summary of characteristics of three strain-gage materials. Reference [7] furnishes additional information.

Wire and foil gages may be manufactured in various ways, but the important point is that the resistance element must be securely bonded to its mounting. It is essential that the bond between the resistance element and the cement joining it to the workpiece be stronger than the resistance wire itself. In this way, the strength of the resistance element is the smaller, and hence the deformation of the entire gage is governed by the deformation of the resistance element. Most wire strain gages employ either a nitrocellulose cement or a phenolic resin for the bonding agent with a thin paper backing to maintain the wire configuration. Such gages may be used up to 300°F. A Bakelite mounting is usually employed for temperatures up to 500°F. Foil gages are manufactured by an etching process and use base materials of paper, Bakelite, and epoxy film. Epoxy cements are also employed for both wire and foil gages.

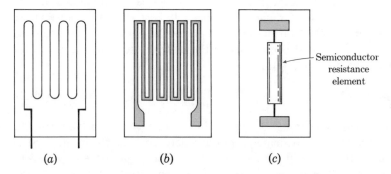

(a)   *(b)*   *(c)*

Fig. 10-11   Three types of resistance strain gages.   *(a)* Wire gage; *(b)* foil gage; *(c)* semiconductor gage.

Table 10-1 Characteristics of some resistance strain-gage materials

Material	Approx. gage factor, F	Approx. resistance, ohms	Temp. coef. of resistance, $°F^{-1} \times 10^6$	Remarks
57% Cu, 43% Ni (Driver-Harris Co., "Advance")	2.0	100	6.0	F constant over wide range of strain; low-temperature use (under 500°F)
Platinum alloys	4.0	50	1200	Suitable for high-temperature use above 1000°F
Silicon semiconductor	-100 to $+150$	200	50,000	Brittle but has high gage factor; not suitable for large strain measurements

When strain gages are mounted on a specimen, two notes of caution should be observed: (1) The surface must be absolutely clean. Cleaning with an emery cloth followed by acetone is usually satisfactory. (2) Sufficient time must be allowed for the cement to dry and harden completely. Even though the cement is dry around the edge of the gage, it may still be wet under the gage. If possible, 24 hr should be allowed for drying at room temperature. Drying time may be reduced for higher temperatures.

Several different cements are available for mounting strain gages. These cements are discussed in Ref. [2] along with rather detailed instructions for mounting the various types of gages. The interested reader should consult this reference for more information as well as the literature of various manufacturers of strain gages.

Problems associated with strain-gage installations generally fall into three categories: (1) temperature effects, (2) moisture effects, and (3) wiring problems. It is assumed that the gage is properly mounted. Temperature problems arise because of differential thermal expansion between the resistance element and the material to which it is bonded. Semiconductor gages offer the advantage that they have a lower expansion coefficient than either wire or foil gages. In addition to the expansion problem, there is a change in resistance of the gage with temperature, which must be adequately compensated for. We shall see how this compensation is performed in a subsequent paragraph. Moisture absorption by the paper and cement can change the electrical resistance between the gage and the ground potential and thus affect the output-resistance readings. Methods of moistureproofing are discussed in Refs. [1] and [2]. Wiring problems are those situations that arise because of faulty connections between the gage-resistance element and the external readout circuit. These problems may develop from poorly soldered connections or from inflexible wiring, which may pull the gage loose from the test specimen or break the gage altogether. Proper wiring practices are discussed in Refs. [2] and [6].

Electrical-resistance strain gages cannot be easily calibrated because once they are attached to a calibration workpiece removal cannot be made without destroying the gage. In practice, then, the gage factor is taken as the value specified by the manufacturer and a semicalibration effected by checking the bridge measurement and readout system.

10-8 MEASUREMENT OF RESISTANCE STRAIN-GAGE OUTPUTS

Consider the bridge circuit of Fig. 4-17. The voltage present at the detector is given by Eq. (4-24) as

$$E_D = E\left(\frac{R_1}{R_1 + R_4} - \frac{R_2}{R_2 + R_3}\right) \tag{10-33}$$

If the bridge is in balance, $E_D = 0$. Let us suppose that the strain gage represents R_1 in this circuit and a high-impedance readout device is employed so that the bridge operates as a voltage-sensitive deflection circuit. We assume that the bridge is balanced at zero-strain conditions and that a strain on the gage of ϵ results in a change in resistance ΔR_1. R_1 will be used to represent the resistance of the gage at zero-strain conditions. We immediately obtain a voltage due to the strain as

$$\frac{\Delta E_D}{E} = \frac{R_1 + \Delta R_1}{R_1 + \Delta R_1 + R_4} - \frac{R_2}{R_2 + R_3} \tag{10-34}$$

Solving for the resistance change gives

$$\frac{\Delta R_1}{R_1} = \frac{(R_4/R_1)[\Delta E_D/E + R_2/(R_2 + R_3)]}{1 - \Delta E_D/E - R_2/(R_2 + R_3)} - 1 \tag{10-35}$$

Equation (10-35) expresses the resistance changes as a function of the voltage unbalance at the detector ΔE_D.

The bridge circuit may also be used as a current-sensitive device. Combining Eqs. (4-22) to (4-24) gives

$$i_g = \frac{E(R_1 R_3 - R_2 R_4)}{R_1 R_2 R_4 + R_1 R_3 R_4 + R_1 R_2 R_3 + R_2 R_3 R_4 + R_g(R_1 + R_4)(R_2 + R_3)} \tag{10-36}$$

Again, we assume the bridge to be balanced under zero-strain conditions and take R_1 as the resistance of the gage under these conditions. Thus,

$$R_1 R_3 = R_2 R_4 \tag{10-37}$$

The galvanometer current ΔI_g is that value which results from a change in resistance ΔR_1 from the balanced condition. It may be shown that the denominator of Eq. (10-36) is not very sensitive to small changes in R_1 and hence is very nearly a constant which we shall designate as C. Thus,

$$\Delta I_g = \frac{E}{C}[(R_1 + \Delta R_1)R_3 - R_2 R_4] \tag{10-38}$$

Applying the balance conditions from Eq. (10-37) gives

$$\Delta I_g = \frac{E}{C} R_2 \, \Delta R_1 \tag{10-39}$$

Introducing the gage factor from Eq. (10-31),

$$\Delta I_g = \frac{E}{C} R_3 R_1 F \epsilon = \text{const} \times \epsilon \tag{10-40}$$

Thus, the deflection current may be taken as a direct indication of the strain imposed on the gage.

10-9 TEMPERATURE COMPENSATION

It is generally not possible to calculate corrections for temperature effects in strain gages. Consequently, compensation is made directly by means of the experimental setup. Such a compensation arrangement is shown in Fig. 10-12. Gage 1 is installed on the test specimen, while gage 2 is installed on a like piece of material that remains unstrained throughout the tests but at the same temperature as the test piece. Any changes in resistance of gage 1 due to temperature are thus cancelled out by similar changes in the resistance of gage 2, and the bridge circuit detects an unbalanced condition resulting only from the strain imposed on gage 1. Of course, care must be exerted to ensure that both gages are installed in exactly the same manner on their respective workpieces.

Example 10-4 A resistance strain gage with $R = 120$ ohms and $F = 2.0$ is placed in an equal-arm bridge in which all resistances are equal to 120 ohms. The battery voltage is 4.0 volts. Calculate the detector current in microamperes per microinch of strain. The galvanometer resistance is 100 ohms.

Solution The denominator of Eq. (10-36) is calculated as

$$C = (4)(120)^2 + (100)(240)^2 = 1.267 \times 10^7$$

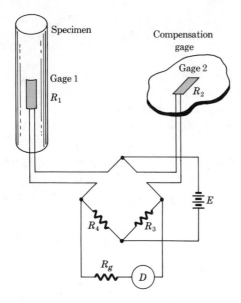

Fig. 10-12 Temperature-compensation arrangement for electrical-resistance strain gages.

We calculate the current sensitivity from Eq. (10-40) as

$$\frac{\Delta I_g}{\epsilon} = \frac{E}{C} R_3 R_1 F = \frac{(4.0)(120)^2(2.0)}{1.267 \times 10^7} = 9.08 \times 10^{-3} \text{ amp/in.}$$

$$= 9.08 \times 10^{-3} \; \mu a/\mu \text{in.}$$

Example 10-5 For the gage and bridge in Example 10-4 calculate the voltage indication for a strain of 1.0 μin./in. when a high-impedance detector is used.

Solution We have $R_1 = R_2 = R_3 = R_4 = 120$ ohms. Under these conditions Eq. (10-34) becomes

$$\frac{\Delta E_D}{E} = \frac{1 + (\Delta R_1/R_1)}{2 + (\Delta R_1/R_1)} - \frac{1}{2} = \frac{\Delta R_1/R_1}{4 + 2(\Delta R_1/R_1)}$$

The resistance change is calculated from Eq. (10-31)

$$\frac{\Delta R_1}{R_1} = F\epsilon = 2.0 \times 10^{-6}$$

so that

$$\frac{\Delta E_D}{E} = \frac{2.0 \times 10^{-6}}{4 + 4 \times 10^{-6}} = 0.5 \times 10^{-6}$$

and $\Delta E_D = (4.0)(0.5 \times 10^{-6}) = 2.0 \; \mu v$

10-10 STRAIN-GAGE ROSETTES

The installation of a strain gage on a bar specimen like that shown in Fig. 10-10 is a useful application of the gage, but it is quite restricted. The strain that is measured in such a situation is a principal strain since we assumed that the bar is operating under only a tensile load.   Obviously, a more general measurement problem will involve strains in more than one direction, and the orientation of the principal stress axes will not be known.   It would, of course, be fortunate if the strain gages were installed on the specimen so that they were oriented exactly with the principal stress axes.   We now consider the methods that may be used to calculate the principal stresses and strains in a material from three strain-gage measurements.   The arrangements for strain gages in such applications are called rosettes.   We shall give only the final relations that are used for calculation purposes.   The interested reader should consult Refs. [2] and [5] for the derivations of these equations.

Consider the rectangular rosette shown in Fig. 10-13.   The three strain gages are oriented as shown, and the three strains measured by these gages are ϵ_1, ϵ_2, and ϵ_3.   The principal strains for this situation are

$$\epsilon_{max}, \epsilon_{min} = \frac{\epsilon_1 + \epsilon_3}{2} \pm \frac{1}{\sqrt{2}} [(\epsilon_1 - \epsilon_2)^2 + (\epsilon_2 - \epsilon_3)^2]^{\frac{1}{2}} \tag{10-41}$$

The principal stresses are

$$\sigma_{max}, \sigma_{min} = \frac{E(\epsilon_1 + \epsilon_3)}{2(1 - \mu)} \pm \frac{E}{\sqrt{2}(1 + \mu)} [(\epsilon_1 - \epsilon_2)^2 + (\epsilon_2 - \epsilon_3)^2]^{\frac{1}{2}}$$

$$\tag{10-42}$$

The maximum shear stress is designated by τ_{max} and calculated from

$$\tau_{max} = \frac{E}{\sqrt{2}(1 + \mu)} [(\epsilon_1 - \epsilon_2)^2 + (\epsilon_2 - \epsilon_3)^2]^{\frac{1}{2}} \tag{10-43}$$

The principal stress axis is located with the angle θ according to

$$\tan 2\theta = \frac{2\epsilon_2 - \epsilon_1 - \epsilon_3}{\epsilon_1 - \epsilon_3} \tag{10-44}$$

This is the axis at which the maximum stress σ_{max} occurs.   A problem arises with the determination of the quadrant for θ since there will be two values obtained as solutions to Eq. (10-44).   The angle θ will lie in the first quadrant

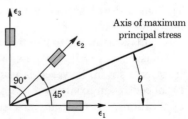

Fig. 10-13   Rectangular strain-gage rosette.

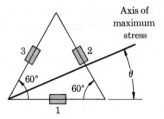

Fig. 10-14   Delta strain-gage rosette.

$(0 < \theta < \pi/2)$ if

$$\epsilon_2 > \frac{\epsilon_1 + \epsilon_3}{2} \tag{10-45}$$

and in the second quadrant if ϵ_2 is less than this value.

Another type of strain-gage rosette in common use is the delta rosette shown in Fig. 10-14.   The principal strains in this instance are given by

$$\epsilon_{max}, \epsilon_{min} = \frac{\epsilon_1 + \epsilon_2 + \epsilon_3}{3} \pm \frac{\sqrt{2}}{3} [(\epsilon_1 - \epsilon_2)^2 + (\epsilon_2 - \epsilon_3)^2 \\ + (\epsilon_3 - \epsilon_1)^2]^{\frac{1}{2}} \tag{10-46}$$

The principal stresses are

$$\sigma_{max}, \sigma_{min} = \frac{E(\epsilon_1 + \epsilon_2 + \epsilon_3)}{3(1 - \mu)} \pm \frac{\sqrt{2}E}{3(1 + \mu)} [(\epsilon_1 - \epsilon_2)^2 + (\epsilon_2 - \epsilon_3)^2 \\ + (\epsilon_3 - \epsilon_1)^2]^{\frac{1}{2}} \tag{10-47}$$

The maximum shear stress is calculated from

$$\tau_{max} = \frac{\sqrt{2}E}{3(1 + \mu)} [(\epsilon_1 - \epsilon_2)^2 + (\epsilon_2 - \epsilon_3)^2 + (\epsilon_3 - \epsilon_1)^2]^{\frac{1}{2}} \tag{10-48}$$

The principal stress axis is located according to

$$\tan 2\theta = \frac{\sqrt{3}(\epsilon_3 - \epsilon_2)}{2\epsilon_1 - \epsilon_2 - \epsilon_3} \tag{10-49}$$

The angle θ will be in the first quadrant when $\epsilon_3 > \epsilon_2$ and in the second quadrant when $\epsilon_2 > \epsilon_3$.

It is worthwhile to mention that resistance strain gages may be sensitive to transverse as well as axial strains.   The resistance change produced by a transverse strain, however, is usually less than 2 or 3 percent of the change produced by an axial strain.   For this reason, it may be neglected in many applications.   If the transverse strain is to be considered, the above rosette formulas must be modified accordingly. We shall not present this modification but refer the reader to Refs. [1] and [5] for more information.

Example 10-6 A rectangular rosette is mounted on a steel plate having $E = 29 \times 10^6$ psi and $\mu = 0.3$. The three strains are measured as

$\epsilon_1 = +500 \ \mu in./in.$

$\epsilon_2 = +400 \ \mu in./in.$

$\epsilon_3 = -100 \ \mu in./in.$

Calculate the principal strains and stresses and the maximum shear stress. Locate the axis of principal stress.

Solution As an intermediate step we calculate the quantities

$$A = \frac{\epsilon_3 + \epsilon_1}{2} = 200 \ \mu in./in.$$

$$B = [(\epsilon_1 - \epsilon_2)^2 + (\epsilon_2 - \epsilon_3)^2]^{\frac{1}{2}} = 510 \ \mu in./in.$$

Then,

$$\epsilon_{max} = A + \frac{1}{\sqrt{2}} B = 561 \ \mu in./in.$$

$$\epsilon_{min} = A - \frac{1}{\sqrt{2}} B = -161 \ \mu in./in.$$

$$\sigma_{max} = \frac{EA}{1 - \mu} + \frac{EB}{\sqrt{2}(1 + \mu)} = \frac{(29 \times 10^6)(200 \times 10^{-6})}{1 - 0.3}$$

$$+ \frac{(29 \times 10^6)(510 \times 10^{-6})}{\sqrt{2}(1 + 0.3)} = 8,280 + 8,050 = 16,330 \ psi$$

$$\sigma_{min} = 8,280 - 8,050 = 230 \ psi$$

$$\tau_{max} = \frac{EB}{\sqrt{2}(1 + \mu)} = 8,050 \ psi$$

We also have

$$\tan 2\theta = \frac{2\epsilon_2 - \epsilon_1 - \epsilon_3}{\epsilon_1 - \epsilon_3} = \frac{(2)(400) - (500) - (-100)}{(500) - (-100)} = 0.667$$

$2\theta = 33.7$ or $213.7°$

$\theta = 16.8$ or $106.8°$

We choose the first quadrant angle ($\theta = 16.8°$) in accordance with Eq. (10-45).

10-11 THE UNBONDED-RESISTANCE STRAIN GAGE

The bonded electrical-resistance strain gage discussed above is the most widely used device for strain measurements. An alternate resistance gage is the unbonded type shown in Fig. 10-15. A springloaded mechanism holds the two plates in a close position while the fine-wire filaments are

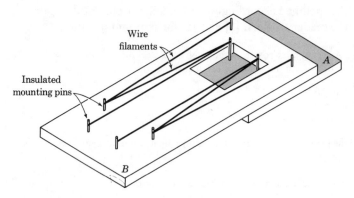

Fig. 10-15   Schematic of unbonded-resistance strain gage.

stretched around the mounting pins as shown.   The mounting pins must be rigid and also serve as electrical insulators.   When plate A moves relative to B, a strain is imposed on these filaments, which may be detected through a measurement of the change in resistance.   The allowable displacement of commercial gages is of the order of ± 0.0015 in., and the wire diameter is usually less than 0.001 in.   The I^2R heating in the unbonded gage can be a problem because the wires have no ready means for heat dissipation other than convection to the surrounding air.   The principle of the unbonded gage has been applied to acceleration and diaphragm pressure transducers with good success.

REVIEW QUESTIONS

1.   Why must a correction be applied for the specific gravity of a sample when used in a mass balance?

2.   How are elastic elements used for force or torque measurements?

3.   What is a proving ring?

4.   What is a Prony brake?

5.   What does a dynamometer measure?

6.   Distinguish between stress and strain.

7.   What is Poisson's ratio?

8.   What is deformation sensitivity?

9.   What is strain sensitivity?

10.   How may a brittle coating be used for strain measurements?

11.   Define the gage factor for strain gages.

12.   What is the main advantage of a semiconductor strain gage?

13.   How are measurements performed with a resistance strain gage?

14.   How may temperature compensation be performed on resistance strain gages?

15.   What is meant by a strain-gage rosette?   How is it used?

PROBLEMS

10-1. Many balances use brass weights ($\rho_s = 530$ lb$_m$/ft^3) in atmospheric air ($\rho_a = 0.075$ lb$_m$/ft^3). Plot the percent error in the balance determination as a function of ρ_u/ρ_s.

10-2. Using the conditions given in Example 10-1, calculate the required angular sensitivity in order that the uncertainty in the weight determination is of the order of 1 part in 10^8.

10-3. An LVDT is used to measure the axial deformation of the simple bar of Fig. 10-3. The uncertainty in this measurement is ± 0.001 in. A steel bar is to be used to measure a force of 250 lb$_f$ with an uncertainty of 1 percent. Determine a suitable diameter and length of rod for this measurement, assuming the rod dimensions will be known exactly.

10-4. The diaphragm of Fig. 6-10c is to be used as a force-measurement device with the load applied at the center boss. Steel is to be used as the diaphragm material, and the maximum displacement is not to exceed one-third the thickness. A load of 1,000 lb$_f$ is to be measured with an uncertainty of 1 percent using an LVDT with an uncertainty of ± 0.0001 in. Determine suitable diaphragm dimensions for this measurement.

10-5. A farmer decides to build a crude weighing device to weigh bags of grain up to 150 lb$_f$. For this purpose he employs a section of 1-in. steel pipe (1.315 in. OD, 0.957 in. ID) as a cantilever beam. He intends to measure the deflection of the beam with a metal carpenter's scale having graduations of $\frac{1}{16}$ in. On the basis of these data what length of pipe would you recommend for the farmer's application?

10-6. A proving ring is constructed of steel ($E = 30 \times 10^6$ psi) having a cross section with 0.250 in. thickness and 1.000 in. depth. The overall diameter of the ring is 6.000 in. A micrometer is used for the deflection measurement, and its uncertainty is ± 0.0001 in. Assuming that the dimensions of the ring are exact, calculate the applied load when the uncertainty in this load is 1.0 percent. Calculate the percent uncertainty in the load when $y = 0.01$ in. and the uncertainties in the dimensions are

$$w_h = w_b = 0.0002 \text{ in.}$$
$$w_d = 0.01 \text{ in.}$$

where b and h are the thickness and depth of the ring, respectively.

10-7. A hollow steel cylinder ($G = 12 \times 10^6$ psi) is subjected to a moment such that ϕ is 1.50°. The dimensions of the cylinder are

$$r_i = 0.500 \pm 0.0003 \text{ in.}$$
$$r_0 = 0.625 \pm 0.0003 \text{ in.}$$
$$L = 5.000 \pm 0.001 \text{ in.}$$

The uncertainty in the angular deflection is $\pm 0.05°$. Calculate the nominal value of the impressed moment and its uncertainty. Also calculate the 45° strains.

10-8. In a bridge circuit like that of Fig. 4-17, $R_2 = R_3 = 100$ ohms. The galvanometer resistance is 50 ohms. The strain-gage resistance at zero strain is 120 ohms, and the value of R_4 is adjusted to bring the bridge into balance at zero-strain conditions. The gage factor is 2.0. Calculate the galvanometer current when $\epsilon = 400$ μin./in. Take the battery voltage as 4.0 volts.

10-9. Calculate the voltage output of the bridge in Prob. 10-8, assuming a detector of very high impedance.

10-10. Show that Eq. (10-41) reduces to

$$\epsilon_{max} = \epsilon_1$$
$$\epsilon_{min} = \epsilon_3$$

when $\theta = 0$. Obtain relations for the principal stresses and maximum shear stress in this instance.

10-11. A delta rosette is placed on a steel plate and indicates the following strains:

$\epsilon_1 = 400 \ \mu\text{in./in.}$

$\epsilon_2 = 84 \ \mu\text{in./in.}$

$\epsilon_2 = -250 \ \mu\text{in./in.}$

Calculate the principal strains and stresses, the maximum shear stress, and the orientation angle for the principal axes.

10-12. Obtain simplified relations for the delta rosette under the conditions that $\epsilon_1 = \epsilon_{max}$; that is, $\theta = 0$.

10-13. Suppose a rectangular rosette is used to measure the same stresses as in Prob. 10-11. The bottom leg of this rosette is placed in the same location as the bottom arm of the delta rosette. Calculate the strains that would be indicated by the rectangular rosette under these conditions.

10-14. Calculate the percent uncertainty in the maximum and minimum stresses resulting from uncertainties of 2 percent in the strains as measured with a rectangular rosette. For this calculation assume E and μ are known exactly.

10-15. Rework Prob. 10-14 for the delta rosette.

10-16. A rectangular rosette is placed on a steel plate and indicates the following strains:

$\epsilon_1 = 563 \ \mu\text{in./in.}$

$\epsilon_2 = -155 \ \mu\text{in./in.}$

$\epsilon_3 = -480 \ \mu\text{in./in.}$

Calculate the principal strains and stresses, the maximum shear stress, and the orientation angle for the principal axes.

10-17. A strip of steel sheet, $\frac{1}{16} \times 2 \times 20$ in., is available for use as a force-measuring elastic element. The strip is to be used by cementing strain gages to its flat surfaces and measuring the deformation under load. The strain gages have a maximum strain of 2,000 $\mu\text{in./in.}$ and a gage factor of 1.90. A battery is available with a voltage of 4.0 volts, and the readout voltmeter has a high impedance and an accuracy of $\pm 1 \ \mu\text{v}$. The nominal resistance of the strain gages is 120 ohms. Calculate the force range for which the measurement system may be applicable.

10-18. A bridge circuit is used to measure the output of a strain gage having a resistance of 110 ohms under zero conditions. $R_2 = R_3 = 100$ ohm, and the galvanometer resistance is 75 ohms. The gage factor is 2.0. Calculate the galvanometer current when $\epsilon = 300 \ \mu\text{in./in.}$ for a battery voltage of 6.0 volts. Also calculate the voltage output of the bridge for a very-high-impedance detector.

10-19. An electrical-resistance strain gage records a strain of 400 $\mu\text{in./in.}$ on a steel-tension member. What is the axial stress?

10-20. A hollow steel cylinder is used as a torque-sensing element. The dimensions are: $r_i = 1$ in., $r_0 = 1.25$ in., $L = 6$ in. Calculate the angular deflection for an applied moment of 200 in./lb. Calculate the strain that would be indicated by gages attached as shown in Fig. 10-7.

10-21. A certain automotive engine is reported to produce a torque of 400 ft-lb$_f$ at 3,000 rpm. What power dynamometer would be necessary to test this engine?

REFERENCES

1. Dean, M. (ed.): "Semiconductor and Conventional Strain Gages," Academic Press, Inc., New York, 1962.
2. Dove, R. C., and P. H. Adams: "Experimental Stress Analysis and Motion Measurement," Charles E. Merrill Books, Inc., Columbus, Ohio, 1964.
3. Durelli, A. J., E. A. Phillips, and C. H. Tsao: "Introduction to the Theoretical and Experimental Analysis of Stress and Strain," McGraw-Hill Book Company, New York, 1958.
4. Harris, C. M., and C. E. Crede (eds.): "Basic Theory and Measurements," vol. 1, "Shock and Vibration Handbook," McGraw-Hill Book Company, New York 1961.
5. Hetenyi, M. (ed.): "Handbook of Experimental Stress Analysis," John Wiley & Sons, Inc., New York, 1950.
6. Perry, C. C., and H. R. Lissner: "The Strain Gage Primer," 2d ed., McGraw-Hill Book Company, New York, 1962.
7. Stein, Peter K.: Material Considerations for Strain Gages, *Inst. & Control Systems*, vol. 37, no. 10, pp. 132–135, October, 1964.
8. *Proc. Soc. Exptl. Stress Anal.*, published by the Society.
9. Dally, J. W., and W. F. Riley: "Experimental Stress Analysis," McGraw-Hill Book Company, New York, 1965.
10. Juvinall, R. C.: "Engineering Considerations of Stress, Strain, and Strength," McGraw-Hill Book Company, New York, 1967.
11. Beckwith, T. G., and N. L. Buck: "Mechanical Measurements," 2d ed., Addison-Wesley Publishing Company, Inc., Reading, Mass., 1969.

11

Motion and Vibration Measurement

11-1 INTRODUCTION

Measurements of motion and vibration parameters are important in many applications. The experimental quantity that is desired may be velocity, acceleration, or vibration amplitude. These quantities may be useful in predicting the fatigue failure of a particular part or machine or may play an important role in analyses which are used to reduce structure vibration or noise level.

The central problem in any type of motion or vibration measurement concerns a determination of the appropriate quantities in reference to some specified state, i.e., velocity, displacement, or acceleration in reference to ground. Ideally, one would like to have a motion or vibration transducer which connects to the body in motion and furnishes an output signal proportional to the vibrational input. The ideal transducer should be independent of its location; i.e., it should function equally well whether it is connected to a vibrating structure on the ground, in an airplane, or in a space vehicle. In this chapter we shall see some of the compromises which are necessary to approach such behavior.

Sound may be classified as a vibratory phenomenon, and we shall indicate some of the important parameters necessary for specification of sound level. The measurement and analysis of sound levels are very specialized subjects, which are becoming increasingly important in modern building and equipment design [8].

11-2 TWO SIMPLE VIBRATION INSTRUMENTS

Consider the simple wedge shown in Fig. 11-1 which is attached to a vibrating wall. When the wall is at rest, the wedge appears as in Fig. 11-1a; when the wall is in motion, the wedge appears as in Fig. 11-1b. An observation is made of the distance x. At this distance the thickness of the wedge is equal to the double amplitude of the motion. In terms of x the amplitude is given by

$$a = x \tan \frac{\theta}{2} \qquad (11\text{-}1)$$

where θ is the total included angle of the wedge. The wedge-measurement device is necessarily limited to rather large amplitude motions, usually for $a > \frac{1}{32}$ in.

Another simple vibration-frequency-measurement device is shown in Fig. 11-2. The small cantilever beam mounted on the block is placed against the vibrating surface, and some appropriate method is provided for varying the beam length. When the beam length is properly adjusted so that its natural frequency is equal to the frequency of the vibrating surface, the resonance condition shown in Fig. 11-2b will result. The natural frequency of such a beam is given by

$$\omega_n = 11.0 \sqrt{\frac{EI}{mL^4}} \qquad (11\text{-}2)$$

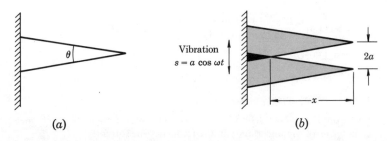

Fig. 11-1 Simple wedge used as device for amplitude-displacement measurements. (a) At rest; (b) in motion.

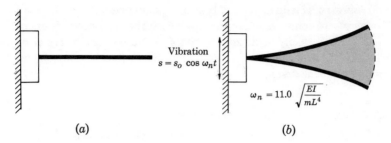

$$s = s_0 \cos \omega_n t$$

Vibration

$$\omega_n = 11.0 \sqrt{\frac{EI}{mL^4}}$$

(a)   (b)

Fig. 11-2   Cantilever beam used as frequency-measurement device.   (a) No vibration or vibration other than ω_n; (b) vibration at ω_n.

where ω_n = natural frequency

E = Young's modulus for the beam material, psi

I = moment of inertia of the beam, in.^{4}

m = mass per unit length, lb_m/in.

L = beam length, in.

Example 11-1   A $\frac{1}{16}$-in.-diam spring-steel rod is to be used for a vibration-frequency measurement as shown in Fig. 11-2.   The length of the rod may be varied between 1 and 4 in.   The density of this material is 489 lb_m/ft^3, and the modulus of elasticity is 28.3×10^6 psi.   Calculate the range of frequencies that may be measured with this device and the allowable uncertainty in L at $L = 4$ in. in order that the uncertainty in the frequency is not greater than 1 percent.   Assume the material properties are known exactly.

Solution   We have

$$E = 28.3 \times 10^6 \text{ psi}$$

$$I = \frac{\pi r^4}{4} = \frac{\pi (1/32)^4}{4} = 7.49 \times 10^{-7} \text{ in.}^4$$

$$m = \rho \pi r^2 = \frac{\pi (489)(1/32)^2}{(144)} \frac{1}{12} = 8.69 \times 10^{-4} \text{ lb}_m/\text{in.}$$

At $x = 1$ in.,

$$\omega_n = 11.0 \left[\frac{(28.3 \times 10^6)(7.49 \times 10^{-7})}{(8.69 \times 10^{-4})(1)^4} \right]^{\frac{1}{2}} = 1{,}720 \text{ rad/sec}$$

$$= 273 \text{ Hz}$$

At $x = 4$ in.,

$$\omega_n = 107.5 \text{ rad/sec} = 17.06 \text{ Hz}$$

We use Eq. (3-2) to determine the allowable uncertainty in the length measurement in terms of the uncertainty in the frequency measurement.

$$\frac{\partial \omega_n}{\partial L} = 11.0 \sqrt{\frac{EI}{m}} \left(\frac{-2}{L^3} \right)$$

$$w_{\omega n} = \left[\left(\frac{\partial \omega_n}{\partial L} \, w_L \right)^2 \right]^{\frac{1}{2}}$$

so that

$$w_L = \frac{w_{\omega n} L^3}{22.0 \sqrt{EI/m}}$$

At $L = 4$ in.,

$$w_{\omega n} = (0.01)(107.5) = 1.075 \text{ rad/sec}$$

$$w_L = \frac{(1.075)(4)^3}{(22.0)(156.2)} = 0.02 \text{ in.}$$

11-3 PRINCIPLES OF THE SEISMIC INSTRUMENT

The seismic instrument is a device that has the functional form of the system shown in Fig. 2-3. A schematic of a typical instrument is shown in Fig. 11-3. The mass is connected through the parallel spring and damper arrangement to the housing frame. This frame is then connected to the vibration source whose characteristics are to be measured. The mass tends to remain fixed in its spatial position so that the vibrational motion is registered as a relative displacement between the mass and the housing frame. This displacement is then sensed and indicated by an appropriate transducer, as shown in the schematic diagram. Of course, the seismic mass does not remain absolutely steady, but for selected frequency ranges it may afford a satisfactory reference position.

The seismic instrument may be used for either displacement or acceleration measurements by proper selection of mass, spring, and damper combinations. In general, a large mass and soft spring are desirable for

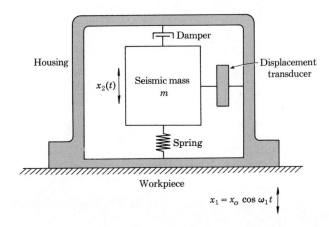

Fig. 11-3 Schematic of typical seismic instrument.

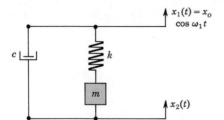

Fig. 11-4 Mechanical system for a seismic instrument.

vibrational displacement measurements, while a relatively small mass and stiff spring are used for acceleration indications. This will be apparent from the theoretical discussion that follows.

Figure 2-3 is reproduced as Fig. 11-4 to show the mechanical system to be analyzed. Using Newton's second law of motion, we have

$$m\frac{d^2x_2}{dt^2} + c\frac{dx_2}{dt} + kx_2 = c\frac{dx_1}{dt} + kx_1 \qquad (11\text{-}3)$$

where it is assumed that the damping force is proportional to velocity. We assume that a harmonic vibratory motion is impressed on the instrument such that

$$x_1 = x_0 \cos \omega_1 t \qquad (11\text{-}4)$$

and wish to obtain an expression for the relative displacement $x_2 - x_1$ in terms of this impressed motion. The relative displacement is that which is detected by the transducer shown in Fig. 11-3. Rewriting Eq. (11-3) and substituting Eq. (11-4) gives

$$\frac{d^2x_2}{dt^2} + \frac{c}{m}\frac{dx_2}{dt} + \frac{k}{m}x_2 = x_0\left(\frac{k}{m}\cos \omega_1 t - \frac{c}{m}\omega_1 \sin \omega_1 t\right) \qquad (11\text{-}5)$$

The solution to Eq. (11-5) is

$$x_2 - x_1 = e^{-(c/2m)t}(A \cos \omega t + B \sin \omega t) + \frac{mx_0\omega_1^2 \cos (\omega_1 t - \phi)}{[(k - m\omega_1^2)^2 + c^2\omega_1^2]^{\frac{1}{2}}} \qquad (11\text{-}6)$$

where the frequency is given by

$$\omega = \left[\left(\frac{c}{2m}\right)^2 - \frac{k}{m}\right]^{\frac{1}{2}} \qquad (11\text{-}7)$$

and the phase angle by

$$\phi = \tan^{-1}\frac{c\omega_1}{k - m\omega_1^2} \qquad (11\text{-}8)$$

A and B are constants of integration determined from the initial or boundary conditions.

Note that Eq. (11-6) is composed of two terms: (1) the transient term involving the exponential function and (2) the steady-state term. This means that after the initial transient has died out a steady-state harmonic motion is established in accordance with the second term. The frequency of this steady-state motion is the same as that of the impressed motion, and its amplitude is

$$(x_2 - x_1)_0 = \frac{x_0(\omega_1/\omega_n)^2}{\{[1 - (\omega_1/\omega_n)^2]^2 + [2(c/c_c)(\omega_1/\omega_n)]^2\}^{\frac{1}{2}}} \tag{11-9}$$

where the natural frequency ω_n and critical damping coefficient c_c are given by

$$\omega_n = \sqrt{\frac{k}{m}} \tag{11-10}$$

$$c_c = 2\sqrt{mk} \tag{11-11}$$

The phase angle may also be written

$$\phi = \tan^{-1} \frac{2(c/c_c)(\omega_1/\omega_n)}{1 - (\omega_1/\omega_n)^2} \tag{11-12}$$

A plot of Eq. (11-9) is given in Fig. 11-5. It may be seen that the output

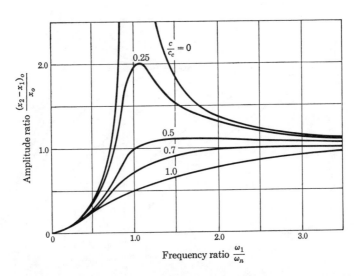

Fig. 11-5 Displacement response of a seismic instrument as given by Eq. (11-9).

amplitude is very nearly equal to the input amplitude when $c/c_c = 0.7$ and $\omega_1/\omega_n > 2$. For low values of the damping ratio the amplitude may become quite large. The output becomes essentially a linear function of input at high-frequency ratios. Thus, a seismic-vibration pickup for measurement of displacement amplitude should be utilized for measurement of frequencies substantially higher than its natural frequency. The instrument constants c/c_c and ω_n should be known or obtained from calibration. The anticipated accuracy of measurement may then be calculated for various frequencies.

The acceleration amplitude of the input vibration is

$$a_0 = \left(\frac{d^2 x_1}{dt^2}\right)_0 = \omega_1^2 x_0 \tag{11-13}$$

We may thus use the measured output of the instrument as a measure of acceleration. There are problems associated with this application, however. In Eq. (11-9) the bracketed term is the one that governs the linearity of the acceleration response since ω_n will be fixed for a given instrument. In Fig. 11-6 we have a plot of

$$\frac{(x_2 - x_1)_0 \omega_n^2}{a_0} \qquad \text{versus} \qquad \frac{\omega_1}{\omega_n}$$

which indicates the response. Generally unsatisfactory performance is

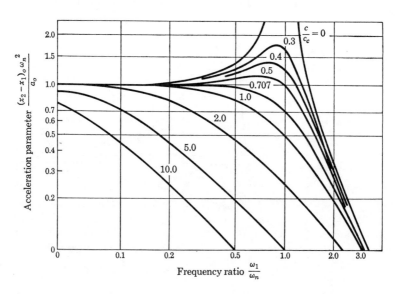

Fig. 11-6 Acceleration response of a seismic instrument as given by Eq. (11-13).

observed at frequency ratios above 0.4. Thus, for acceleration measurements we wish to operate at frequencies much lower than the natural frequency, in contrast to the desirable region of operation for amplitude measurements. From the standpoint of instrument construction this means that we wish to have a low natural frequency (soft spring, large mass) for amplitude measurements and a high natural frequency (stiff spring, small mass) for acceleration measurements in order to be able to operate over a wide range of frequencies and still enjoy linear response.

The seismic instrument may also be used for vibration-velocity measurements by employing a variable-reluctance magnetic pickup as the sensing transducer. The output of such a pickup will be proportional to the relative velocity amplitude, i.e., the quantity

$$\left[\frac{d}{dt}(x_2 - x_1)\right]_0$$

From the above discussion it may be seen that the seismic instrument is a very versatile device that may be used for measurement of a variety of vibration parameters. For this reason, many commercial vibration and acceleration pickups operate on the seismic-instrument principle. Calibration methods for the seismic instrument are discussed in Refs. [1], [4], [6], and [7].

The transient response of the seismic instrument is governed partially by the exponential decay term in Eq. (11-6). The time constant for this term could be taken as

$$T = \frac{2m}{c} \tag{11-14}$$

or, in terms of the natural frequency and critical damping ratio,

$$T = \frac{1}{\omega_n(c/c_c)} \tag{11-15}$$

The specific transient response of the seismic-instrument system is also a function of the type of input signal, i.e., whether it is a step function, harmonic function, ramp function, etc. The linearity of a vibration transducer is thus influenced by the frequency-ratio requirements that are necessary to give linear response as indicated by Eqs. (11-6) and (11-9). The design of a transducer for particular response characteristics must involve a compromise between these two effects, combined with a consideration of the sensitivity of the displacement-sensing transducer and its transient response characteristics. The response of the seismic instrument to different types of inputs is discussed by Dove and Adams [4].

Example 11-2   Using Eq. (11-9) and $c/c_c = 0.7$, calculate the value of ω_1/ω_n such that $(x_2 - x_1)_0/x_0 = 0.99$; that is, the error is 1 percent.

Solution   We have

$$0.99 = \frac{(\omega_1/\omega_n)^2}{\{[1 - (\omega_1/\omega_n)^2]^2 + [2(0.7)(\omega_1/\omega_n)]^2\}^{\frac{1}{2}}}$$

Rearranging this equation gives the quadratic relation

$$\left[\left(\frac{1}{0.99}\right)^2 - 1\right]\left(\frac{\omega_1}{\omega_n}\right)^4 + 0.04\left(\frac{\omega_1}{\omega_n}\right)^2 - 1 = 0$$

which yields

$$\frac{\omega_1}{\omega_n} = 2.47$$

Example 11-3   A small seismic instrument is to be used for measurement of linear acceleration. It has $\omega_n = 100$ rad/sec and a displacement-sensing transducer which detects a maximum of ± 0.1 in. The uncertainty in the displacement measurement is ± 0.001 in. Calculate the maximum acceleration that may be measured with this instrument and the uncertainty in the measurement, assuming ω_n is known exactly.

Solution   From Eq. (11-9) with $\omega_1/\omega_n = 0$, we have

$$a_0 = \omega_n^2 (x_2 - x_1)_0$$
$$= 100^2(0.1/12) = 83.3 \text{ ft/sec}^2$$

The uncertainty is calculated from

$$w_a = \left[\left(\frac{\partial a}{\partial x} w_x\right)^2\right]^{\frac{1}{2}} = \omega_n^2 w_x = (100)^2(0.001/12) = 0.833 \text{ ft/sec}^2$$

or an uncertainty of 1 percent. It may be noted that larger uncertainties would be present when smaller accelerations were measured.

Example 11-4   Calculate the frequency ratio for which the error in acceleration measurement is 1 percent, with $c/c_c = 0.7$; that is, $[(x_2 - x_1)_0 w_n^2]/a_0 = 0.99$.

Solution   We have

$$0.99 = \frac{1}{\{[1 - (\omega_1/\omega_n)^2]^2 + [2(0.7)(\omega_1/\omega_n)^2]\}^{\frac{1}{2}}}$$

This gives the quadratic relation

$$\left(\frac{\omega_1}{\omega_n}\right)^4 - 0.04\left(\frac{\omega_1}{\omega_n}\right)^2 + 1 - (1/0.99)^2 = 0$$

which yields

$$\frac{\omega_1}{\omega_n} = 0.306$$

Example 11-5 Calculate the time required for the exponential transient term in Eq. (11-6) to decrease by 99 percent with $\omega_n = 100$ rad/sec and $c/c_c = 0.7$. Compare this time with the period of an acceptable frequency ω_1 for displacement and acceleration amplitude measurements as calculated from the 99 percent relationships of Examples 11-2 and 11-4.

Solution   We have

$$e^{-(c/2m)t} = 0.01 \quad \text{and} \quad \frac{ct}{2m} = 10.06$$

$$10.06 = \omega_n\left(\frac{c}{c_c}\right)t$$

$$t = 0.1438 \text{ sec}$$

From Example 11-2, $\omega_1/\omega_n = 2.47$ and the corresponding period for ω_1 in the displacement measurement is

$$P = \frac{2\pi}{\omega_1} = \frac{2\pi}{247} = 0.0255 \text{ sec}$$

From Example 11-4, $\omega_1/\omega_n = 0.306$ and the corresponding period for ω_1 in the acceleration measurement is

$$P = \frac{2\pi}{\omega_1} = \frac{2\pi}{30.6} = 0.205 \text{ sec}$$

No specific conclusions may be drawn from this comparison because the transient response depends on the manner in which the instrument is subjected to a change.

11-4 PRACTICAL CONSIDERATIONS FOR SEISMIC INSTRUMENTS

The previous paragraphs have shown the basic response characteristics of seismic instruments to an impressed harmonic vibrational motion. Let us now consider some of the ways that these instruments might be constructed in practice.

In Fig. 11-7 a seismic instrument is illustrated that uses a voltage-divider potentiometer for sensing the relative displacement between the frame

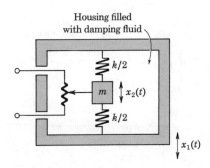

Fig. 11-7 Seismic instrument using a voltage-divider potentiometer for sensing relative displacement.

and the seismic mass.   To provide the damping for the system, the case of the instrument might be filled with a viscous liquid which would interact continuously with the frame and the mass.   Because of the relatively large mass of the potentiometer, such systems have rather low natural frequencies (less than about 100 Hz) and, as such, are limited to acceleration measurements at frequencies less than about 50 Hz.   Such acceleration transducers may weigh a pound or more.

The linear variable differential transformer (LVDT) described in Sec. 4-13 offers another convenient means for measurement of the relative displacement between the seismic mass and accelerometer housing.   Such devices have somewhat higher natural frequencies than potentiometer devices (270–300 Hz) but are still restricted to applications with low frequency-response requirements.   The differential transformer, however, has a much lower resistance to motion than the potentiometer and is capable of much better resolution.   In addition, the seismic accelerometer using an LVDT can be considerably lighter in construction than one with a potentiometer. Table 4-2 gives a comparison of the characteristics of the LVDT and potentiometer.

The electrical-resistance strain gage discussed in Chap. 10 may also be used for a displacement-sensing device in a seismic instrument.   Consider the schematic in Fig. 11-8.   The seismic mass is mounted on a cantilever beam.   On each side of the beam a resistance strain gage is mounted to sense the strain in the beam resulting from the vibrational displacement of the mass. Damping for the system is provided by the viscous liquid, which fills the housing.   The outputs of the strain gages are connected to an appropriate bridge circuit, which is used to indicate the relative displacement between the mass and housing frame.   The natural frequencies of such systems are fairly low and roughly comparable to those for the LVDT systems.   The low natural frequencies result from the fact that the cantilever beam must be sufficiently large to accommodate the mounting of the resistance strain gages.

For high-frequency measurements the seismic instrument frequently

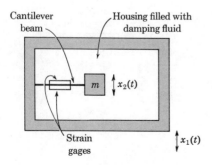

Fig. 11-8 Seismic instrument utilizing an electrical-resistance strain gage for sensing relative displacement.

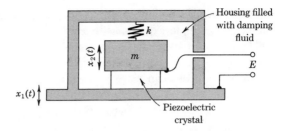

Fig. 11-9 Seismic instrument utilizing a piezoelectric transducer.

employs a piezoelectric transducer (Sec. 4-15), as shown in Fig. 11-9. The natural frequency of such instruments may be as high as 100 kHz, and the entire instrument may be quite small and light in weight. A total weight of 1 oz for a piezoelectric accelerometer is not uncommon. Piezoelectric devices have rather high electric outputs but are not generally suitable for measurements at frequencies below about 10 Hz. Electrical impedance matching between the transducer and readout circuitry is usually a critical matter requiring careful design considerations.

Seismic vibration instruments may be influenced by temperature effects. Devices employing variable-resistance displacement sensors will require a correction to account for resistance changes due to temperature. The damping of the instrument is affected by changes in viscosity of the fluid surrounding the seismic mass. Silicone oils are frequently employed in such applications, and their viscosity is strongly dependent on temperature. It would not be uncommon for such oils to experience a change in viscosity by a factor of 2 for a temperature change of only 50°F. The viscosity-temperature effect could be eliminated, of course, if a fluid were used with a small change in viscosity with temperature. Unfortunately, fluids which have this desirable behavior, viz., gases, have low viscosities as well, so that they are generally unsuitable for use as damping agents in the seismic instrument. One way of alleviating the viscosity-temperature problem is to install an electrical-resistance heater in the fluid to maintain the temperature at a constant value regardless of the surrounding temperature.

Obviously, vibration measurements are not always performed on large, massive pieces of equipment. It is frequently of interest to measure the vibration of a thin plate or other small, low-mass structures. The question immediately arises about the effect of the presence of the vibration instrument in such measurements. If the mass of the instrument is not small in comparison with the mass of the workpiece on which it is to be installed, the vibrational characteristics may be altered appreciably. In such cases a correct interpretation of the output of the vibrational instrument may only be obtained by analyses beyond the scope of our discussion, and the interested reader should consult Ref. [6] for more information about this subject. A simple experimental technique, however, will tell the engineer whether the presence of the instrument has altered the vibrational characteristics of the structure under test. First, a measurement is made with the vibration instrument in place. Then, an additional mass is mounted on the structure along with the instrument and a second measurement taken under the same conditions as the first. If there is no appreciable difference between the two measurements, it may be assumed that the presence of the vibrometer has not altered the vibrational characteristics. If there is an appreciable difference in the measurements, an analysis will be necessary to determine the "free" vibrational characteristics in terms of the experimental measurements.

11-5 SOUND MEASUREMENTS

Sound waves are a vibratory phenomenon and for this reason are appropriately discussed in this chapter. They might also be discussed in the chapter on pressure measurements because acoustic effects are usually measured in terms of harmonic pressure fluctuations that they produce in a liquid or gaseous medium.

It is standard practice in acoustic measurements to relate sound intensity and sound pressure to certain reference values I_0 and p_0, which correspond to the intensity and mean pressure fluctuations of the faintest audible sound at a frequency of 1,000 Hz. These reference levels are:

$$I_0 = 10^{-16} \text{ watt/cm}^2 \tag{11-16}$$
$$p_0 = 2 \times 10^{-4} \text{ dyne/cm}^2$$

$$= 2 \times 10^{-5} \text{ newton/m}^2 \tag{11-17}$$
$$= 2.9 \times 10^{-9} \text{ psi}$$

Intensity and pressure levels are measured in decibels. Thus,

$$\text{Intensity level (db)} = 10 \log \frac{I}{I_0} \tag{11-18}$$

$$\text{Pressure level (db)} = 20 \log \frac{p}{p_0} \tag{11-19}$$

When pressure fluctuations and particle displacements are in phase such as in a plane acoustic wave, these levels are equal, $10 \log I/I_0 = 20 \log p/p_0$.

The magnitudes of the particle velocity and pressure fluctuations created by a sound wave are small. For example, a plane sound wave having an

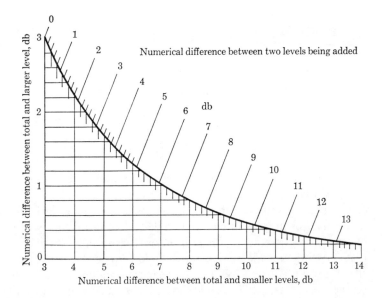

Fig. 11-10 Calculation aid for adding or subtracting sound levels measured in decibels.

intensity of 140 db generates a maximum oscillation velocity of approximately 2.4 ft/sec and a root-mean-square pressure fluctuation of about 0.029 psi. A sound intensity of 90 db is considered the maximum permissible level for extended human exposure.

In many circumstances we shall be interested in the sound intensity that results from several sound sources. This calculation could, of course, be performed with Eqs. (11-18) and (11-19), but the chart in Fig. 11-10 is frequently used to advantage in such computations. The following examples illustrate the use of the figure.

Example 11-6   Calculate the total sound intensity resulting from two sound sources at 60 and 70 db.

Solution   The numerical difference between the two levels being added is 10 db. Consulting Fig. 11-10 we find that the numerical difference between the total and larger level is about 0.4 db. Therefore, the total sound intensity is 70 + 0.4 = 70.4 db.

Example 11-7   Calculate the total sound intensity resulting from three sound sources at 50, 55, and 60 db.

Solution   We first combine the two lower levels to find their equivalent intensity. This intensity is then combined with the highest level to obtain the total sound level. Figure 11-10 is used for the calculations.

50- and 55-db levels

Difference = 5 db

Amount to be added to larger level = 1.2 db

Equivalent level = 55 + 1.2 = 56.2 db

56.2- and 60-db combination

Difference = 3.8 db

Amount to be added to larger level = 1.5 db

Total sound level = 60 + 1.5 + 61.5 db

The human ear responds differently to different sound frequencies and to certain groups of frequencies, which are called "noise." The acoustical engineer is responsible for designing rooms or systems such that unwanted frequencies or noise is filtered out while still allowing clarity of transmission of certain other sounds. An auditorium, for example, should be designed so that very little outside noise is transmitted through the walls, while the inside structure and fixtures should allow clear transmission of audio performances on the stage. The walls should have good sound insulation, while the inside surfaces should be properly designed so that unwanted reverberations are eliminated without causing undue attenuation of sound from the stage. Almost all acoustical design problems involve the exercise of considerable judgment based on experience as well as a knowledge of some basic sound-level measurements.

Sound-level measurements are performed with some type of microphone, which may be considered a type of seismic vibration instrument. The electric output of the microphone is proportional to sound-pressure level, which may be used to calculate the sound intensity according to Eq. (11-18). Appropriate power amplifiers and readout meters or recorders indicate the sound level. In general, the microphone must be calibrated in a test facility with a source of known frequency and intensity. Even with careful calibration, accuracies of better than ± 1 db may not be expected in sound-pressure-level measurements. The characteristics of some typical microphones are given in Table 11-1.

A typical practical application of sound-level measurements may call for an analysis of the noise spectrum in a certain sound source. For this purpose a noise analyzer with bandpass-filter circuits is used. Several sound-pressure-level measurements are taken to determine the sound intensity in various wavelength bands. Based on experience, the acoustical engineer may then take steps to attenuate whatever frequencies he feels are necessary to achieve the design objective. For many commercial noise and vibration meters calibration may be achieved with a simple whistle or tuning fork in a quiet room. Sound-level readings will generally follow the inverse square law.

Table 11-1 Characteristics of microphones according to Ref. [9]

Mechanism	Frequency range	Approximate open-circuit sensitivity, db, below 1 volt/dyne/ cm²	Approximate impedance, ohms	Application and remarks
Carbon microphone	Up to 5 kHz	−40	100	Telephones
Electrodynamic	Up to 20 kHz	−85	10	Field measurements, communications, etc.
Electrostatic	Up to 50 kHz	−50	500,000	Precision measurements, standards
Piezoelectric:				
Rochelle	Audio region ultrasonic	−50	100,000	Temperature-dependent Hygroscopic
ADP crystal		−50		Use in underwater sound, solids, etc.
Quartz	Mainly ultrasonic	−90 to −100	High	
Barium titanate		−90	Low	High-intensity work in air; water; measurements with small probes
Magnetostrictive	Mainly ultrasonic	−100	Low	Underwater sound
Ribbon microphone	Audio region	−100	1	Directive

Example 11-8 A certain microphone has an open-circuit sensitivity of −80 db referenced to 1 volt for a sound-pressure excitation of 1 dyne/cm². Calculate the voltage output when exposed to a sound field of (a) 100 db and (b) 50 db.

Solution We first calculate the sound pressure from Eq. (11-19). We have, for the 100-db level

$$100 = \frac{20 \log p}{2 \times 10^{-4}}$$

$$p = 20 \text{ dyne/cm}^2$$

The reference voltage for an excitation of 1 dyne/cm² is calculated as

$$-80 \text{ db} = \frac{20 \log E}{1.0}$$

$$E = 10^{-4} \text{ volts}$$

Since we assume the output voltage varies in a linear manner with the impressed sound field, the output voltage at 100-db sound level is

$$E = 20 \times 10^{-4} = 2 \text{ mv}$$

For a sound level of 50 db the sound pressure is obtained from

$$50 \text{ db} = \frac{20 \log p}{2 \times 10^{-4}}$$

$$p = 6.24 \times 10^{-2} \text{ dyne/cm}^2$$

The output voltage is therefore

$$E = (6.24 \times 10^{-2})(10^{-4}) = 6.24 \times 10^{-6} \text{ volts}$$

A good summary of acoustical principles and materials for use in sound-attenuation applications is given in Refs. [2], [3], and [8]. The following paragraph presents a summary discussion of some of the types of acoustical phenomena that are encountered in practice.

TRAVELING PLANE WAVES

A plane acoustic wave may be described in terms of the displacement of a particle at a distance x from the wave source and at a time t.   The analytical relation expressing the particle displacement as a function of distance and time for a harmonic wave is

$$\xi = \xi_m \cos \frac{2\pi}{\lambda} (ct - x) \tag{11-20}$$

where ξ is the particle displacement, ξ_m is the particle-displacement amplitude, c represents the sound velocity, and λ is the wavelength.   An observer located at some particular distance x_1 from a plane wave source will experience a periodic variation of the particle displacement and velocity.   Another observer located at a distance x_2 from the source will experience exactly the same relative periodic motion; however, the absolute time at which the maximum and minimum points in the motion occur will depend on the distance $(x_2 - x_1)$ and the wave velocity c.   The amplitude of the oscillatory motion will decrease with the distance from the wave source owing to viscous dissipation in the fluid.   The plane wave described by Eq. (11-20) is called a traveling wave since the particle displacement is dependent on time and the distance from the wave source.   Traveling waves may be described in a similar manner for cylindrical wave propagation.

The fluctuations in pressure due to the passage of a plane sound wave may be described in terms of the amplitude of the particle displacement through the relation

$$p = \beta \frac{\partial \xi}{\partial x} = -\beta \frac{2\pi}{\lambda} \xi_m \sin \frac{2\pi}{\lambda} (ct - x) \tag{11-21}$$

where β is the adiabatic bulk modulus of the fluid defined by

$$\beta = -V\frac{dp}{dV} \tag{11-22}$$

The intensity of a sound wave is defined as the flux of energy per unit time and per unit area. Several equivalent expressions for the intensity of a plane wave are given in Eq. (11-23).

$$
\begin{aligned}
I &= \tfrac{1}{2}\rho_a c(\dot{\xi}_m)^2 \\
&= \tfrac{1}{2}\rho_a c\omega^2 \xi_m{}^2 \\
&= \rho_a c(\dot{\xi}_{rms})^2 \\
&= p_{rms}\dot{\xi}_{rms} \\
&= \frac{p_{rms}^2}{\rho_a c} \tag{11-23}
\end{aligned}
$$

where ρ_a = air density

ω = frequency

$\dot{\xi}_m$ = velocity amplitude of the wave

$\dot{\xi}_{rms}$ = rms value of the particle velocity

p_{rms} = rms pressure fluctuation

It is worthwhile to note that the intensity expressions containing the pressure terms do not apply for the case of cylindrical or spherical waves unless the radius of curvature of these waves is large enough that the wave front may be approximated by a plane.

STANDING WAVES

Standing waves, like traveling waves, are described in terms of periodic particle displacements and the corresponding periodic particle velocities. For the case of traveling waves the amplitude of the particle displacement is the same regardless of the position x provided that viscous dissipation effects are neglected. For standing waves the amplitude of the particle displacement will follow a periodic variation with the distance x from the sound source. At even quarter-wavelengths the amplitude will be zero, and at odd quarter-wavelengths the particle-displacement amplitude will take on its maximum value. Consequently, the term "standing wave" is derived from the fact that the amplitude of the particle displacement has a periodic variation which is independent of time. Thus, we have the idea that the wave "stands" in a certain position. The particle motion in a standing wave may be represented by the relation

$$\xi = 2\xi_m \sin\left(\frac{2\pi x}{\lambda}\right) \cos\left(\frac{2\pi ct}{\lambda} + \text{const}\right) \tag{11-24}$$

CONSTANT-PRESSURE SOUND FIELD

Constant-sound-pressure waves are distinctly different from either traveling waves or standing waves and are encountered in reverberant chambers whose walls reflect the major portion of the wave energy striking them. For the ideal case, the energy of the sound source is distributed uniformly throughout the reverberant room so that the fluid will experience a uniform compression and expansion in all directions.

Traveling waves and standing waves represent a vector field; i.e., they depend on the direction in which the wave is propagating. Reverberant waves, for the ideal case, are independent of either distance or direction from the wave source; hence, they represent a scalar field. There is evidently no simple correlation between the pressure fluctuations and the particle displacements in the free space of a reverberant room.

PSYCHOACOUSTIC FACTORS

The sound-pressure scale for typical sound sources is indicated in Fig. 11-11, ranging from the threshold of hearing to the intensive sound of a rocket engine. We have already noted that the human ear responds differently to different frequencies; therefore, in order to interpret the response of the ear to a particular sound source we must know the distribution of the sound over the entire frequency spectrum. The nonlinear response of the ear has been determined, resulting in the Fletcher-Munson [11] curves for equal loudness shown in Fig. 11-12. According to these curves, the ear is much more sensitive to midfrequencies than to low or high frequencies. Based on the above concepts, it is possible to define the *loudness level* as a relative measure of sound strengths judged by an average human listener. The unit of loudness level is the *phon*, which is defined so that the loudness in phons is numerically equal to the sound-pressure level at a frequency of 1,000 Hz. The loudness levels are noted on the curves of Fig. 11-12.

MEASUREMENT SYSTEMS

The typical sound-measurement system is shown in Fig. 11-13. An appropriate microphone is used to sense the sound wave. Before measurements begin, an acoustic calibrator is placed over the microphone to effect a calibration. The calibrator consists of a special cavity for the particular microphone and a transducer that will produce a known sound-pressure level inside the cavity for given driving voltage inputs. The response of the

	180	Rocket engine
	160	Supersonic boom
	140	Threshold of pain
Hydraulic press (3 ft)	130	
Large pneumatic riveter (4 ft)		
		Boiler shop (maximum level)
Pneumatic chipper (5 ft)		
	120	
Overhead jet aircraft, four-engine (500 ft)		
		Jet-engine test-control room
Unmuffled motorcycle		
	110	Construction noise (compressors and
Chipping hammer (3 ft)		hammers) (10 ft)
		Woodworking shop
Annealing furnace (4 ft)		
		Loud power mower
Rock and roll band	100	
Subway train (20 ft)		
Heavy trucks (20 ft)		Inside subway car
Train whistles (500 ft)		Food blender
	90	
10-hp outboard (50 ft)		Inside sedan in city traffic
Small trucks accelerating (30 ft)		
		Heavy traffic (25 to 50 ft)
	80	Office with tabulating machines
Light trucks in city (20 ft)		
Automobiles (20 ft)		
	70	
Dishwashers		Average traffic (100 ft)
		Accounting office
Conversational speech (3 ft)		
	60	
	50	Private business office
		Light traffic (100 ft)
		Average residence
	40	
	30	
		Broadcasting studio (music)
	20	
	10	
	0	

Fig. 11-11 Typical overall sound levels; decibels RE 0.0002; microbar.

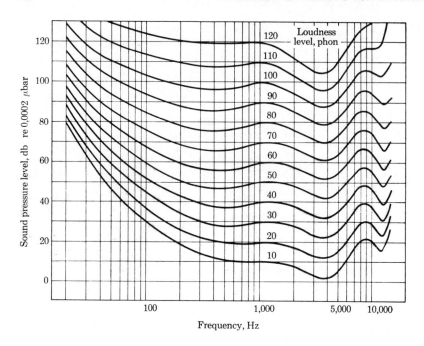

Fig. 11-12 Fletcher-Munson [11] curves for equal-loudness contours for pure tones.

microphone to the calibration signals is noted on the sound-level meter and thereby furnishes reference levels for use in comparing the signals produced by the unknown sound field.

In practice, the microphone may be mounted directly on the sound-level meter or separated with a connecting cable in order to accommodate some particular measurement objective. The sound-level meter incorporates

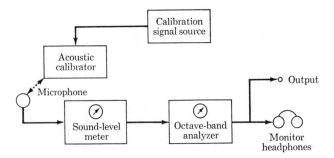

Fig. 11-13 Schematic of instrument arrangement for sound measurements.

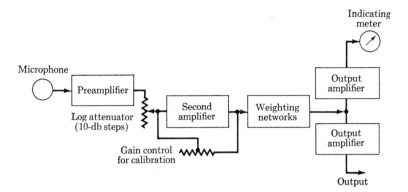

Fig. 11-14 Schematic of sound-level meter.

the functions shown in Fig. 11-14. An input preamplifier amplifies the attenuator switch, which may be varied in 10-db steps. The signal is then amplified further, and three standardized weighting networks modify the signal in accordance with approximations to the standard loudness contours. The frequency-response characteristics of these networks are shown in Fig. 11-15. The *A* scale approximates the 40-phon contour and is used for levels below 55 db. The *B* scale approximates the 70-phon contour and is used for levels between 55 and 85 db. Finally, the *C* scale is used above 85 db and provides a fairly flat response curve. The output from the weighting networks is either observed on the self-contained indicating meter

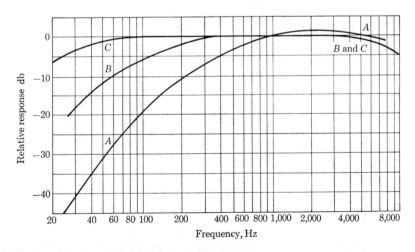

Fig. 11-15 Standardized weighting networks for sound-level meters for random sound incidence upon the microphone. (*American Standard Specification for General Purpose Sound-Level Meters SI-4-1961.*)

or with some type of external recording device such as an oscillograph or tape recorder.

The entire sound spectrum is of importance because of the influence of different frequencies on human response. For this reason the output of the sound-level meter is usually fed to an octave-band analyzer, which is an instrument containing amplifiers, standard filter circuits, and an indicating-meter circuit to measure the signal strength in each filter passband. The standard octave passbands are given in Table 11-2. Facilities are provided to measure the output signal of the octave-band analyzer with standard recording equipment. In general, a large variety of instruments are available for measurement and analysis of sound level, and new products appear on the market regularly. The reader is urged to consult well-disposed manu-facturers' representatives for information on the instruments that are cur-rently available and that may be most suitable to his application.

ACOUSTICAL PROPERTIES OF MATERIALS

One of the important objectives of acoustic measurements is a specification of the sound-absorption properties of materials. With such information available, the acoustical engineer may then choose the proper material to accomplish his environmental objectives in noise control or architectural design. There are many complicating factors to be considered in the selection of sound-insulation materials, and we shall discuss only two parameters that are of common interest. For further information the reader should consult Refs. [3] and [10].

The sound-absorption coefficient $\bar{\alpha}$ is defined by

$$\bar{\alpha} = \frac{\text{sound energy absorbed}}{\text{sound energy incident upon surface}}$$

The absorption coefficient is strongly frequency-dependent and influenced by such factors as porosity of the surface, thickness of material, rigidity of material, etc. Figure 11-16 illustrates the frequency dependence for a

**Table 11-2   Standard
octave-band-
analyzer filters**

Low pass to 75 Hz
75–150 Hz
150–300 Hz
300–600 Hz
600–1,200 Hz
1,200–2,400 Hz
2,400–4,800 Hz
High pass from 4,800 Hz

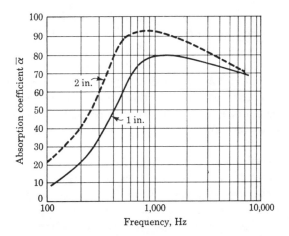

Fig. 11-16 Increase in sound absorption resulting from a doubling in thickness of a typical homogeneous porous sound-absorbing material, according to Ref. [3].

typical porous sound-absorbing material. The material is clearly most effective in absorbing sound fields at mid- and high frequencies. The absorption characteristics of some different types of materials are shown in Fig. 11-17.

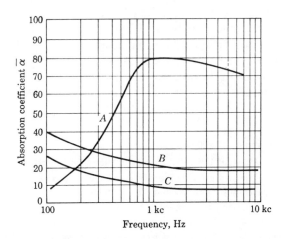

Fig. 11-17 Absorption coefficients for: (*a*) 1-in.-thick porous acoustical material; (*b*) two sheets of ⅛-in.-thick plywood, separated by tufts of randomly glued felt causing ⅛-in. spacing between sheets; (*c*) one sheet of ⅜-in.-thick randomly braced plywood, according to Ref. [3].

The more rigid wood panels are substantially less effective as absorbers than the porous material. In general, rigid materials are poor sound absorbers because they have the ability to vibrate and reemit sound. A highly damped, nonrigid material like the porous sheet or, to a lesser degree, the assembly represented by curve B in Fig. 11-17 is more effective in absorbing an incident sound wave. A thin sheet of steel mounted in an undamped fashion would be a very poor absorber because of the ease with which it may be set into vibration.

NOISE-REDUCTION COEFFICIENT

Since the sound-absorption characteristics of materials are so strongly frequency-dependent, some average parameter must be selected which represents the material behavior over the entire frequency spectrum. One could choose an integrated average of the absorption coefficient, but the accepted practice is to define a noise-reduction coefficient (NRC) as

$$\text{NRC} =$$
$$\frac{\text{sound-absorption coefficients of material at 250, 500, 1,000, and 2,000 Hz}}{4}$$

Thus the NRC may be determined by measuring the absorption coefficients at the four test frequencies and taking their arithmetic average. It should be noted that this definition does not take into account the absorption characteristics at a low frequency of, say, 125 Hz. For this reason, comparison of materials on the basis of NRC alone may be misleading if the absorption characteristics at low frequencies are of particular interest. Certain inconsistencies are present in published data of noise-reduction coefficients because of the influence of sample mounting and test procedures on the final reported measurements. For further information the interested reader should consult Refs. [3] and [10].

REVIEW QUESTIONS

1. Describe the basic concept of the seismic instrument.
2. Under what conditions is a seismic instrument suitable for (a) amplitude measurements and (b) acceleration measurements?
3. Describe some alternative constructions of a seismic instrument. When would each construction be used?
4. When would a piezoelectric transducer be used to advantage in a seismic instrument?
5. What are the reference-energy flux and pressure levels for sound fields?
6. What is the maximum permissible sound level for human exposure?
7. What is meant by the open-circuit sensitivity of a microphone?
8. Distinguish between traveling, standing, and constant-pressure sound fields.
9. Why is an octave-band analyzer used in sound measurements?

10. What is meant by loudness level?

11. Is the ear more or less sensitive to low frequencies than to midfrequencies (that is, 1,000 Hz)?

12. How is a sound-level meter calibrated?

13. Define the sound-absorption coefficient.

14. Define the noise-reduction coefficient (NRC).

PROBLEMS

11-1. The vibrating wedge shown in Fig. 11-1 is used for an amplitude measurement. The length of the wedge is 6.0 ± 0.02 in., and the thickness is 1.0 ± 0.01 in. The x distance is measured as $x = 2.2 \pm 0.05$ in. Calculate the vibration amplitude and its uncertainty in percent.

11-2. A small cantilever vibrometer is available for measurement of vibration frequency, but the specification sheet is lost so that the properties of the device are not known. The instrument is calibrated by placing it on a large compressor in the laboratory which is rotating at 300 ± 2.0 rpm. The measured length for resonance conditions is 2.2 ± 0.01 in. Calculate the frequency that the instrument will indicate when $L = 4.0 \pm 0.02$ in. Also calculate the uncertainty in the measurement at this length.

11-3. Consider the cantilever beam of Example 11-1. Suppose the uncertainties in the material properties and dimensions are

$$w_E = 2\%$$
$$w_r = 0.0005 \text{ in.}$$
$$w_\rho = 5 \text{ lb}_m/\text{ft}^3$$
$$w_L = 0.01 \text{ in.}$$

Calculate the resulting uncertainty in ω_n at $L = 1.0$ and $L = 4.0$ in.

11-4. A cantilever beam is to be used for the measurement of frequencies between 100 and 1,000 Hz. The beam is to be made from a rod of spring steel having the properties given in Example 11-1. The maximum length of the rod is to be 5 in., and the uncertainties in material dimensions and properties are those given in Prob. 11-3. Calculate the nominal diameter of the rod and the uncertainty in the frequency at 100 and 1,000 Hz.

11-5. A seismic accelerometer is to be used to measure linear acceleration over a range from 100 to 1,000 ft/sec². The natural frequency of the instrument is 200 Hz, and this value may vary by ± 2 Hz owing to temperature fluctuations. Calculate the allowable uncertainty in the relative displacement measurement in order to ensure an uncertainty of no more than 5 percent in the acceleration measurement.

11-6. A seismic instrument is to be used to measure velocity. Show that the input velocity amplitude is

$$v_0 = \left(\frac{dx_1}{dt}\right)_0 = \omega_1 x_0$$

Subsequently show that this velocity amplitude may be expressed in terms of the steady-state relative displacement as

$$(x_2 - x_1)_0 = \frac{v_0(\omega_1/\omega_n)}{\omega_n\{[1 - (\omega_1/\omega_n)^2]^2 + [2(c/c_c)(\omega_1/\omega_n)]^2\}^{\frac{1}{2}}}$$

11-7. Calculate the value of the time constant for the instrument in Prob. 11-5 if $c/c_c = 0.65$.

11-8. Plot the error in acceleration measurement of a seismic instrument for $c/c_c = 0.7$ versus frequency ratio, i.e.,

$$1 - \frac{(x_2 - x_1)_0 \omega_n^2}{a_0} \qquad \text{versus} \qquad \frac{\omega_1}{\omega_n}$$

11-9. A large seismic instrument is constructed so that $m = 100$ lb_m and $c/c_c = 0.707$. A spring with $k = 200$ lb_f/ft is used so that the instrument will be relatively insensitive to low-frequency signals for displacement measurements and relatively insensitive to high-frequency signals for acceleration measurements. Calculate the value of linear acceleration that will produce a relative displacement of 0.1 in. on the instrument. Calculate the value of ω_1/ω_n such that

$$\frac{(x_2 - x_1)_0}{x_0} = 0.99$$

11-10. A seismic accelerometer is to be designed so that the time constant calculated from Eq. (11-15) is equal to the period of the maximum acceptable frequency for a 1 percent error in measurement; that is, $(x_2 - x_1)_0/a_0 = 0.99$. Plot ω_n versus c/c_c in accordance with this condition.

11-11. Calculate the phase angle ϕ for the conditions of Example 11-2.

11-12. Calculate the phase angle ϕ for the conditions of Example 11-3.

11-13. Calculate the energy flux and rms pressure variation corresponding to a sound intensity of 100 db.

11-14. The estimated accuracy of a sound-intensity measurement is usually expressed in decibels. Calculate the uncertainties in intensity in watts per square centimeter corresponding to ± 1, ± 2, and ± 3 db. Express these as percentage values at sound levels of 50, 80, and 120 db.

11-15. A standing sound wave is created in room air ($70°F$, 14.7 psia) with a frequency of 1,000 Hz. The peak sound intensity occurring in the wave is 120 db. Plot the sound intensity and pressure fluctuations as functions of x for one wavelength. Over what fractional portion of a wavelength could the intensity be considered constant, consistent with an estimated uncertainty of ± 1 db in the intensity measurement?

11-16. Calculate the equivalent sound intensity of a combination of two, three, and four 50-db sources.

11-17. Calculate the total sound intensity of a combination of 63- and 73-db sources.

11-18. Four sound sources of 62, 58, 65, and 70 db are added together. Calculate the total sound intensity in two different ways showing that the same result is obtained.

11-19. A microphone has an open-circuit sensitivity of -55 db. Calculate the voltage output for sound levels of 60 and 80 db.

11-20. Two microphones having sensitivities of -55 and -70 db are available. What amplification factor must be applied to the signal of the least sensitive unit in order to make its output the same as the other microphone?

11-21. Calculate the noise-reduction coefficient (NRC) for the materials shown in Figs. 11-16 and 11-17.

11-22. Calculate the attenuation in decibels for the 2-in. porous material of Fig. 11-16 within each octave band of Table 11-2. For this calculation choose some average absorption coefficient for each range of interest. Assume that all energy not absorbed is transmitted through the material.

11-23. A 1,000-Hz tone is generated with an intensity of 70 db. What must be the intensities of tones at 50, 100, and 10,000 Hz in order to be detected with the same loudness by the ear?

11-24. Suppose that the playback of a certain musical passage in a high-fidelity system produces an intensity of 80 db at 1,000 Hz when the reproduction is judged to be most authentic to the original source. For various reasons, a person may wish to play the music at a lower level. When the 1,000-Hz tones are reduced to 45 db, how much must the 60- and 100-Hz tones be boosted in order to retain the same sense of authenticity to the original source?

11-25. Compare the sound-level-meter weighting-characteristic curves with the loudness curves of 40, 70, and 90 phon, corresponding to the *A*, *B*, and *C* scales.

REFERENCES

1. Beckwith, T. G., and N. L. Buck: "Mechanical Measurements," 2d ed., Addison-Wesley Publishing Company, Inc., Reading, Mass., 1969.
2. Beranek, L. L.: "Acoustics," McGraw-Hill Book Company, New York, 1954.
3. Beranek, L. L.: "Noise Reduction," McGraw-Hill Book Company, New York, 1960.
4. Dove, R. C., and P. H. Adams: "Experimental Stress Analysis and Motion Measurement," Merrill, Columbus, Ohio, 1964.
5. Gross, E. E.: Noise Measuring and Sound Control, *Refrig. Eng.*, vol. 66, p. 49, 1957.
6. Harris, C. M., and C. M. Crede (eds.): "Basic Theory and Measurements," vol. 1, "Shock and Vibration Handbook," McGraw-Hill Book Company, New York, 1961.
7. Hetenyi, M.: "Handbook of Experimental Stress Analysis," John Wiley & Sons, Inc., New York, 1950.
8. "ASHRAE Guide and Data Book," Amer. Soc. Heating, Refrig., and Air Cond., 1963.
9. Condon, E. U. (ed): "Handbook of Physics," McGraw-Hill Book Company, New York, 1958.
10. Harris, C. M. (ed.): "Handbook of Noise Control," McGraw-Hill Book Company, New York, 1957.
11. Fletcher, H., and W. A. Munson: Loudness, Its Definition, Measurement, and Calculation, *J. Acous. Soc. Am.*, vol. 5, pp. 82–108, 1933.

12
Thermal- and Nuclear-radiation Measurements

12-1 INTRODUCTION

In Chap. 8 we have already mentioned the importance of thermal-radiation measurements in temperature determinations. In this chapter we shall examine the physical principles and operating characteristics of some of the more important thermal-radiation detectors and indicate their range of applicability. For a rather complete survey of the field of infrared engineering and thermal-radiation detectors, the reader is referred to the monograph by Hackforth [7].

Radioactivity measurements comprise a broad field of activity by engineers and are becoming increasingly important in many applications. In this chapter we shall discuss some of the measurements that find wide application and indicate the devices used for performing these measurements. Our discussion is, of necessity, a cursory one, and the interested reader should consult the book by Price [10] for detailed information on detection of nuclear radiation.

12-2 DETECTION OF THERMAL RADIATION

The measurement of thermal radiation is basically a measurement of radiant energy flux. The detection of this energy flux may be accomplished through

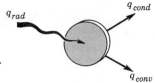

Fig. 12-1 Schematic of the general thermal-radiation detector.

a measurement of the temperature of a thin metal strip exposed to the radiation. The strip is usually blackened to absorb most of the radiation incident upon it and is constructed as thin as possible to minimize the heat capacity and thereby bring about the most desirable transient characteristics. A schematic of the general thermal-radiation detector is given in Fig. 12-1. The temperature attained by the element is not only a function of the radiant energy absorbed, but is also dependent on the convection losses to the surroundings and conduction to the mounting fixtures. Convection losses from the element may be reduced by enclosing the detector in an evacuated system with an appropriate window for transmission of the radiation. The infrared-transmission characteristics of several substances employed as window materials are given in Fig. 12-2 according to Ref. [1]. Conduction losses may be reduced with suitable insulating materials.

Either thermocouples or thermopiles may be used for detecting the temperature of the blackened radiation-sensitive element. Thermopiles offer the advantage that they produce a higher voltage output. Many ingenious methods of construction have been devised for such thermopiles, a commercial example of which is given in Fig. 12-3. The expanded view of the thermopile shows the blackened junction pairs surrounded by an annular ring of mica that serves as both electrical and thermal insulation. The thermopile registers the difference in temperature between the hot junctions and the ambient temperature surrounding the detector. The lens at the front of the device focuses the radiation on the thermopile junctions. Special circuitries are used to provide for compensation of ambient temperatures between 50 and 250°F.

Thermal radiation may also be sensed by a metal bolometer, which consists of a thin strip of blackened metal foil such as platinum. The temperature of the foil is indicated through its change in resistance with temperature. An appropriate bridge circuit is used to measure the resistance.

Thermistors are widely used as thermal-radiation detectors, and a cutaway drawing of a commercial detection device is given in Fig. 12-4. Two thermistors are enclosed in the detector case, which is covered by an appropriate glass window having satisfactory transmission characteristics. One thermistor element is exposed to the incoming radiation that is to be measured, while the other element is shielded from this radiation. The shielded element is connected in the circuit so that it furnishes a continuous compensation for the temperature of the detector enclosure. A schematic of a commercial radiometer using thermistor detectors is given in Fig. 12-5. The incoming radiation is focused by the mirror system onto the thermistor detector. A motor-driven chopper periodically interrupts the radiation so that an alternating signal is produced which may be amplified more easily. The amplified signal is subsequently rectified to produce an output voltage proportional to the radiant flux incident on the thermistor detector. In this system mirrors may be adjusted so that the optical system can be focused on an area as small as 1 mm². The movable mirror and focusing lamp are used for this purpose. The system is exceedingly sensitive and may even be used for detection of radiation from sources near room temperatures.

Thermal-radiation detectors are calibrated directly by obtaining the

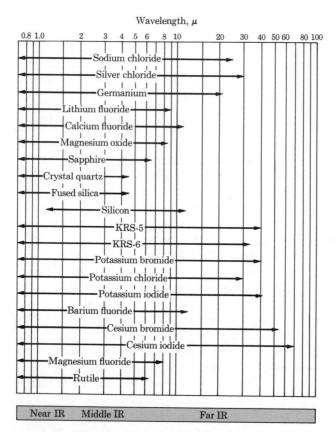

Fig. 12-2 Infrared transmission characteristics of several optical materials according to Ref. [1] as arranged by Hackforth [7]. Sample thickness = 2 mm, long-wavelength cutoff at 10 percent transmittance.

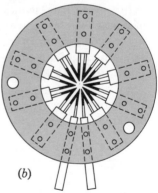

Fig. 12-3 A commercial radiometer utilizing a thermopile sensor. (*a*) Cutaway of instrument; (*b*) detail of thermopile sensor. (*Courtesy Minneapolis-Honeywell Co.*)

output as a function of the known radiation from a blackbody source at various temperatures. A typical blackbody source is constructed as shown in Fig. 12-6. The conical cavity is constructed of some high-conductivity material such as aluminum or copper, and the inside surface is blackened. An electric heater maintains the cavity at a desired temperature, which is

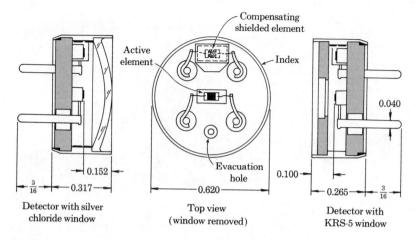

Fig. 12-4 A commercial thermistor-radiation detector. (*Courtesy Barnes Engineering Co., Stamford, Conn.*)

indicated and controlled through a sensitive electric-resistance thermometer or thermistor. Baffles near the opening prevent stray radiation from the surroundings from influencing the radiation output of the cavity. The detailed construction of blackbody standards is discussed by Marcus [9]. It may be noted that cavities like the one discussed above may give effective emissivities within 1 percent of blackbody conditions.

A comparison of the approximate transient and energy threshold characteristics of three types of thermal-radiation receivers is given in Table 12-1 according to Ref. [2]. The energy-threshold values are the minimum energy which may adequately be detected by such a device.

Photoconductive-radiation detectors have been discussed in Sec. 4-17 and response characteristics of a number of practical sensors noted   The

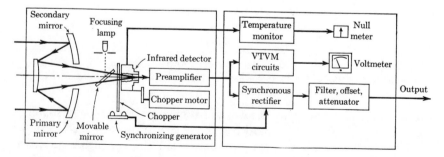

Fig. 12-5 Schematic of a commercial radiometer utilizing a thermistor detector. (*Courtesy Barnes Engineering Co., Stamford, Conn.*)

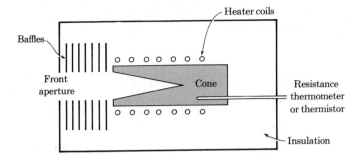

Fig. 12-6 Typical construction of a blackbody source.

lead-sulfide cell is widely used for detection of radiation in the 1- to 4-μ range.

Example 12-1 A certain thermistor radiation bolometer detector has characteristics such that the minimum input power necessary to produce a signal-to-noise ratio of unity is 10^{-8} watt. Suppose such a detector is located 3 ft from a blackbody radiation source and is perfectly black. Calculate the temperature of the source necessary to produce a signal-to-noise ratio of 10, assuming that the detector and surroundings are maintained at 70°F. The source is a black sphere, 2 in. in diameter, and the detector area is 1 mm².

Solution   The net energy absorbed by the detector is

$$q = \sigma A (T_s^4 - 530^4)(\tfrac{1}{36})^2$$

where A is the detector area. For a signal-to-noise ratio of 10 the input power must be $10 \times 10^{-8} = 10^{-7}$ watt $= 3.413 \times 10^{-7}$ Btu/hr $= q$. Thus,

$$T_s^4 - 530^4 = \frac{(3.413 \times 10^{-7})(36)^2}{(0.1714 \times 10^{-8})(0.01)(0.001076)}$$

$$= 240 \times 10^8$$

$$T_s = 566°\text{R} = 106°\text{F}$$

Table 12-1   Characteristics of thermal receivers†

Type	Response time	Energy threshold, watts
Thermocouple	0.2 sec	3.0×10^{-6}
Thermistor bolometer	3.0 msec	7.2×10^{-8}
Metal bolometer	4.0 msec	3.3×10^{-8}

† According to Ref. [2].

For a signal-to-noise ratio of 100 the corresponding value of T_s would be 291°F. From the results of this example we see why optical systems like those shown in Figs. 12-3 and 12-5 are used to collect the radiation over a fairly large area and focus it on the small surface area of the detector.

12-3 MEASUREMENT OF EMISSIVITY

An apparatus for the measurement of total normal emissivity has been described by Snyder, Gier, and Dunkle [11].   The apparatus uses a thermo-pile radiometer and is constructed as shown in Fig. 12-7.   An electric heater is used to maintain the temperature of the sample, while thermocouples embedded in the sample furnish an indication of its temperature.   A detailed drawing of the thermopile receiver is shown in Fig. 12-7b.   It is constructed of 160 junctions of silver constantan mounted in a cylindrical housing which is blackened on the inside.   Two blackened aluminum-foil strips are attached to the junctions.   The rear shield has a narrow slot which allows exposure of the hot-junction strip to the radiant flux from the sample while exposing

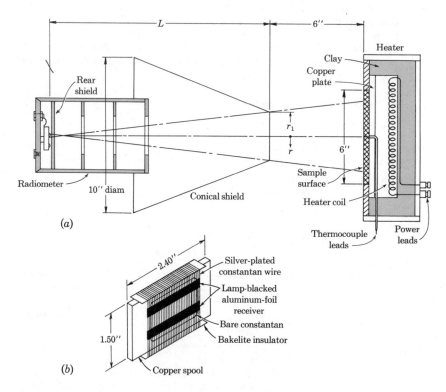

Fig. 12-7 Apparatus for measurement of normal emissivity according to Ref. [11]. (*a*) Schematic; (*b*) detail of thermopile construction.

the cold-junction strip to the temperature of the thermopile enclosure. The temperature difference between the hot and cold junctions is thus taken as an indication of the radiant energy flux, which, in turn, is related to the emissivity of the sample. The conical shield on the front of the device ensures a proper concentration of energy from the sample. According to Ref. [11], the device furnishes essentially a linear output of 0.1275 mv per Btu/(hr)(ft²) radiant flux. If the temperatures of the hot and cold junctions are assumed to be essentially the same as the inside temperature of the radiometer (this apparently is a good assumption because of the small total energy absorbed in the radiometer) and the sample is assumed gray, the total normal emissivity of the sample is

$$\epsilon_s = \frac{7.84\,E}{F_{ts}\sigma(T_s^4 - T_R^4)} \tag{12-1}$$

where E = voltage output of the thermopile, mv

T_s = sample temperature

T_R = radiometer temperature

The view factor F_{ts} is calculated from

$$F_{ts} = \frac{r_1^2}{r_1^2 + L^2} \tag{12-2}$$

In order for the measurement to be valid the temperature of the radiometer must be maintained at very nearly that of the surroundings.

Glaser and Blau [6] have presented a method for measurement of spectral emissivities, and the interested reader should consult their paper for additional information on spectral characteristics of surfaces.

Example 12-2 A measurement of total emissivity is made with the apparatus shown in Fig. 12-7. $L = 14.25$ in. and $r_1 = 1.50$ in. The output is 0.823 ± 0.005 mv, $T_s = 703 \pm 1.0°F$, and $T_R = 70 \pm 0.5°F$. Calculate the nominal value of the emissivity, and estimate the uncertainty in the measurement.

Solution We first calculate the view factor F_{ts} as

$$F_{ts} = \frac{(1.50)^2}{(1.50)^2 + (14.25)^2} = 0.011$$

The emissivity is then calculated with the use of Eq. (12-1).

$$\epsilon_s = \frac{(7.84)(0.823)}{(0.011)(0.1714 \times 10^{-8})(1163^4 - 530^4)}$$

$$= 0.195$$

We use Eq. (3-2) to calculate the uncertainty. The appropriate parameters are

$$\frac{\partial \epsilon}{\partial E} = \frac{7.84}{F_{ts}\sigma(T_s^4 - T_R^4)} = 0.237 \text{ mv}^{-1}$$

$w_E = 0.005 \text{ mv}$

$$\frac{\partial \epsilon}{\partial T_s} = \frac{-(7.84)(4)T_s^3 E}{F_{ts}\sigma(T_s^4 - T_R^4)^2} = -7.0 \times 10^{-4} \text{ }^{\circ}\text{R}^{-1}$$

$w_{T_s} = 1.0 \text{ }^{\circ}\text{R}$

$$\frac{\partial \epsilon}{\partial T_R} = \frac{(7.84)(4)T_R^3 E}{F_{ts}\sigma(T_s^4 - T_R^4)^2} = 0.663 \times 10^{-5} \text{ }^{\circ}\text{R}^{-1}$$

$w_{T_R} = 0.5 \text{ }^{\circ}\text{R}$

and the uncertainty in the emissivity is calculated as

$$w_\epsilon = \{[(0.237)(0.005)]^2 + [(-7.0 \times 10^{-4})(1.0)]^2 + [(0.663 \times 10^{-5})(0.5)]^2\}^{\frac{1}{2}}$$
$$= 0.00137 \text{ or } 0.7\%$$

If the uncertainty in the source temperature had been $\pm 5^{\circ}\text{F}$, the resulting uncertainty in the emissivity would be 0.0037 or 1.9 percent.

12-4 REFLECTIVITY AND TRANSMISSIVITY MEASUREMENTS

In Chap. 8 we have already described the nature of thermal radiation and the properties which describe surface characteristics.   Emissivity is one important property, and Sec. 12-3 has described one method for its measurement. Reflectivity and transmissivity are other important surface properties, and we shall now describe a technique for their measurement through the use of an integrating sphere.   For completeness we shall describe the technique for determining monochromatic properties, i.e., properties over a range of thermal-radiation wavelengths.

Let us assume that a suitable *monochromer* is available to produce monochromatic radiation.   This radiation might be produced by some kind of prism arrangement or through the use of narrow-band optical filters.   We wish to determine the fraction of some incident radiation that is either reflected or transmitted by a surface.   For this purpose the integrating sphere of Fig. 12-8 is employed.   The monochromatic radiation is chopped to produce an ac source to a detector that may be easily amplified.   Consider first the measurement of reflectivity.   The pivoted mirror permits a divergence of the monochromatic beam either directly to the specimen inside the sphere or to some location on the inner surface of the sphere.   The inside of the sphere is coated with a thick layer (about 2 mm) of magnesium oxide to produce a very high and diffuse reflectance of about 0.99.   Thus, radiation incident on the inner surface will be diffusely reflected throughout the sphere, producing a uniform irradiation of the inner surface.   Around the surface are placed radiation detectors to sense the reflected radiation.   The general

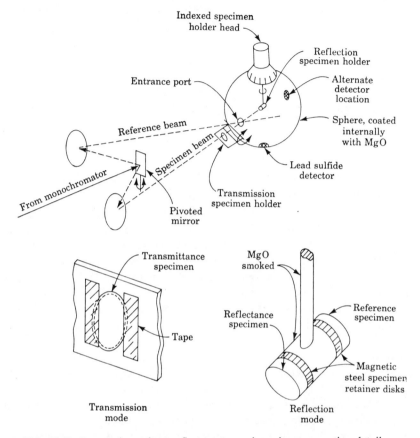

Fig. 12-8 Integrating-sphere reflectometer and specimen mounting details.

measurement technique consists of a comparison of the detector signal for direct impingement of the reference beam with that which results from reflected radiation from the sample.   The ratio of these two signals is the reflectivity of the sample.   It must be noted that a chopped (ac) input beam is essential for this measurement technique because the detectors in the side of the sphere sense all radiation incident upon them, including the continuous signal resulting from thermal emission of the wall of the sphere.   Appropriate electronic circuits filter out all but the chopped signals so that the output will be indicative of the input chopped radiation.

A measurement of transmissivity is made by placing a sample over the input port to the sphere.   After transmission through the specimen, the signal is then compared to that obtained for direct impingement on the inner surface of the sphere.

The integrating-sphere technique is widely regarded as the most reliable method for obtaining reflectance and transmissivity measurements and may properly be described as an absolute measurement standard. It is applicable only up to about 3-μ wavelength because of the spurious properties of magnesium oxide at higher wavelengths. The use of the integrating sphere for radiation measurements has been given a very complete exposition in the literature, and the interested reader should consult Refs. [13], [15], and [5] for additional information. A good discussion of radiation-property measurements in both gases and solids is given in Refs. [14] and [16] to [20]. A complete survey of thermal-radiation-property data is available in Ref. [21].

12-5 NUCLEAR RADIATION

In the following sections we shall be concerned with methods for detecting nuclear radiation. A detailed consideration of the origins of such radiation is quite beyond the scope of this discussion, but a few introductory remarks are appropriate in order to categorize the types of nuclear radiation and indicate some of their particular characteristics. A detailed discussion of the principles of nuclear radiation is given by Kaplan [8]; Price [10] presents a comprehensive appraisal of the subject of nuclear-radiation detection.

We shall be concerned with the detection of four kinds of nuclear radiation: (1) alpha (α) particles, (2) beta (β) particles, (3) gamma (γ) rays, and (4) neutrons. An alpha particle is a helium nucleus that has a positive charge of 2 and a relative mass of 4. A beta particle is a negatively charged electron having a relative mass of 0.000549. The neutron has a zero charge and a relative mass of unity. Gamma rays are high-energy electromagnetic waves that result from nuclear transformations and, as such, do not have mass or charge in the classical sense. They may, however, produce ionizing effects in their interaction with matter. Alpha particles are absorbed rather readily in many materials, beta particles are usually more penetrating, and neutrons and gamma rays are the most penetrating types of radiation because they usually have higher energies and do not interact with coulomb force fields when entering a material. The interaction of these nuclear radiations with particular materials is a very complicated subject and forms the basis for nuclear-shielding applications. Our concern is with the detection of these types of radiation.

12-6 DETECTION OF NUCLEAR RADIATION

Nuclear radiation is detected through an interaction of the radiation with the detecting device which produces an ionization process. The degree of ionization may be measured with appropriate electronic circuitry. Two types of detection operations are normally performed: (1) a measurement of the *number* of interactions of nuclear radiation with the detector and (2) a measurement of the total effect of the radiation. The first type of operation is a counting process, while the second operation may be characterized as a mean-level measurement. The counting operation frequently ignores the energy level of the radiation, while the mean-level measurement is used for determining the energy level of the irradiation. The popular Geiger-Müller counter is typically used for nuclear counting operations, while ionization chambers and photographic plates are used for energy-level measurements. Scintillation detectors may be used for both counting and energy-level measurements. We shall discuss these different types of detectors and indicate their range of applicability.

12-7 THE GEIGER-MÜLLER COUNTER

A typical cylindrical-tube arrangement for a Geiger-Müller tube is shown in Fig. 12-9. The anode is a tungsten or platinum wire, while the cylindrical tube forms the cathode for the circuit. The tube is filled with argon with perhaps a small concentration of alcohol or some other hydrocarbon gas. The gas pressure is slightly below atmospheric. The ionizing particle or radiation is transmitted through the cathode material, or some window material installed therein, and through interaction with the gas molecules produces an ionization of the gas. If the voltage E is sufficiently high, each particle will produce a voltage pulse. The counting performance of the tube is indicated in Fig. 12-10. The plateau region typically slopes slightly upward at a rate of from 1 to 10 percent per 100 volts. The tube must be operated in the plateau region, which has a width of approximately 200 volts for commercial tubes. When the particle causes a discharge or pulse, there is a time delay before the tube can detect another particle and register another pulse. This delay is roughly the time required to recharge the anode and cathode system, i.e., to establish a new space charge in the gas. The counting rate of the G-M tube is thus limited by this delay time. The maximum counting rates are of the order of 10^4 counts/sec.

An end-type Geiger-counter tube may be constructed as shown in Fig.

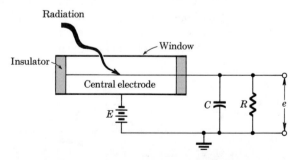

Fig. 12-9 Typical cylindrical-tube arrangement for a Geiger-Müller counter.

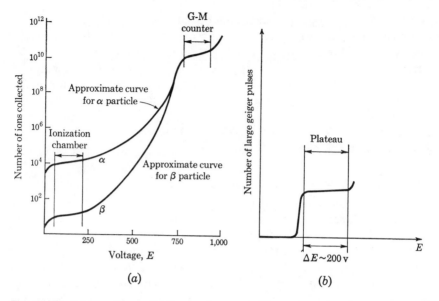

Fig. 12-10 (*a*) Counting performance of system in Fig. 12-9 as a function of impressed voltage; (*b*) detail of Geiger-counter region.

12-11.   This type of tube is used for counting α and β particles and low-energy γ rays.   For these low-energy radiations the window is usually a thin sheet of mica or Mylar.

12-8 IONIZATION CHAMBERS

An ionization chamber may be constructed in basically the same way as the Geiger counter shown in Fig. 12-9 except that the tube is operated at a much lower voltage.   This region of operation is indicated in Fig. 12-10.   The arrangement may be modified, however, to accommodate specific applications.   A schematic of a typical parallel-plate ionization chamber is shown in Fig. 12-12 according to Price [10].   In operation, the chamber is charged with a voltage that is high enough to ensure that electrons produced by the ionizing radiation will be collected on the anode.   The voltage is not so high

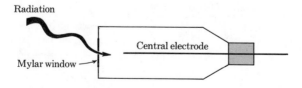

Fig. 12-11   Construction of an end-type Geiger counter.

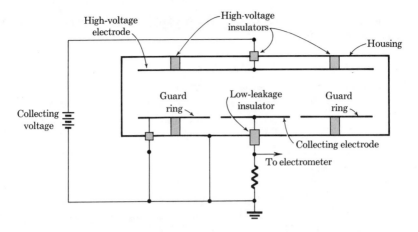

Fig. 12-12 Parallel-plate ionization chamber according to Ref. [10].

that it triggers excitation of other molecules by electrons produced in the ionization process.   This is in contrast to the operation of the G-M counter, which has a sufficiently high voltage that once ionization is effected by the incoming radiation, subsequent ionization of other molecules also takes place in an avalanche fashion, which produces the voltage pulse used for counting.   To measure the energy level of the incoming radiation, a measurement is made of the output current of the device or a determination is made of the charge released in the chamber over a period of time.   The ionization chamber may also be used as a pulse-type device for the measurement of the number and energies of high-energy alpha particles.   In this application a measurement of the pulse height (strength) is made in order to determine the energy levels of the alpha particles.

12-9 PHOTOGRAPHIC-DETECTION METHODS

When certain types of photographic films are exposed to nuclear radiation and subsequently developed, the opacity of the print may be taken as an indication of the total amount of radiation incident on the film during the time of exposure.   Many specialized films are available for nuclear-radiation measurement, and this method is the primary one used for measuring the total radiation exposure for workers in atomic-energy installations.   The photographic film badge may be used for detecting α and β particles, γ rays, and neutrons.   In order to use a single badge for measurement of the exposure rates to the different types of radiation, several openings, or windows, may be constructed in a single badge.   A different type of filter is placed over each window so that only one type of radiation is permitted to strike the photographic emulsion under each window.   Thus, the opacity of the developed film under each of the windows gives an indication of the different radiation-exposure rates.

12-10 THE SCINTILLATION COUNTER

The scintillation counter is the most versatile of the conventional devices for the measurement of nuclear radiation. A schematic of the device is shown in Fig. 12-13. The ionizing radiation strikes the scintillation crystal, which is enclosed in the thin aluminum housing and produces a flash of light. The light flashes are detected by the photomultiplier tube, the output of which is fed to appropriate electronic circuits. If the number of pulses is to be counted, the output circuit will include appropriate electronic counters. The scintillation counter may also be used for measuring the energy of the incoming radiation since, for certain crystals, the intensity of the flash of light is proportional to the energy of the radiation.

A number of scintillation materials are used, including both solids and liquids. The selection of the proper type of material for detection of specific radiations is discussed by Price [10] and Birks [3].

12-11 NEUTRON DETECTION

The three detection methods described above depend on an ionization process caused by an interaction of the α, β, or γ radiation with a gas or scintillation material. These detectors may not be used for neutrons because neutrons do not produce an ionizing effect. For this reason, the measurement of neutron flux is usually an indirect process, which involves the utilization of an intermediate reaction to produce some type of ionizing radiation (α, β, or γ rays), which may then be used to indicate the incoming flux. A variety of neutron-measurement devices are available, and the type which is used in practice depends strongly on the energy level of the neutrons to be measured and the total neutron flux. We shall discuss the principles of operation of only one type of detector which is applicable for the measurement of thermal neutron fluxes ($E \sim 0.025$ ev).

A typical reaction which is used for neutron detection is the interaction of neutrons with B^{10} to produce Li^7 and an alpha particle. The alpha particles may subsequently produce ionizations in a gas and corresponding

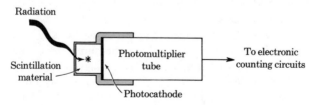

Fig. 12-13 Schematic of a scintillation counter.

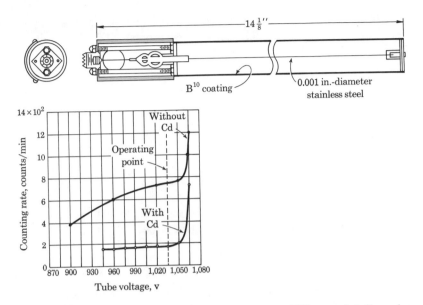

Fig. 12-14 Diagram of a B^{10}-lined neutron counter. Filling gas is helium plus 5 percent ether at 10 cm Hg pressure. (*From Ref.* [*12*].)

voltage pulses. The reaction is utilized in two ways: (1) the boron is present in the form of boron-trifluoride (BF_3) gas and placed in a chamber very similar to the G-M counter shown in Fig. 12-9. The potential difference between the central wire and the cylindrical shell is about 1,300 volts. The wire diameter is about 0.002 in., while the cylinder diameter is about 1 in., with a length of about 6 in. (2) An alternate method of utilizing the B^{10} (n, α) reaction is shown in Fig. 12-14. The inside surface of the cylinder is coated with B^{10}, while the volume is filled with helium and 5 percent ether at 10 cm Hg pressure. The sensitivity of a typical BF_3 detector is about 1.0 counts/sec per unit neutron flux. The sensitivity of the boron-coated chamber is about 5 to 10 counts/sec per unit neutron flux. Neutron flux is measured in neutrons/cm^2-sec.

If sufficiently high neutron fluxes are encountered, the detector may be operated as a current-sensitive device. The current output is then proportional to the incoming flux. Currents of about 0.1 ma may be obtained for neutron fluxes of 10^{10} neutrons/cm^2-sec.

12-12 STATISTICS OF COUNTING

We have already seen that many nuclear-radiation measurements rely on a counting operation. In this section we wish to consider the statistics associated with such operations and indicate the possible errors involved.

In a given sample of radioactive material there are a large number of atoms, and the probability that each of these atoms will decay is quite small. The probability that n atoms will decay is given by the Poisson distribution [Eq. (3-9)] as

$$p(n) = \frac{n^n e^{-n}}{n!} \tag{12-3}$$

where $Np = n$

     N = total number of atoms

     p = probability that each atom will decay

Equation (12-3) is generally valid for $N > 100$ and $p < 0.01$, and the standard deviation of the distribution may be shown to be†

$$\sigma = \sqrt{n} \tag{12-4}$$

The quantity n represents the average number of radioactive decays recorded in a series of observations. The radioactivity of a certain sample of material might be studied by observing the number of particles or γ rays given off in specified increments of time. The counting operation could be performed with a G-M counter or other device. In general, the number of counts in a fixed interval of time, say, 1 min, might vary considerably. Several counts would be taken, and the arithmetic average of the number of counts in each time interval is given by $\bar{n}$. The standard deviation of the measurement would be given by Eq. (12-4). In the above and following equations, n is taken to represent the average number of counts.

The parameter which is of interest in an experimental determination is the *counting rate* defined by

$$r = \frac{n}{t} \tag{12-5}$$

where t is the elapsed time for the average number of counts n. The accuracy of the determination of the counting rate depends on the number of counts and the time t.

When a counting measurement is made to determine the counting rate, the detector senses not only the signal from the radioactive source but also some background signal which is the result of spurious radioactivity of the environment. A separate measurement must be made of the background counting rate and this information used to determine the counting rate resulting from the radioactive source. The counting rate and its associated

† See Ref. [10], p. 56.

uncertainty are written as

$$r = r \pm w_r \tag{12-6}$$

The counting rate for the source is

$$r_s = r_T - r_b \tag{12-7}$$

where r_T is the total counting rate and r_b is the counting rate due to the background. The uncertainty in the source counting rate is thus

$$w_s = (w_T{}^2 + w_b{}^2)^{\frac{1}{2}} \tag{12-8}$$

where w_T and w_b are the uncertainties in the total and background counting rates, respectively. The uncertainties for these counting rates may be estimated through the use of the standard deviation for the Poisson distribution given by Eq. (12-4) if the nuclear radiation satisfies the conditions for this distribution. The accuracy of the counting-rate measurement is clearly dependent on the length of time available for the measurement. A longer time affords the opportunity for observation of a larger number of counts and hence the establishment of a better average counting rate. Example 12-5 illustrates the use of the above relations for determining counting rates.

Example 12-3 A certain radioactive sample is observed, and the total and background counting rates are approximately 800 and 35 counts/min, respectively. If a total time of 45 min is available for both counting measurements, estimate the apportionment of this total time such that there is minimum uncertainty in the source counting rate.

Solution We may assume that both the source and background counts follow the Poisson distribution so that the uncertainty is proportional to the standard deviation; i.e.,

$$\sigma_s = n_s{}^{\frac{1}{2}}$$
$$\sigma_b = n_b{}^{\frac{1}{2}}$$
$$\sigma_T = n_T{}^{\frac{1}{2}}$$

Thus,

$$w_s = \frac{n_s{}^{\frac{1}{2}}}{t_s} \qquad w_b = \frac{n_b{}^{\frac{1}{2}}}{t_b} \qquad w_T = \frac{n_T{}^{\frac{1}{2}}}{t_T} \tag{a}$$

where t_s, t_b, and t_T are the counting times for the source, background, and total, respectively. We have also

$$n_s = r_s t_s \qquad n_b = r_b t_b \qquad n_T = r_T t_T \tag{b}$$

so that

$$w_s{}^2 = \frac{r_s}{t_s} = \frac{r_b}{t_b} + \frac{r_T}{t_T} \tag{c}$$

in accordance with Eq. (12-8).   We wish to optimize w_s by proportionment of the times t_b and t_T in accordance with

$$t_b + t_T = \text{const} = 45 \text{ min} \tag{d}$$

Thus, we optimize w_s by setting $dw_s/dt_b = 0$ and obtain

$$2w_s \frac{dw_s}{dt_b} = 0 = -\frac{r_b}{t_b^2} - \frac{r_T}{t_T^2} \frac{dt_T}{dt_b} \tag{e}$$

From Eq. (d),

$$\frac{dt_T}{dt_b} = -1$$

so that the optimum condition is

$$\frac{t_b}{t_T} = \left(\frac{r_b}{r_T}\right)^{\frac{1}{2}} \tag{f}$$

For the given counting rates,

$$t_b + t_T = 45$$

$$\frac{t_b}{t_T} = (\tfrac{35}{800})^{\frac{1}{2}} = 0.2091$$

and $t_T = 37.2$ min; $t_b = 7.8$ min.   In the above calculations it has been assumed that the time interval is accurately determined and that the uncertainty for this determination is zero.

Example 12-4   For the estimated counting rates given in Example 12-3 calculate the optimum total time necessary to give a standard deviation of the source counting rate of 1 percent.   Repeat for total and background counting rates of 400 and 100, respectively.

Solution   The nominal value of the source counting rate is

$$r_s = r_T - r_b = 800 - 35 = 765 \text{ counts/min}$$

We have

$$\sigma_s^2 = \frac{r_b}{t_b} + \frac{r_T}{t_T} = [(0.01)(765)]^2 = 58.5$$

Also, from relation (f) of Example 12-3,

$$t_b = t_T\left(\frac{r_b}{r_T}\right)^{\frac{1}{2}}$$

so that

$$t_T = \frac{1}{\sigma_s^2}\left(r_b^{\frac{1}{2}}r_T^{\frac{1}{2}} + r_T\right)$$

$$= \frac{1}{58.5}\left[(35)^{\frac{1}{2}}(800)^{\frac{1}{2}} + 800\right] = 16.5 \text{ min}$$

The total counting time is $t_b + t_T$, or

$$t_{tot} = (1.209)(16.5) = 20 \text{ min}$$

For total and background counting rates of 400 and 100,

$$r_s = 400 - 100 = 300 \text{ counts/min}$$
$$\sigma_s^2 = [(0.01)(300)]^2 = 9$$
$$t_T = \tfrac{1}{9}[(100)^{\frac{1}{2}}(400)^{\frac{1}{2}} + 400] = 66.7 \text{ min}$$

The total counting time is

$$t_{tot} = 66.7[(\tfrac{100}{400})^{\frac{1}{2}} + 1] = 100 \text{ min}$$

Example 12-5 A radioactive sample is observed over a period of 30 min, and a total of 21,552 counts is recorded. The sample is then removed, and the background radiation is observed for another 30 min. The background produces 1,850 counts. Calculate the source counting rate and its uncertainty.

Solution We have

$$r_T = \frac{21,552}{30} = 715.1 \text{ counts/min}$$

$$r_b = \frac{1,850}{30} = 61.7 \text{ counts/min}$$

so that

$$r_s = 715 - 62 = 653 \text{ counts/min}$$

The standard deviation of the source counting rate is calculated from Eq. (*c*) of Example 12-3.

$$w_s = \sigma_s = \left(\frac{r_b}{t_b} + \frac{r_T}{t_T}\right)^{\frac{1}{2}} = 5.08 \text{ counts/min}$$

REVIEW QUESTIONS

1. How is thermal radiation distinguished from other EM radiation?

2. How is thermal radiation detected?

3. Why is it sometimes necessary to chop thermal radiation before it strikes the detector?

4. What is an integrating sphere? How is it used for reflectivity and transmissivity measurements?

5. What are the four types of nuclear radiation that are normally measured?

6. Describe the principle of operation of the Geiger counter.

7. How does an ionization chamber work?

8. How does a scintillation counter operate?

9. What types of instruments are used for detection of (*a*) alpha particles, (*b*) beta particles, (*c*) gamma rays, and (*d*) neutrons?

10. Describe the techniques for measuring neutron irradiation.

11. Why is a Poisson distribution an adequate representation of nuclear counting statistics?

PROBLEMS

12-1. Consider the thermistor detector and radiation source of Example 12-1. Instead of the bare detector an optical system collects the radiation over a circular area of radius r and focuses it on the detector which is maintained at 70°F. Plot the source temperature necessary to maintain a signal-to-noise ratio of 200 as a function of the radius of the collecting optical system. Assume that the distance between the source and detector is 3 ft, as in Example 12-1. Neglect losses in the optical system for this calculation.

12-2. Repeat Prob. 12-1 for distances of 6 and 12 ft between the source and detector.

12-3. A thermocouple detector has characteristics such that the minimum input power necessary to produce a signal-to-noise ratio of unity is 3×10^{-6} watt. A 2-in.-diam optical system collects the input thermal radiation,and focuses it on the thermocouple. The optical-thermocouple system is to be used to measure the temperature of a 2-in. black sphere 6 ft away. The thermocouple bead is maintained at a temperature of 70°F. Estimate and plot the uncertainty in the temperature measurement as a function of the temperature of the sphere.

12-4. The apparatus of Example 12-2 is used for an emissivity determination between the temperature limits of 300 and 1000°F. Plot the percent uncertainty in emissivity as a function of T_s for nominal emissivities of 0.2 and 0.8. Take the uncertainties in radiometer temperature and output voltage as

$$w_{T_R} = \pm 0.5°F$$
$$w_E = 0.005 \text{ mv}$$

The uncertainty in the source temperature is taken as $\pm 1.0°F$.

12-5. A radioactive sample is observed, and the total and background counting rates are 500 and 100 counts/min, respectively. Estimate the total time necessary to produce a standard deviation of 2 percent in the nominal source counting rate. Repeat for counting rates of 100 and 22.

12-6. The following counting data are collected on a radioactive sample. Each of the number of counts is taken in an exact time interval of 1 min. The background counting rate is known accurately as 23 counts/min. Calculate the nominal value of the source counting rate and the uncertainty in this rate. Apply Chauvenet's criterion to eliminate some of the data if necessary. Assume that the Poisson distribution applies.

Interval number	No. of counts
1	523
2	410
3	342
4	595
5	490
6	611
7	547
8	512

12-7. After a set of radioactive counting measurements has been made, it is found that the timing device has been spurious in operation. A set of calibration measurements on the timer has given the standard deviation of the time-interval measurements as σ_t. Using the approach taken in Example 12-3, derive an expression for the standard deviation of the source counting rate in terms of the standard deviations of the number of background and total counts and the standard deviation of the time interval.

12-8. Derive an expression for the optimum time t_T under the conditions that $r_b = r_s$. Express in terms of σ_s and r_s.

12-9. Plot the optimum counting time from Prob. 12-8 as

$$\frac{(t_T)_{\text{opt}}\sigma_{.}^2}{r_s} \qquad \text{versus} \qquad \frac{r_b}{r_s}$$

12-10. Under the condition that the total number of counts for both the background and sample measurements is limited, i.e.,

$$n_b + n_T = K = \text{const}$$

show that the minimum error condition is

$$\frac{n_b}{n_T} = \frac{r_b}{r_T}$$

and the total counting time for the background and sample measurements is

$$t_b + t_T = \frac{2K}{2r_b + r_s}$$

Plot

$$\frac{(t_b + t_T)r_s}{K} \qquad \text{versus} \qquad \frac{r_b}{r_s}$$

REFERENCES

1. Ballard, S., and W. Wolfe: Optical Materials in Equipment Design, A Critique, ONR, *Proc. Infrared Inform. Symposia*, vol. 4, no. 1, p. 185, March, 1959.
2. Billings, B. H., E. E. Barr, and W. L. Hyde: An Investigation of the Properties of Evaporated Metal Bolometers, *J. Opt. Soc. Am.*, vol. 37, p. 123, 1947.
3. Birks, J. B.: "Scintillation Counters," McGraw-Hill Book Company, New York, 1953.
4. Fuson, N.: The Infrared Sensitivity of Superconducting Bolometers, *J. Opt. Soc. Am.*, vol. 38, p. 845, 1948.
5. Gier, J. T., R. V. Dunkle, and J. T. Bevans: Measurement of Absolute Spectral Reflectivity from 1.0 to 15 Microns, *J. Opt. Soc. Am.*, vol. 44, p. 558, 1954.
6. Glaser, P. E., and H. H. Blau: A New Technique for Measuring the Spectral Emissivity of Solids at High Temperatures, *Trans. ASME*, vol. 81C, p. 92, 1959.
7. Hackforth, H. L.: "Infrared Radiation," McGraw-Hill Book Company, New York, 1960.
8. Kaplan, I.: "Nuclear Physics," 2d ed., Addison-Wesley Publishing Company, Inc., Reading, Mass., 1963.
9. Marcus, N.: A Blackbody Standard, *Instr. and Automation*, vol. 28, March, 1955.
10. Price, W. J.: "Nuclear Radiation Detection," 2d ed., McGraw-Hill Book Company, New York, 1964.
11. Snyder, N. W., J. T. Gier, and R. V. Dunkle: Total Normal Emissivity

Measurements on Aircraft Materials between 100 and 800°F, *Trans. ASME*, vol. 77, p. 1011, 1955.

12. "Reactor Handbook," vol. 2, p. 951, U.S.A.E.C. Document, McGraw-Hill Book Company, New York, 1955.

13. Edwards, D. K., J. T. Gier, K. E. Nelson, and R. D. Roddick: Integrating Sphere for Imperfectly Diffuse Samples, *J. Opt. Soc. Am.*, vol. 51, no. 11, pp. 1279–1288, November, 1961.

14. Cox, R. L.: Emittance of Translucent Materials: A Qualitative Evaluation of Governing Fundamental Properties and Preliminary Design of an Experimental Apparatus to Measure These Properties to 6000°F, M.S. thesis, Southern Methodist University, Dallas, Tex., 1961.

15. Birkebak, R. C., and S. H. Cho: Integrating Sphere Reflectometer Center-Mounted Blockage Effects, ASME Paper 67-HT-54, August, 1967.

16. "Precision Measurement and Calibration: Optics, Metrology, and Radiation," Natl. Bur. Std., Handbook 77, vol. 3, 1961.

17. Richmond, J. C.: Measurement of Thermal Radiation Properties of Solids, NASA Sp. 31, 1963.

18. Katzoff, S.: Symposium on Thermal Radiation of Solids, NASA Sp. 55, 1965.

19. Smith, R. A., F. E. Jones, and R. P. Chasmar: "The Detection and Measurement of Infra-red Radiation," Oxford University Press, New York, 1957.

20. Richmond, J. C., S. T. Dunn, D. P. Dewitt, and W. D. Hayes: Procedures for the Precise Determination of Thermal Radiation Properties, *Air Force Materials Lab., Tech. Doc. Rept.* ML-TDR-64-257, pt. 2, April, 1965.

21. Gubareff, G. G., J. E. Jansen, and R. H. Torborg: "Thermal Radiation Properties Survey," Honeywell Research Center, Honeywell, Inc., Minneapolis, Minn. 1960.

13

Data Acquisition and Processing

13-1 INTRODUCTION

All the previous chapters have illustrated, in one way or another, the manner in which an experiment can be planned to collect meaningful data. Selection of proper instrumentation and analyses of experimental uncertainties are all part of the planning process. But nothing has been said about the manner of collecting data and later processing of these data to produce the desired results of the experiments. The acquisition of data might consist simply of several people (or perhaps only one person) reading a number of instruments and recording the observations on a data sheet. The processing of data could be done in many ways, from simple slide-rule calculations to a complicated digital-computer routine.

There are systems available today for rapidly collecting a great bulk of data, processing it, and displaying the desired results in printed form. The purpose of this chapter is to present a brief qualitative description of such systems and the functions of the elements that go together to make up an overall data-acquisition and processing installation.

13-2 THE GENERAL ACQUISITION SYSTEM

The essential element in a modern data-acquisition system is the instrument transducer, which furnishes an electric signal that is indicative of the physical variable being measured. The signal may be voltage, current, resistance, frequency, or electric pulses. Our discussion assumes that suitable transducers are available to convert the physical variables of interest into electric signals. As examples of such transducers we may recall that a thermocouple gives a voltage representation of temperature, a strain gage gives a resistance representation of strain, and so on.

The object of a data-acquisition system is to collect the data and record them in a form suitable for processing or presentation. Thus, a recording potentiometer is a simple data-acquisition system that may be used for collecting temperature data from thermocouples. In this case, the data points must be read from the recorder chart. A more complicated system might convert the voltage signal from the thermocouple to a digital signal, which could be used to operate a printing recorder so that the numerical value of the temperature is printed on a sheet of paper. Such a system is much more complicated than the simple voltage recorder because of the digital-conversion process. It is easy to see, however, that the digital output has many advantages.

The general data-acquisition system has three stages:

1. The input stage, which consists of appropriate transducers and signal-conditioning circuits (amplifiers, filters, etc.).
2. A signal-conversion stage, which converts the input signal to a voltage and subsequently expresses this voltage in digital form. The conversion to digital form is made by converting the voltage to frequency, which, in turn, may be used to drive electronic counting circuits. The conversion is essentially a frequency-modulation process.
3. An output stage, which takes the digital signal and expresses it in printed form on a sheet of paper, plots the data on graph paper, punches the data on cards, or stores the data on magnetic or punched tape. The output stage must include suitable coupling circuits to express the digital signal in a form for driving a printer, card punch, magnetic tape, etc.

A schematic of this general system is shown in Fig. 13-1. The analog-to-digital converter is the conversion stage.

It is a rare circumstance when data in only one variable are to be collected. The data-acquisition system, then, should include provisions for collecting multiple channels of data inputs. This collection process could be accomplished by having a channel like the one shown in Fig. 13-1 for each variable to be studied. The cost of such a system, however, would be quite

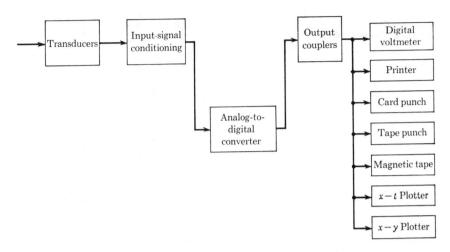

Fig. 13-1 Schematic of a general data-acquisition system.

high so that a scanner/programmer is normally employed for multiple-channel work. The scanner is a device that samples the data channels in rapid sequence so that only one conversion and output stage is necessary. Commercial devices are constructed so that any particular sequencing of 1 to 100 channels may be used at the discretion of the laboratory personnel. Thus, the system may be *programmed* to collect any desired range of variables, and the device is called a scanner/programmer. The scanner may be considered as part of the input section of the general data-acquisition system.

Many experimental programs involve the collection of data at regular time intervals or in some particular time sequence. The acquisition system may perform this timing function automatically by incorporating a digital clock and time standard in the scanner and/or converter stages.

Finally, it may be advantageous to apply signal conditioning to the output of the scanner/programmer. This conditioning might be amplification, voltage-to-frequency conversion for only a few of the channels, filtering, distortion or harmonic analysis of the waveform of some of the signals, etc. When all the above elements are combined, a very flexible data-acquisition system results, as shown in Fig. 13-2. It may be noted that the programmable feature of the scanner is normally used in the converter stage as well. This is essential since some channels may require signal conditioning while others may not.

The use of a flexible data-acquisition system like the one described above depends on many factors, not the least of which is cost. Such systems are expensive. The expense may be entirely justified, however, in an installation where large quantities of data are to be collected.

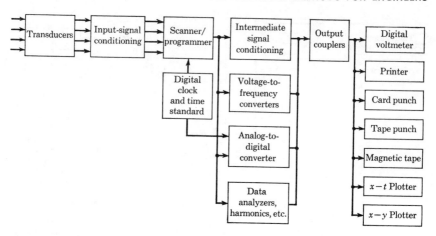

Fig. 13-2 Schematic of a multichannel data-acquisition system.

13-3 DATA PROCESSING

The digital computer is undoubtedly the most suitable means for processing large quantities of data. By processing, we mean the calculation of results from the raw data and/or analysis of these data for consistency using statistical methods or other means. The computer must be programmed to perform the desired calculations and print out or feed the results to some suitable output device.

In the above section we have seen that a flexible data-acquisition system can collect and present data in many output forms. The punched card, punched tape, and magnetic-tape forms are all suitable for use as inputs to a digital computer, provided that the systems use coding which is compatible with the computer installation. It is easy to see that a complete data-acquisition and processing installation might be organized something like the schematic shown in Fig. 13-3. Such large-scale systems are used, of course, in many industrial applications.

13-4 SUMMARY

What are the advantages of a flexible data-acquisition system like the one described above? There are several.

1. The output of the acquisition system can be obtained in a form which is ready for immediate input to a digital computer.
2. A large amount of data can be collected very quickly, thereby freeing personnel and instrumentation for other projects.
3. The multichannel system may be able to collect data for some particular transient process which could not be obtained otherwise.

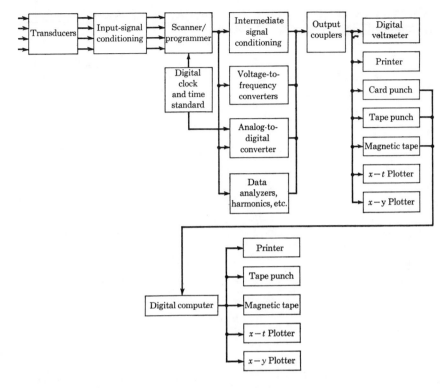

Fig. 13-3 Combined data-acquisition and processing system.

4. Since the collection of data can be accomplished very rapidly with such systems, instrument transducers may enjoy more varied use in a laboratory because they are used for a shorter period of time on each project.

5. The rapid collection of data coupled with rapid data processing by means of a computer can speed the overall experimental program so that men, money, and machines are not wasted because of excessive waiting periods between data collection, processing, and analysis.

Offsetting these advantages are the high cost of the data-acquisition system and the absolute necessity that all measurements be made with electric transducers; e.g., we cannot use a simple manometer for a differential pressure measurement; some type of transducer must be employed.

The decision of whether or not to purchase a data-acquisition system is clearly dependent on the economics of the situation. The author has no desire to belabor this point, but a statement that has been made by many people in the past years in regard to digital computers may also be applicable to data-acquisition systems.   The statement is:

> You can never economically justify the cost of a digital computer *before* you buy one.   After you buy it you cannot get along without it because you find yourself solving problems you were never able to solve before.

Appendix

Table A-1 The International System of Units

The following set of units was officially adopted in a resolution of the Eleventh General Conference on Weights and Measures in Paris, 1960. The system is designated by SI (for Système Internationale d'Units) and is increasingly accepted as the worldwide standard for presentation of research results. Although the system is only in modest use in engineering, it will undoubtedly increase in importance in the future and thus merits the attention of all those engaged in experimental work.

Quantity	Unit	Symbol	
Elemental units			
Length	meter	m	
Mass	kilogram	kg	
Time	second	s	
Electric current	ampere	A	
Temperature	degree Kelvin	°K	
Luminous intensity	candela	cd	
Supplementary units			
Plane angle	radian	rad	
Solid angle	steradian	sr	
Derived units			
Area	square meter	m^2	
Volume	cubic meter	m^3	
Frequency	hertz	Hz	(s^{-1})
Density	kilogram per cubic meter	kg/m^3	
Velocity	meter per second	m/s	
Angular velocity	radian per second	rad/s	
Acceleration	meter per second squared	m/s^2	
Angular acceleration	radian per second squared	rad/s^2	
Force	newton	N	$(kg \cdot m/s^2)$
Pressure	newton per sq meter	N/m^2	
Kinematic viscosity	sq meter per second	m^2/s	
Dynamic viscosity	newton-second per sq meter	$N \cdot s/m^2$	
Work, energy, quantity of heat	joule	J	$(N \cdot m)$
Power	watt	W	(J/s)
Electric charge	coulomb	C	$(A \cdot s)$
Voltage, potential difference, electromotive force	volt	V	(W/A)
Electric field strength	volt per meter	V/m	
Electric resistance	ohm	Ω	(V/A)
Electric capacitance	farad	F	$(A \cdot s/V)$
Magnetic flux	weber	Wb	$(V \cdot s)$
Inductance	henry	H	$(V \cdot s/A)$
Magnetic flux density	tesla	T	(Wb/m^2)
Magnetic field strength	ampere per meter	A/m	
Magnetomotive force	ampere	A	
Luminous flux	lumen	lm	$(cd \cdot sr)$
Luminance	candela per sq meter	cd/m^2	
Illumination	lux	lx	(lm/m^2)

Table A-2 Standard prefixes and multiples

Multiples and submultiples	Prefixes	Symbols	Pronunciations
10^{12}	tera	T	tĕr'å
10^9	giga	G	jĭ' gå
10^6	mega	M	mĕg'å
10^3	kilo	k	kĭl'ô
10^2	hecto	h	hek'tô
10	deka	da	dĕk'å
10^{-1}	deci	d	dĕs'ĭ
10^{-2}	centi	c	sĕn'tĭ
10^{-3}	milli	m	mĭl ĭ
10^{-6}	micro	μ	mĭ' krô
10^{-9}	nano	n	năn'ô
10^{-12}	pico	p	pĕ' cô
10^{-15}	femto	f	fĕm' tô
10^{-18}	atto	a	ăt' tô

Table A-3   Conversion factors

Length
  12 in. = 1 ft
  2.54 cm = 1 in.
  $1\mu = 10^{-6}$ m $= 10^{-4}$ cm

Mass
  1 kg $= 2.205$ lb_m
  1 slug $= 32.16$ lb_m
  454 g $= 1$ lb_m

Force
  1 dyne $= 2.248 \times 10^{-6}$ lb_f
  1 $lb_f = 4.448$ newtons
  10^5 dynes $= 1$ newton

Energy
  1 ft-$lb_f = 1.356$ joules
  1 kwhr $= 3413$ Btu
  1 hp-hr $= 2545$ Btu
  1 Btu $= 252$ cal
  1 Btu $= 778$ ft-lb_f

Pressure
  1 atm $= 14.696$ lb_f/in.$^2 = 2,116$ lb_f/ft^2
  1 atm $= 1.0132 \times 10^5$ newtons/m^2
  1 in. Hg $= 70.73$ lb_f/ft^2

Viscosity
  1 centipoise $= 2.42$ lb_m/hr-ft
  1 lb_f-sec/ft$^2 = 32.16$ lb_m/sec-ft

Thermal conductivity
  1 cal/sec-cm-°C $= 242$ Btu/(hr)(ft)(°F)
  1 watt/cm-°C $= 57.79$ Btu/(hr)(ft)(°F)

Table A-4   Properties of metals at 70°F

Metal	Density, lb_m/ft^3	Specific heat C, $Btu/lb_m\text{-}°F$	Linear coef. of expansion, $°F^{-1} \times 10^4$	Electrical conductivity relative to copper	Thermal conductivity k, $Btu/(hr)(ft)(°F)$
Aluminum, pure	167	0.22	0.13	0.66	118
Red brass: 85% Cu, 9% Sn, 6% Zn	544	0.092	0.10		35
Copper, pure	555	0.095	0.09	1.00	223
Iron, pure	493	0.108	0.065	0.18	42
Lead	710	0.031	0.16	0.08	20
Magnesium	109	0.242	0.145	0.40	99
Nickel	556	0.106	0.07	0.25	52
Platinum	1,335	0.032	0.05	0.18	40
Silver	657	0.056	0.107	1.15	235
Stainless steel: 18% Cr, 8% Ni	488	0.11	0.003	0.0234	9.4
Steel, structural	485	0.11	0.07	0.12	33
Tin	456	0.054	0.11	0.16	37
Tungsten	1,208	0.032	0.02	0.33	94
Zinc	446	0.092	0.18	0.31	37

Table A-5   Thermal properties of some nonmetals at 70°F

Material	Density, lb_m/ft^3	Specific heat, $Btu/lb_m\text{-}°F$	Thermal conductivity k, $Btu/(hr)(ft)(°F)$
Asbestos sheet	60	0.2	0.083
Brick, common	110	0.22	0.042
Concrete	140	0.16	1.0
Fiberboard	15	0.5	0.027
Fiberglass	6	0.5	0.022
Glass, window	160	0.16	0.50
85% magnesia pipe covering	15	0.2	0.042
Plaster	95	0.25	0.27

Table A-6   Properties of some saturated liquids at 68°F

Fluid	Chemical formula	Density, lb_m/ft^3	Specific heat C, $Btu/lb_m\text{-}°F$	Kinematic viscosity v, ft^2/sec	Thermal conductivity k, $Btu/(hr)(ft)(°F)$	Prandtl no. Pr, $c_p\mu/k$
Ammonia	NH_3	38.2	1.15	0.386×10^{-5}	0.301	2.02
Carbon dioxide	CO_2	48.2	1.2	0.98×10^{-5}	0.0504	4.10
Freon-12	CCl_2F_2	83.0	0.231	0.213×10^{-5}	0.042	2.17
Glycerine	$C_2H_3(OH)_3$	78.9	0.57	0.0127	0.165	12.5
Mercury	Hg	847.7	0.033	0.123×10^{-5}	5.02	0.025
Motor oil (typical)		55.0	0.45	0.010	0.08	10.000
Water	H_2O	62.4	1.0	1.08×10^{-5}	0.345	7.02

Table A-7 Properties of gases at atmospheric pressure and 68°F

Gas	Chemical formula	Molecular weight	c_p, Btu/lb_m-$°F$	$\gamma = \dfrac{c_p}{c_v}$	Thermal conductivity k, $Btu/(hr)(ft)(°F)$	Viscosity, centipoise
Air	...	28.95	0.24	1.40	0.025	0.018
Acetylene	C_2H_2	26.02	0.38	1.25	0.0124	0.010
Ammonia	NH_3	17.03	0.52	1.30	0.0142	0.010
Carbon dioxide	CO_2	44.00	0.20	1.30	0.0095	0.015
Freon-12	CCl_2F_2	120.9	0.15	1.14	0.0053	0.013
Hydrogen	H_2	2.01	3.4	1.40	0.103	0.0089
Methane	CH_4	16.03	0.54	1.30	0.0196	0.011
Nitrogen	N_2	28.02	0.245	1.40	0.0150	0.018
Oxygen	O_2	32.00	0.215	1.40	0.0153	0.020

Table A-8 Properties of dry air at atmospheric pressure†

T, $°F$	μ, $\dfrac{lb_m}{hr\text{-}ft}$	k, $\dfrac{Btu}{(hr)(ft)(°F)}$	c_p, $\dfrac{Btu}{lb_m\text{-}°F}$	Pr
−100	0.0319	0.0104	0.239	0.739
−50	0.0358	0.0118	0.239	0.729
0	0.0394	0.0131	0.240	0.718
50	0.0427	0.0143	0.240	0.712
100	0.0459	0.0157	0.240	0.706
150	0.0484	0.0167	0.241	0.699
200	0.0519	0.0181	0.241	0.693
250	0.0547	0.0192	0.242	0.690
300	0.0574	0.0203	0.243	0.686
400	0.0626	0.0225	0.245	0.681
500	0.0675	0.0246	0.248	0.680
600	0.0721	0.0265	0.250	0.680
700	0.0765	0.0284	0.254	0.682
800	0.0806	0.0303	0.257	0.684
900	0.0846	0.0320	0.260	0.687
1000	0.0884	0.0337	0.263	0.690

† From: *Natl. Bur. Std. (U.S.), Circ.* 564, 1955.

Table A-9   Properties of water [saturated liquid]†

$°F$	$c_p, \dfrac{Btu}{lb_m\text{-}°F}$	$\rho, \dfrac{lb_m}{ft^3}$	$\mu, \dfrac{lb_m}{ft\text{-}hr}$	$k, \dfrac{Btu}{(hr)(ft)(°F)}$	$Pr, \dfrac{c_p\mu}{k}$
32	1.009	62.42	4.33	0.327	13.35
40	1.005	62.42	3.75	0.332	11.35
50	1.002	62.38	3.17	0.338	9.40
60	1.000	62.34	2.71	0.344	7.88
70	0.998	62.27	2.37	0.349	6.78
80	0.998	62.17	2.08	0.355	5.85
90	0.997	62.11	1.85	0.360	5.12
100	0.997	61.99	1.65	0.364	4.53
110	0.997	61.84	1.49	0.368	4.04
120	0.997	61.73	1.36	0.372	3.64
130	0.998	61.54	1.24	0.375	3.30
140	0.998	61.39	1.14	0.378	3.01
150	0.999	61.20	1.04	0.381	2.73
160	1.000	61.01	0.97	0.384	2.53
170	1.001	60.79	0.90	0.386	2.33
180	1.002	60.57	0.84	0.389	2.16
190	1.003	60.35	0.79	0.390	2.03
200	1.004	60.13	0.74	0.392	1.90
220	1.007	59.63	0.65	0.395	1.66
240	1.010	59.10	0.59	0.396	1.51
260	1.015	58.51	0.53	0.396	1.36
280	1.020	57.94	0.48	0.396	1.24
300	1.026	57.31	0.45	0.395	1.17
350	1.044	55.59	0.38	0.391	1.02
400	1.067	53.65	0.33	0.384	1.00
450	1.095	51.55	0.29	0.373	0.85
500	1.130	49.02	0.26	0.356	0.83
550	1.200	45.92	0.23		
600	1.362	42.37	0.21		

† From A. I. Brown and S. M. Marco: "Introduction to Heat Transfer," 3d ed., McGraw-Hill Book Company, New York, 1958.

Table A-10 Diffusion coefficients of some gases and vapors in air at 25°C and 1 atm†

Substance	D, cm^2/sec
Ammonia	0.28
Benzene	0.088
Carbon dioxide	0.164
Hydrogen	0.410
Methanol	0.159
Water	0.256

† From R. H. Perry (ed.): "Chemical Engineers' Handbook," 3d ed., McGraw-Hill Book Company, New York, 1950.

Table A-11 Approximate total normal emissivities of various surfaces at 70°F

Surface	Emissivity
Aluminum, heavily oxidized	0.20–0.30
Aluminum, polished	0.09
Asbestos sheet	0.93–0.96
Brass, polished	0.06
Brick	0.93
Carbon, lampblack	0.96
Copper, heavily oxidized	0.75
Copper, polished	0.03
Enamel, white fused on iron	0.90
Iron, oxidized	0.74
Paint, aluminum	0.27–0.67
Paint, flat black lacquer	0.97
Steel, polished	0.07
Steel plate, rough	0.95
Tin, bright	0.06
Water	0.95

Table A-12 Area moments of inertia for some common geometric shapes

Rectangle

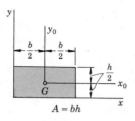

$$A = bh$$

$$I_{y0} = \frac{bh^3}{12}$$

$$I_{y0} = \frac{hb^3}{12}$$

Right triangle

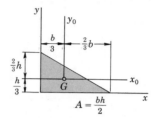

$$A = \frac{bh}{2}$$

$$I_{x0} = \frac{bh^3}{36}$$

$$I_{y0} = \frac{hb^3}{36}$$

Quadrant of circle

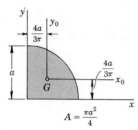

$$A = \frac{\pi a^2}{4}$$

$$I_x = I_y = \frac{\pi a^4}{16}$$

$$I_{x0} = I_{y0} = 0.054a^4$$

Circle

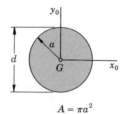

$$A = \pi a^2$$

$$I_{x0} = I_{y0} = \frac{\pi a^4}{4} = \frac{\pi d^4}{64}$$

$$\text{(Polar)} \quad J = \frac{\pi a^4}{4} = \frac{\pi d^4}{32}$$

Quadrant of ellipse

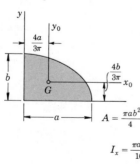

$$A = \frac{\pi ab^2}{4}$$

$$I_x = \frac{\pi ab^3}{16}$$

$$I_y = \frac{\pi a^3 b}{16}$$

$$I_{x0} = 0.054ab^3$$

$$I_{y0} = 0.054a^3 b$$

Ellipse

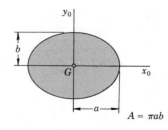

$$A = \pi ab$$

$$I_{x0} = \frac{\pi ab^3}{4}$$

$$I_{y0} = \frac{\pi a^3 b}{4}$$

Index

Absolute performance, 24
Accuracy, general discussion, 7
Acoustic calibrator, 366
Air, properties of, 411
Alpha particle detection, 388
Alphatron vacuum gage, 180
Alternating current:
  high-frequency measurement, 85
  measurement, 82
  rms value, 84
Amplification ratio, 103
Amplifiers:
  cathode-follower, 100
  charge, 101
  emitter-follower, 100
  impedance-transforming, 100
  voltage, 100

Amplitude response, definition, 19
Analog-to-digital converter in data
    acquisition, 402
Analysis of data, 34–44
  graphical, 70–72
  statistical, 44–53
    chi-square test, 59
    Gaussian distribution, 49
    method of least squares, 65
    probability distribution, 47
    standard deviation, 44
Analytical balance, 323
  influence of buoyancy forces, 325
  sensitivity, 324
Apparent blackbody temperature, 263
Area measurement:
  graphical methods, 155

Area measurement:
    numerical methods, 156
    planimeter, 152
Arithmetic mean, definition, 44

Ballast circuit, 88
Base length for strain gage, 333
Bellows gage, 172
Beta particle detection, 388
Bimetallic strip thermometer, 242
Binomial distribution, 49
Blackbody, 259
    apparent temperature, 263
    source, 381
Bomb calorimeter, 307
Boron used for neutron detection, 391
Bourdon-tube pressure gage, 170
Bridges, 92
    ac, 96
    Hay, 97
    Maxwell, 97
    Owen, 96
    resonance, 97
    Wheatstone, 92, 96
    Wien, 96
Bridgman gage, 174

Calibration:
    for acoustic devices, 366
    general discussion, 7
Calorimetry, 303
Cantilever beam:
    as force transducer, 327
    vibrating, 350
Capacitance pressure gage, 172
Capacitive transducer, 127
Carrier's equation, 312
Cathode-follower amplifier, 100
Cathode-ray oscilloscope, 112
Celsius temperature scale, 239
Centigrade temperature scale, 239
Charge amplifier, 101
Chauvenet's criterion, 56
Chi-square test, 59

Cigars, as smoke generators, 223
Commonsense analysis, 36
Conduction error in temperature
    measurement, 266
Conduction heat-flux sensors, 315
    sensitivity of, 316
Consistency, importance, 35
Constant-pressure sound field, 366
Contrast in flow visualization, 216
Control, relation to measurement
    process, 3
Convection heat-transfer formulas,
    summary, 272
Convection measurements:
    forced convection, 308
    free convection, 309
Conversion factors, 409
Corrugated-disk diaphragm, 171
Counting rate, 392
Counting statistics, 391
Current measurement:
    ac, 84
    dc, 79
    high-frequency, 85
Current-sensitive input circuit, 87
Curve fitting, 69

D'Arsonval movement, 81
Data-acquisition system, general, 401
    advantages of, 403
Data analysis:
    correlation, 73
    effect of consistency, 72
    general considerations, 72
    uncertainty analysis, 73
Data collection, 25, 401
Dead-weight tester, 167
Decibels:
    calculation aid for sound
        levels, 361
    related to gain, 105
    sound wave, 360
Deformation sensitivity, 333
Delay, definition, 19
Density related to interferometer, 219
Detectivity, 132

Deviation:
  definition, 44
  standard, definition, 44
Dew-point temperature, 313
Dial indicators, 145
Diaphragms:
  corrugated-disk, 171
  deflection characteristics, 171
  natural frequency, 173
  pressure gage, 172
Differential transformer:
  principles, 124
  used with pressure measurements,
    172
Diffusion coefficient:
  definition, 299
  measurement, 300
  tabulated values, 413
Digital voltmeter, 110
Dimensional measurements:
  definition, 143
  optical methods, 146
Discharge coefficient, definition, 189
  (See also Flowmeters)
Displacement measurement:
  capacitive transducer, 127
  definition, 143
  ionization transducer, 135
  LVDT, 124
  pneumatic gage, 149
  resistance transducer, 124
Distortion related to frequency and
    phase-shift response, 20
Drag flowmeters, 203
Dry-bulb temperature, 311
Dynamometer, 331

Elastic elements, as force
    transducers, 326
Electrical filters (see Filters,
    electrical)
Electrical resistance, variation with
    temperature, 245
Electrical-resistance strain gage, 334
  bridge circuits, 338
  gage factor, 335

Electrical-resistance strain gage:
  installation, 337
  materials, 336
  temperature compensation, 339
Electrical-resistance thermometer,
    244
  correction of lead resistance, 245
Electrical standards, 9
Electrodynamometer movement, 84
Electronic counter, 120
Electronic switch, 114
Emissivity:
  definition, 261
  measurement, 382
  spectral, 383
  tabulated values, 413
Emitter-follower amplifier, 100
Errors:
  conduction, 267
  fixed, 35
  radiation, 269
  random, 35
  related to uncertainty, 35
  systematic, 35
Expansion factor:
  charts, 197
  definition, 191
Experiment planning:
  general discussion, 23
  use of uncertainty analysis, 25, 39
Experimentation, relation to theoretical
    work, 2

Fahrenheit temperature scale, 238
Fick's law, 299
Filter, electrical:
    basic discussion, 102
    design formulas, 104
    passive networks, 105
Fixed errors, 35
Fletcher-Munson curves for
    loudness, 368
Flow calorimeter, 303
Flow measurement (see Flowmeters)
Flow visualization:
    basic optical effect, 212

Flow visualization:
hydrogen-bubble method, 213
interferometer, 217
laser anemometer, 219
schlieren, 215
shadowgraph, 213
smoke method, 222
Flowmeters:
drag, 203
flow nozzle, 190
lobed-impeller, 187
magnetic, 210
nutating-disk, 186
obstruction, 188
orifice, 189
positive-displacement, 186
rotameter, 203
rotary-vane, 187
sonic-nozzle, 201
summary, 230
turbine, 205
venturi, 189
Fluid-expansion thermometer, 243
Fog juice, 223
Foil-type heat flux-meter, Gardon, 314
Frequency, relation to probability, 46
Frequency measurement:
electronic counter, 120
oscilloscope, 115
Frequency response, definition, 16
Fringes for interferometer, 219

Gage blocks, 145
Gage factor, 335
Gain, 104
Galvanometer:
principles, 81
tabulation, 82
as used in bridge circuits, 95
Gamma rays, detection, 387
Gardon foil-type heat-flux meter, 314
Gases, properties of, 411
Gaussian distribution, 49
values of, 51
Geiger-Müller counter, 387
counting performance, 389

Generalized measurement system, 14
Graphical analysis, 69
probability distribution, 57
Gray body, 261
Guarded hot-plate apparatus, 288

Hall coefficients, 138
Hall-effect transducers, 137
Heat-flux meters, 313
Heise gage, 169
Helmholtz cavity, resonant frequency,
164
High-speed temperature probe, 278
Hot wire for turbulence measurements,
209
Hot-wire probe, 208
Humidity measurements, 310
Humidity ratio, 311
Humidity transducer, 312
Hydrogen-bubble method, 213
Hysteresis, 7

Ideal-gas flow, 190
Ideal-gas thermometer, 239
Illumination measurement, 131
Impedance matching, general discus-
sion, 21
Impedance-transforming amplifier,
100
Incompressible fluid, flow, 189
Index of refraction, 212
Indium antimonide detector, 132
Input circuitry, 87
Integrating sphere, 385
Integrating voltmeter, 110
Intensity:
of sound wave, 360
of turbulence, 209
Interference principle, 146
Interferometer:
density related to, 219
for dimensional measurements, 147
as flow-visualization device, 217
International Temperature Scale, 10
Ionization chambers, 388

Ionization displacement transducer, 135
Ionization vacuum gage, 179
Iron-vane instrument, 83

Junker's calorimeter, 304

Kelvin temperature scale, 238
Kiel probe, 226
Knudsen gage, 178

Laser anemometer, 219
  sensitivity of, 222
Lead sulphide cell, as used for thermal radiation detection, 381
Lead telluride detector, 132
Least count, 6
Least squares, method of, 65
Length standards, 8
Linear variable-differential transformer, 124
  characteristics of, 126
Linearity, definition, 17
Liquid-in-glass thermometer, 241
Liquid-level measurement, 128
Liquids, properties of, 410
Lissajous diagrams, 115
Loading error, 90
Lobed-impeller flowmeter, 187
Loschmidt apparatus, 300

McLeod gage, 174
Magnetic field measurements:
  Hall effect, 137
  search coil, 136
Magnetic flowmeters, 210
Magnetometer search coil, 136
Manometers, 165
Mass balance, 322
Mass standards, 8
Mean free path, 162
Mean molecular velocity, 161
Mean solar day, 9

Measurement system:
  dynamic, 14
  generalized, 12
  response, 16
Metals, properties of, 410
Micrometer calipers, 145
Microphones, characteristics of, 363
Moments of inertia, 414
Monochrometer, 384
Multisample data, 34

Natural frequency of tube cavity, 163
  (See also Seismic instrument)
Neutron detection, 372
Newton-Cotes integration formulas, 156
Noise, 362
Noise equivalent power, 132
Noise reduction coefficient, 372
Nonmetals, properties of, 410
Normal error curve (see Gaussian distribution)
Nozzle, sonic, 201
Nozzle flowmeter, 190, 193
Nuclear radiation measurement, 386
Numerical integration:
  plane areas, 156
  surface areas, 157
Nutating-disk meter, 186

Obstruction flowmeter, 188
Octave bands, standard, 370
Optical pyrometer, 263
Orifice flowmeter, 190, 195
Oscillograph, 116
Oscilloscope, 112
  dual-beam, 113
  measurement: frequency, 115
    phase-shift, 114
  used with electronic switch, 114

Performance testing distinguished from research, 4

Phase shift:
  definition, 16
  measurement: electronic counter,
    121
    oscilloscope, 114
  in seismic instrument, 362
Photoconductive transducer, 131
  bandwidth of, 132
  characteristics of, 133
  spectral response, 133
Photoelectric transducer, 130
Photographic methods for nuclear
    radiation detection, 389
Photovoltaic cells, 134
Piezoelectric constants, 130
Piezoelectric transducers, 129
Pirani gage, 176
Pitot tube, 226
Plan for general experiment, 23
Planck distribution, 260
Planimeter:
  polar, 152
    zero circle of, 153
  roller, 154
Pneumatic displacement gage, 149
  as control device, 151
Poisson distribution:
  definition, 49
  standard deviation, 392
  used with nuclear radiation, 392
Poisson's ratio, 332
Positive-displacement meter, 186
Potentiometer:
  with battery standardization, 91
  simple, 90
Potentiometric voltmeter, 110
Power measurement, 330
Precision, 7
Pressure:
  absolute, 160
  gage, 160
  units, 161
Pressure gage, dynamic response, 162
Pressure measurements:
  bellows gage, 172
  bourdon gage, 169
  diaphragm gage, 169

Pressure measurements:
  dynamic response, 162
  for high pressures, 174
  for low pressures, 174
Pressure probes:
  basic relations, 224
  effect of yaw angle, 225
  Mach number effects, 228
Probability:
  definition, 46
  relation to frequency, 46
  rules for calculating, 46
Probability distributions:
  binomial, 49
  definition, 48
  Gaussian, 49
  Poisson, 49
Probability graph paper, 57
Programming in data acquisition, 401
Prony brake, 330
Proving ring, 327
Psychoacoustic factors, 366
Psychrometer, sling, 312
Pulse counters in nuclear measure-
    ments, 388

Quartz-crystal thermometer, 258

Radiation error in temperature mea-
    surement, 269
Radiation shield in temperature mea-
    surements, 270
Radiometer:
  with thermistor, 380
  with thermopile, 379
Random errors, 35
Rankine temperature scale, 238
Ratio arms in bridge circuits, 93
Readability, 6
Recovery factor, 278
Recovery temperature, 278
Reflectivity, measurement of, 384
Refractive index, 212
Relative humidity, 311
Report writing, 27

Research distinguished from performance testing, 4
Resistance thermometer, 244
Resistance transducer, 124
Responsivity, detector, 132
Reynolds number for flowmeters, 198
Rise time, definition, 19
Rosettes for strain gages, 340
Rotameter:
    influence of density, 205
    principles, 204
Rotary-vane flowmeter, 187

Saybolt seconds, 298
Saybolt viscosimeter, 297
Scanner programmer, 401
Schlieren, 215
    knife edge in, 216
Scintillation counter, 390
Seismic instrument:
    acceleration response, 354
    displacement response, 353
    phase shift, 352
    practical instruments, 357
    principles, 351
    temperature effects, 359
    transient response, 355
Sensitivity, 7
Shadowgraph, 213
Simpson's rule, 156
Single-sample data, 34
Slug heat-flux sensor, 313
Smoke flow visualization, 222
Sonic-nozzle flowmeter, 201
Sound, psychoacoustic factors, 366
Sound-absorption coefficient, 370
    curves, 371
Sound intensity, 360
Sound-level meter, 369
    scales on, 369
Sound levels, typical, 367
Sound measurement, 360–372
Sound-pressure level, 360
Specific humidity, 311
Stagnation pressure probes, 225
    for supersonic flow, 229

Standard cell, as used in potentiometer, 91
Standard deviation:
    definition, 44
    of mean, 69
Standard prefixes and multiples, 408
Standards:
    electrical units, 9
    length, 8
    mass, 8
    temperature, 10
    time, 9
Standing sound wave, 365
Static pressure probes, 227
Statistics of counting, 391
Stefan-Boltzmann law, 264
Strain, definition, 332
Strain-gage rosettes, 340
Strain measurement, 333
    electrical-resistance strain gage, 334
    unbonded-resistance strain gage, 343
Strain sensitivity, 333
Systematic errors, 35
Système International d'Units, 407

Temperature:
    dry-bulb, 311
    thermodynamic meaning, 237
    wet-bulb, 311
Temperature compensation in strain gages, 340
Temperature error in radiation measurement, 264
Temperature-Mach number relation, 278
Temperature measurement:
    effect of heat transfer, 265
    electrical, 244
    mechanical, 240
    by radiation, 259
    (See also Thermometers)
Temperature recovery factor, 278
Temperature scales, 238
Temperature standards, 10

Thermal conductivity:
  definition, 289
  for gases: at high temperatures,
    293
  using concentric cylinders, 292
  measurement: dual-cylinders, 289
  guarded hot-plate, 288
Thermal radiation flux, measurement
    of, 377
Thermal radiation receivers, charac-
    teristics, 381
Thermal radiation transmission char-
    acteristics of materials, 378
Thermistors, 246
  transient response of, 248
Thermocouple current meter, 85
Thermocouples:
  compensation for transient re-
    sponse, 273
  emf-temperature relations, 252
  operating range, 253
  tables, 252
Thermoelectric effects, 249
  constants, 253
  law of intermediate metals, 249
  law of intermediate temperatures,
    250
Thermoelectric sensitivity, 251
Thermometers:
  electrical-resistance, 244
  fluid expansion, 243
  for high-speed flow, 277
  ideal-gas, 239
  liquid-in-glass, 241
  optical, 263
  quartz-crystal, 258
  summary of characteristics, 279
  thermistor, 246
  thermocouple, 249
  transient response of, 273
Thermopile, 256
  used in radiometer, 379
Time constant:
  definition, 20
  for thermocouples, 274
Time-interval measurements, 121
Time standards, 9

Titanium tetrachloride, as smoke gen-
    erator, 223
Torque measurement, 329
Transducers:
  capacitive, 127
  differential (LVDT), 124
  general definition, 12
  Hall-effect, 137
  ionization, 135
  photoconductive, 131
  photoelectric, 130
  piezoelectric, 129
  summary of characteristics, 122
  variable-resistance, 124
Transmissivity, measurement of, 384
Trapezoidal rule, 156
Traveling sound wave, 364
Tube-cavity arrangement, natural fre-
    quency, 163
Turbine flowmeter, 207
Turbulence intensity, 209
Turbulence measurements, 209

Unbonded strain gage, 343
Uncertainty:
  calculation, 38
  general discussion, 35
  in nuclear radiation measurements,
    393
Uncertainty analysis:
  related to experiment planning, 25,
    39
  related to instruments selection, 42
Unit strain:
  axial, 332
  transverse, 332
Units, International System of, 407

Vacuum, definition, 160
Vacuum gages:
  Alphatron, 180
  ionization, 179
  Knudsen, 178
  McLeod, 174
  Pirani, 176

Vacuum-tube voltmeter, 107
Variance, definition, 45
Velocity of sound for air, 163
Venturi flowmeter, 190, 194
Vernier calipers, 144
Vibrating cantilever beam, 350
Vibrating wedge, 349
Vibration measurements (*see* Seismic
   instrument)
Viscosity:
  definition, 293
  related to fluid shear, 294
Viscosity measurement:
  capillary, 296
  rotating cylinder, 295
  Saybolt method, 297
Voltage amplifier, 100
Voltage-divider circuit, 89
Voltage-sensitive input circuit, 88
Voltmeter:
  D'Arsonval, 81
  digital, 109

Voltmeter:
  electrostatic, 86
  integrating, 110
  potentiometric, 110
  vacuum-tube, 107

Water, properties of, 412
Wet-bulb temperature, 311
Wheatstone bridge:
  basic circuit, 92
  deflection analysis, 98
Wien's law, 261

Yaw angle, influence on pressure
   probes, 225
Young's modulus, 326

Zero circle of planimeter, 153